# SECURING CYBERSPACE

*International and Asian Perspectives*

# SECURING CYBERSPACE

*International and Asian Perspectives*

*Editors*

Cherian Samuel

Munish Sharma

idsa

INSTITUTE FOR DEFENCE STUDIES & ANALYSES
NEW DELHI

PENTAGON PRESS

**Securing Cyberspace: International and Asian Perspectives**
***Editors: Cherian Samuel and Munish Sharma***

First Published in 2016



ISBN 978-81-8274-918-4



*Published by*
PENTAGON PRESS
206, Peacock Lane, Shahpur Jat,
New Delhi-110049
Phones: 011-64706243, 26491568
Telefax: 011-26490600
email: rajan@pentagonpress.in
website: www.pentagonpress.in

*In association with*
Institute for Defence Studies and Analyses
No. 1, Development Enclave,
New Delhi-110010
Phone: +91-11-26717983
Website: www.idsa.in

Printed at Aegean Offset Printers, Greater Noida, U.P.

# Contents

# Foreword

This volume consists of the presentations made at the 18th Asian Security Conference at the Institute for Defence Studies and Analyses (IDSA) in February 2016. Since its inception in 1999, the Conference has provided a forum for frank exchanges among analysts, scholars and practitioners from across Asia and the world. 'Securing Cyberspace: Asian and International Perspectives', was chosen as the theme for the last Conference since cyberspace has become an arena for co-operation, competition, as well as conflict. Asian internet users, currently at 48 per cent globally, are growing faster than in any other continent. Although, Asian countries will be highly impacted by developments in cyberspace, they play a marginal role in the management of cyberspace. With the shift of global economic activity towards the south and the east, and a concomitant rise in data flows towards Asia, Asian representation is critical to the management of international institutions and regimes, including in the domain of cyberspace.

The Conference papers cover six different facets of cybersecurity – the global cybersecurity environment, the international and regional responses to cybersecurity challenges, non-state actors and cyberspace, securing strategic critical infrastructure, cybersecurity and the digital economy, and the role of the military in cybersecurity.

In a broad sense, and in a language comprehensible to non-technical readers, cyberspace is where information technology and the electromagnetic spectrum come together – its superstructure layered over by the sub-structure of cables, computers, and sea, land and space-based communications networks, and energised by the use of information technology. India's cybersecurity policy defines cyberspace as a complex environment comprising interaction between people, software and services, based on a world-wide platform of information and communication technology devices and networks. At the strategic level, cyberspace and cyber technologies have become key components in the formulation and execution of public policies. They have added a new variable to the components of comprehensive national power, creating both new synergies and new sources of vulnerabilities. Cyber technologies connect across all key bases of national power and act as a force multiplier, sometimes with disruptive effects.

India has a substantial stake in the stability and security of cyberspace, with a growing number of users from across the spectrum, ranging from government agencies to private enterprises, besides individual users – there are now an estimated 462 million internet users in India. Moreover, the Indian government has invested in a number of programmes to leverage information and communication technologies for the benefit of its citizens. This accelerated capacity building has implications for the country's national security.

Cyber has added a new dimension to conflicts and wars. Cyber operations have the capacity not just to augment conventional military operations, but to subvert them. Attacks in cyberspace can be fast, silent, inexpensive to mount, and potentially, devastating in their impact. While boots on the ground are not going to be replaced by cyber armies operating in a virtual battlefield in the near future, information dominance in the battlefield may well make the difference between victory and defeat. The crossover between cyber and space, and the role of cyber in nuclear deterrence, add to the challenges nations face in maintaining cybersecurity.

These issues are not unique to India. In efforts to forestall increasing national, regional and global vulnerability, the international community has been seeking norms and conventions for securing cyberspace. India has been at the forefront of many such efforts, in different fora. It is a member of the newly constituted UN Group of Governmental Experts on Developments in the Field of Information and Telecommunications in the Context of International Security. Later this year, Hyderabad, also known as Cyberabad, is hosting ICANN57, the second such public meeting of the Internet Corporation for assigned Names and Numbers (ICANN) to be held in India. India has been a strong and consistent votary of an open, global, and secure cyberspace, cognisant that this goal can only be achieved through international co-operation.

The Institute looks forward to continuing its contribution to the national and global discussions on cybersecurity through its research and interaction with partner institutions. Its Task Force Report on "India's Cyber Security Challenge" appeared in March 2012, and was a precursor to India's National Cybersecurity Policy, unveiled on July 2, 2013. We hope that dissemination of the ideas considered at the 2016 Asian Security Conference to a wider audience would inspire further suggestions on how India and the international community could devise cooperative measures on cybersecurity.

**Jayant Prasad**
Director General
Institute for Defence Studies and Analyses

# List of Abbreviations

| | |
|---|---|
| ADF | Australian Defence Force |
| APCERT | Asia Pacific Computer Emergency Response Team |
| APEC | Asia Pacific Economic Cooperation |
| APT | Advanced Persistent Threat |
| ARF | ASEAN Regional Forum |
| ARPANET | Advanced Research Projects Agency Network |
| ASEAN | Association of Southeast Asian Nations |
| ASPI | Australian Strategic Policy Institute |
| ATC | Air Traffic Control |
| AU | African Union |
| BFSR | Battle Field Surveillance Radars |
| BGP | Border Gateway Protocol |
| BIS | Bureau of Industry and Security |
| BTS | Base Transceiver Stations |
| C4ISR | Command, Control, Communication, Computer, Intelligence, Surveillance and Reconnaissance |
| CBMs | Confidence Building Measures |
| CCDCOE | Cooperative Cyber Defence Centre of Excellence |
| CD | Conference on Disarmament |
| CEO | Cyber Effect Operations |
| CEPTOAR | Capabilities for Engineering Protection, Technical Operation, Analyses and Response |
| CERT | Computer Emergency Response Team |
| CII | Critical Information Infrastructure |
| CMF | Cyber Mission Forces |

| | |
|---|---|
| CNE | Computer Network Exploitation |
| CNI | Critical National Infrastructure |
| COMESA | Common Market for Eastern and Southern Africa |
| CSCAP | Council for Security Cooperation in the Asia Pacific |
| CSIRT | Computer Security Incident Response Teams |
| DCS | Distributed Control Systems |
| DDoS | Distributed Denial of Service |
| DNS | Domain Name System |
| DoD | Department of Defense |
| DRDO | Defence Research and Development Organisation |
| DSCI | Data Security Council of India |
| ECOWAS | Economic Community of West African States |
| EOS | Earth Observation System |
| FIRST | Forum of Incident Response and Security Teams |
| FSC | Financial Services Commission |
| FTA | Free Trade Agreement |
| GATT | General Agreement on Tariffs and Trade |
| GDP | Gross Domestic Product |
| GFCE | Global Forum on Cyber Expertise |
| GGE | Group of Governmental Experts |
| GNSS | Global Navigation Satellite System |
| GPS | Global Positioning System |
| HADR | Humanitarian Assistance and Disaster Relief |
| HSPD | Homeland Security Presidential Directive |
| IAEA | International Atomic Energy Agency |
| ICANN | Internet Corporation for Assigned Names and Numbers |
| ICS-CERT | Industrial Control Systems Cyber Emergency Response Team |
| ICT | Information and Communication Technology |
| IDSA | Institute for Defence Studies and Analyses |
| IETF | Internet Engineering Task Force |
| IHL | International Humanitarian Law |
| IISS | International Institute for Strategic Studies |
| INEW | Integrated Network Electronic Warfare |
| INL | Idaho National Laboratory |
| IoT | Internet of Things |
| IP | Internet Protocol |

| | |
|---|---|
| ISAC | Information Sharing and Analysis Centre |
| ISIS | Islamic State of Iraq and Syria |
| ISO | International Organisation for Standardisation |
| ISP | Internet Service Provider |
| ISRO | Indian Space Research Organisation |
| IT | Information Technology |
| ITU | International Telecommunication Union |
| KHNP | Korea Hydro and Nuclear Power |
| MAD | Mutually Assured Destruction |
| MND | Ministry of National Defense |
| MoD | Ministry of Defence |
| NADI | Network of ASEAN Defence and Security Institutions |
| NASSCOM | National Association of Software and Services Companies |
| NATO | North Atlantic Treaty Organisation |
| NCRB | National Crime Records Bureau |
| NCW | Network Centric Warfare |
| NEO | Network Enabled Operations |
| NOAA | National Oceanographic and Atmospheric Administration |
| NPT | Treaty on Non-Proliferation of Nuclear Weapons |
| NRRC | Nuclear Risk Reduction Centre |
| NSA | National Security Agency |
| NSB | National Security Bureau |
| NSCS | National Security Council Secretariat |
| OECD | Organisation for Economic Cooperation and Development |
| OIC-CERT | Organisation of Islamic Cooperation CERT |
| OSCE | Organisation for Security and Cooperation in Europe |
| OSI | Open System Interconnection |
| OSPF | Open Shortest Path First |
| PAROS | Prevention of an Arms Race in Outer Space |
| PLA | People's Liberation Army |
| PLC | Programmable Logic Controllers |
| PPP | Public Private Partnership |
| PPWT | Treaty on Prevention of the Placement of Weapons in Outer Space |
| PSTN | Public Switched Telephone Network |
| R&D | Research and Development |
| S&T | Science and Technology |

| | |
|---|---|
| SAARC | South Asian Association for Regional Cooperation |
| SADC | Southern African Development Community |
| SCADA | Supervisory Control And Data Acquisition |
| SCO | Shanghai Cooperation Organisation |
| SSL | Secure Socket Layer |
| TCBM | Transparency and Confidence Building Measures |
| TCP | Transmission Control Protocol |
| TPP | Trans Pacific Partnership |
| TLS | Transport Layer Security |
| TRIPS | Trade Related Aspects of Intellectual Property Rights |
| TT&C | Telemetry, Tracking and Command |
| UDHR | Universal Declaration of Human Rights |
| UN | United Nations |
| UNCOPUOS | United Nations Committee for the Peaceful Uses of Outer Space |
| UNSC | United Nations Security Council |
| UPU | Universal Postal Union |
| USAFA | United States Air Force Academy |
| USB | Universal Serial Bus |
| USCYBERCOM | US Cyber Command |
| VPN | Virtual Private Network |
| WEF | World Economic Forum |
| WSIS | World Summit on the Information Society |

# INTRODUCTION

The use of cyberspace by governments, businesses and individuals to ease and accelerate all kinds of activities has led to the global expansion of cyber-enabled networks in a relatively short period of time. While cyber experts have repeatedly warned that the many inherent and existing vulnerabilities in devices and networks have neither been resolved nor can be adequately managed to ensure security of the networks, these have largely been ignored or downplayed. The escalation in the number and magnitude of attacks has meant that most policymakers are now cognisant of the wide gamut of issues associated with cybersecurity. The varying perspectives of different countries on cyber issues and the sheer complexity of issues have made cybersecurity a concern for not just national but also international security: the geo-political overtones have increased the cybersecurity challenges.

Even as the connected systems and networks have grown more intertwined and complex, cyberspace is being used by malicious actors for a variety of nefarious purposes – from cyberespionage, both for commercial and security interests, to cybercrime and cyberterrorism. A Distributed Denial of Service (DDoS) attack can disrupt business operations or cause severe outages, having a direct impact on revenue and reputation. Companies also face the risk of losing trade secrets or intellectual property rights. Moreover, a massive data breach for companies or governance portals storing data of customers or citizens compromises personal information. A cyberattack on entities that are part of critical infrastructure can have a debilitating impact on national security. The risk increases manifold for electricity grids, nuclear installations, and telemetry/command and control network of space assets. Surprisingly, social media, as a threat vector, has become a channel of least resistance to conduct reconnaissance, steal identities and gather information on employees, projects, systems and infrastructure, besides spreading hateful propaganda and enticing impressionable youth to follow extremist ideologies.

Over the years, cyberspace has also become an intricate constituent of national power. The strategies for the development of cyberspace are not just restricted to civilian purposes; rather, this domain now falls well under the ambit of the armed forces. With the advent of Network Centricity in military operations and Revolution in Military Affairs, armed forces are at elevated risk of cyber incidents. The integrated use of land, air, maritime and space assets for enhanced domain awareness or real-time information access warrants the armed forces to build expertise in both defensive and offensive cyber operations. Nation states have documented their Cyber Strategies and executed them in the form of Cyber Commands. The military dimension has seen cyberspace witnessing the beginnings of a race for the development and deployment of cyberweapons. An arms control regime, the Wassenaar Arrangement has enlarged its controls list in consonance with the way cyberspace has altered the present-day security landscape. The development of cyberweapons and their potential usage against high-value targets has been one of the major security concerns for nation states.

The threats in cyberspace are varying in nature and intensity. Leading companies operating in the energy, telecommunications, finance and transportation sectors are targets of Advanced Persistent Threats (APTs). Non-state actors, such as terrorist organisations and criminal syndicates, have become tech-savvy, employing human resources to develop malware. These tools are used extensively in committing cybercrime. Terrorist organisations leverage the benefits of cyberspace, harnessing it for ideology propagation, recruitment and communication. The Islamic State of Iraq and Syria (ISIS; also called ISIL/IS/Da'esh), for example, is a prime case study. It leverages its tremendous presence on social media to spread propaganda and recruit sympathisers, from every corner of the world, as fighters. Al-Qaeda is also reported to have developed encryption software to secure its communication in cyberspace.

As the extent of commerce transacted over cyberspace grows, along with the dependence on information technology to gain cost-efficiencies, the risk to enterprises increases manifold. And as Asia continues to grow its share of the global trade and commerce, the threats from cyberattacks are expected to increase proportionately.

Cyberattacks, like many of the new security challenges, are transnational in origin and nature, and no nation can combat them alone. Despite variations in ethnic, economic and governance systems, Asian countries need robust security architecture to resolve the issues specific to the geographical region as well as international issues detrimental to Asia's economic and societal growth.

With the shift in power towards Asia, its representation in international governance mechanisms and its inputs towards creating a secure cyberspace are critical to international politics, the world economy and for the credibility of international institutions and cybersecurity regimes. Within Asia, cyberthreats have altered the security perceptions of institutions and government systems.

Against this backdrop, the papers presented by strategic experts, academicians, domain specialists and policymakers at the 18th Asian Security Conference attempt to examine a range of issues – global cybersecurity environment, non-state actors and cyberspace, securing critical infrastructure and the role of military in cybersecurity, to name a few.

Thus, the chapters in this Volume not only provide an outline of the journey so far, but more importantly, give indicators of future trends in cybersecurity from the vantage point of the respective experts.

## Section I: International Perspectives on Cybersecurity

Arvind Gupta sets the stage with a broad overview of the state of play in cybersecurity. He touches upon every aspect of concern for the international community as well as India, conforming that cybersecurity is now an international security concern. The chapter opens with a brisk overview of important events related to cybersecurity in 2015 and their connotations: Chinese President Xi Jinping's visit to the US in September and the subsequent agreement between the two countries; China and Russia signing a comprehensive agreement on cybersecurity; and the efforts of the United Nations Group of Governmental Experts (UNGGE) concerning international law.

Gupta delves into the challenges before states, such as defending their critical military and civilian infrastructure, tackling cybercrime – theft of personal information and intellectual property – and the blurring distinction between state and non-state actors in cyberspace. He questions whether "cyber deterrence" can work analogous to nuclear deterrence, and declares that an effective cyber deterrence strategy must, nevertheless, include deterrence by denial as well as by punishment; and that the state would need to clearly indicate its cyber thresholds. The chapter also looks at the developments in the *Tallinn Manual*. In addition, Gupta outlines the steps taken by India in the recent past to strengthen its cyber defensive capabilities, pressing on the needs of suitable response measures including the capability to conduct cyber operations, if required, and closely studying the idea of cyber deterrence and spawning strong cyber diplomacy. Ultimately, he draws attention to the dearth of consensus within the international community on issues such as definition of a cyberweapon, means of verification and attributability.

Greg Austin dissects the response by middle powers to the emerging centrality of cyberspace in the conduct of future war, which he assesses to be slow and fragmented. The chapter calls out major trends in the policy settings of two pacesetter countries: China and the US. Both regard military dominance in cyberspace as one of the primary determinants of success in war. Austin argues that the two major powers are placing considerable attention on disabling enemy cyber systems in the early stages of hostilities, or even on a pre-emptive basis. Unlike the US and China, few governments among the middle powers have been prepared to canvas in public the centrality of cyber-enabled warfare and craft policies and doctrines accordingly.

Austin previews trends in the technologies and characteristics of cyber-enabled war (attack technologies and defensive systems) and complex cyber-enabled war scenarios. These trends are moving in a direction that will present almost insurmountable challenges to the security of most middle powers. He infers that middle power defence forces will need to maintain distinct capabilities for cyberwarfare at the strategic level. The capabilities need to be unified in both policy and doctrinal terms in a way that lays a clear pathway for mobilisation in a very short time to fight a medium intensity, cyber-enabled hot war. This will require new technologies of decision-making that do not yet exist in most middle powers. He prescribes that, in the next two decades, the war-fighting needs of middle powers in cyberspace, as for their counterterrorism needs, cannot be met without considerable dilution of pre-existing alliances and blocs and more effective bridging of the big geopolitical divides. He asserts that collective security in cyberspace may be the only answer for middle powers.

Information is a strategic enabler or force-multiplier that has revolutionised the contemporary age to an extent that this age is often termed as the information age; nevertheless, it has also induced a strategic vulnerability in our critical assets which in current scenarios is exploited by a new form of warfare called strategic cyberwarfare, and Amit Sharma meticulously analyses the subject. He draws parallels from Clausewitz's Trinitarian Warfare, postulating that strategic paralytic effect can be induced onto the victim nation if the Warfare is performed in cyberspace. The chapter attempts to define strategy through the Triad Theory of Cyberwarfare, which aims at destroying the elusive Clausewitzian Trinity in cyberspace by performing parallel warfare in cyberspace. He also analyses the critical components to be targeted in waging strategic cyberwarfare and provides satisfactory reasons for the failure of contemporary cyberattacks to generate a strategic effect. Further, Sharma presents a framework of strategic cyberwarfare involving the Triad Theory of Cyberwarfare: its operational cyber campaign plan and formation of a 'Known' and 'Credible' cyber deterrence to generate a scenario of Mutually Assured Destruction in cyberspace. Establishing the strategic aspect of the cyberwarfare, he concludes by providing various recommendations for developing cyber deterrence capabilities, especially embroiling the notion of Prepare, Pursue, Protect and Prevent in cyberspace.

Alexandra Kulikova notes that efforts have been underway for over a decade to develop norms and conventions, and taken together, these efforts and initiatives form a rich ecosystem of tools and standards to ensure cyber stability serving global, regional or local goals. However, this diversity of needs, goals and agendas make reconciliation as well as implementation and operability a major challenge. The landscape consists of multiple stakeholders as well as disruptors. The human dimension of cyber politics, she notes, will certainly be further monitored by civil society and academia expert groups to ensure end users' rights are protected online as well as offline.

Further, Kulikova stresses that rather than congruence, there will be increasing competition among the various fora to be the standard setter and to push its vision of cybersecurity forward. She notes that state efforts in the area of cybersecurity, and especially the protection of critical infrastructure, depend on collaboration with the private sector; integrating business interests and profit models into the agreements achieved is crucially important. She underlines the possibility of a joint effort for the protection of critical infrastructure such as financial, nuclear, water facilities and their cyber interfaces as well as the 'public core of the Internet' – the root servers. Thus, it would be critical for the states to establish a workable format of interaction with the private sector to implement the commitments they make.

Sico van der Meer considers the foreign policy instruments available to nation states in order to secure their states from cyberattacks. Deterrence by denial and deterrence by retaliation are probably the most 'simple' counter-measures to international cyber aggression but while they look like promising policies, they face many difficulties in practice. In the case of the former, it is expensive and complex and requires continuous investment since technology changes at a rapid rate. Even if such investment is available, people are the weakest link in the chain and easily exploitable by malicious actors, as repeated intrusions into even the most safeguarded networks have amply shown. On the other hand, deterrence by retaliation requires very good forensic capabilities since misrepresentation and mis-direction are very easy in cyberspace. Deterrence based on the possibility of retaliation also only works if the party seeking to deter communicates clearly about the retaliatory measures that may be taken in the event of a cyberattack. Defence and deterrence are not able to create long-term cybersecurity and stability, but may instead even create further escalation and uncertainties.

Van der Meer suggests that a more viable option, therefore, is creating sustained diplomatic efforts centred around Confidence Building Measures (CBMs) and internationally accepted norms regarding cyberthreats to actively address the core problems of international cyber aggression. While such efforts are on-going, they have not had much success since they are yet to reach critical mass in terms of support by a large number of states. Nonetheless, such multilateral diplomatic efforts are crucial for long-term cybersecurity and stability. Instead of an on-going 'cyber arms race', efforts could better be focussed on building mutual confidence and respect as well.

An ambiguity remains over the description of cyberattacks on a nation's critical infrastructure, whether it is a subversive action or an act of war, or rather a proxy war. A. Vinod Kumar constructs his argument on this ambiguity, delving into the dimension of warfare. He argues that the numerous calibrated attacks in recent years on critical infrastructure (including strategic assets) of various countries underline the new dimensions and frontiers of warfare, involving 'adversarial forces' that blur conventional combatant identities. While the deployment of non-state actors for transnational cyberattacks highlight the element of proxy conflicts in this spectrum, that states are forming techno-military groups (Cyber Commands) reveals the truism of cyberspace evolving into the new battleground. The chapter conceptualises this battlefield and the threat environment, while pointing to its numerous complexities in terms of objectives of the attack, defining the nature of this neo-warfare, attacks on nuclear infrastructure and destruction and instability of states. Kumar concludes that conventional deterrence doctrines may not suffice as cyberattacks run short of triggering a full-fledged war. In this domain, norms are non-existent and state actors use plausible deniability and virtual camouflage to endow cyberwarfare with the same attributes or flexibility of terrorism.

Liina Areng elaborates the military dimension of cybersecurity, underlining the evolving phenomenon of cyber capabilities being used as a tool to gain leverage in international security. It is further challenging the traditional military capabilities and doctrines. She is apprehensive about cyberattacks against national critical infrastructure, which could have a cascading effect on the economy, society and government in ways difficult to understand, model or predict. These man-made disasters might have serious national security implications, yet the response to cyberthreats cannot be conceived purely in terms of classical warfare. She analyses the need to defend the society as an "ecosystem" by orchestrating military planning and preparation for civil emergencies. Moreover, while nations are developing their arsenals of defensive and offensive cyberweapons, most cyber problems remain in the "grey area". Areng concludes that although hostile cyber activities will continue to flourish, cyberspace will also play an important role in future military conflicts, response to which would not only need proper preparedness and resilience, but also adequate international containment and de-escalation mechanisms. She advocates for militaries to focus on the defence of their networks and infrastructure, and insists that their role in cybersecurity is limited even if governments acknowledge it as a national security concern.

Madan M. Oberoi underscores the importance of understanding how technology is impacting crime and the criminal justice system, and its stakeholders, namely the law enforcement agencies, prosecutors, judiciary and also the world within which the criminal justice system operates. Technology can be exploited by both law enforcement and the criminals, and currently the criminals have the upper hand since they are able to adapt to technological trends much faster than the law enforcement universe. His chapter attempts to understand some of the current technological trends and their impact on a myriad of law enforcement activities from intelligence gathering, investigation, police information management, to training. He notes that while the perception that the criminals are far ahead of the law enforcement agencies may be over-hyped, but the threat is real, and few agencies/organisations/countries are in a state of readiness to fully combat this threat.

This in itself leads to changes in strategy – from a prosecution-based strategy to a disruption-based strategy. However, this approach does not target the criminal actors, and only leads to criminals shifting the infrastructure in case of a successful disruption. The victims are also turning to private agencies in the face of their seeming inability to respond to these challenges, thus eventually leading to the marginalisation of state actors in this domain. Oberoi calls for a recalibration of law enforcement strategy and shift towards a multi-stakeholder model involving private sector, academia, research bodies, inter-governmental bodies, civil society and law enforcement agencies.

Accepting that most of the nation's critical infrastructure (energy, transport, finance, medicine, etc.) now lies in the hands of the private sector, Kah-Kin Ho deliberates on the evolving role of government in cybersecurity, particularly in regard to attacks on critical infrastructure by adversaries ranging from non-state actors, such as terrorist groups, hacktivist groups and organised criminals to state actors, and the high degree of interconnectedness across the globe. He analyses the specifics of challenges in the shadow of divergent interests, between the private and public sectors, and deducts that, on the one hand, the private sector's primary focus is corporate efficiency, in terms of implementing the bare minimum level of security, while, on the other hand, the government is principally concerned with achieving social order, national security and economic prosperity for its population. Ho argues that the role of governments as the legitimate providers of security has diminished, and that it will continue to weaken. However, at present, it is critical, and their remit must transcend what the historical regulatory role has typically entailed. To achieve that, he asserts the Regulate, Facilitate, Collaborate (RFC) framework, through which governments must strategise and draw upon analogous lessons learnt from past strategies geared towards other areas of threat, such as pandemics and terrorism.

Illuminating another dimension of cybersecurity, Jana Robinson analyses the security issues lying at the intersection of space and cyberspace. Concerning the critical role of cyber and space domains for national security and war-fighting, Robinson points out that decision-makers must configure proper crisis management mechanisms, including proportionate responses to any hostile or disruptive actions by nations. She stresses that understanding the cross-domain parallels as well as differences is essential to the successful safeguarding of both of these domains, in addition to physical and technical properties. Accordingly, she examines how the cyber and space domains interact (e.g. the transmission of cyber signals by satellite communications systems), the vulnerabilities associated with how they connect (e.g. cyberattacks on space systems), and possible risk-mitigation strategies. She also emphasises on active involvement of both government and private sector actors to formulate realistic norms and guidelines for responsible behaviour in these two domains. Beyond norms development, Robinson traverses to Transparency and Confidence Building Measures (TCBMs), and deems them to be critical, especially when traditional arms control methods cannot be easily employed.

Caroline Baylon meticulously analyses the case of nuclear facilities, which, even among critical infrastructures, have been at persistent risk from cyberattacks. She affirms that most of the operators in the sector do not fully understand the risk, and therefore, a key first step is to develop guidelines to assess and measure this risk as accurately as possible. This would help them to understand the economic rationale of investment in cybersecurity. Baylon draws from her extensive research on cybersecurity of nuclear facilities to delve into the development of cyber insurance and cyber risk guidelines, disclosure and information-sharing measures, improved communication between the Information Technology (IT) and Operations Technology (OT) teams to bridge cultural divides and the implementation of technical solutions.

Ted G. Lewis brings in the technologists' perspective and proposes a technical solution to address the various policy and technical challenges arising from the spread of malware in a computer network. Delving into the technical aspects of architecture, Lewis justifies his argument with illustrations that the *monoculture* design and *scale-free structure* challenges posed by the Autonomous System (AS)-level Internet promote widespread contamination even under very small probability or vulnerability to an exploit. On the other hand, he explains, the scale-free structure of the AS-level Internet can be turned to the defender's advantage by relating resilience to vulnerability and self-organisation as measured by the *spectral radius* of the network. Furthermore, Ted analyses the AS13579 Internet and identifies 944 (7 per cent) of the AS-level nodes as blocking nodes. This builds his argument that, by blocking these nodes, spectral radius diminishes dramatically, with corresponding increase in resilience and protecting 13.5 per cent of the global Internet's AS-level nodes can eliminate nearly all malware.

Sanjeev Relia seeks for a clear delineation between the different non-state actors. In the first instance, there are the good non-state actors, like companies and non-profits who are largely responsible for the creation and maintenance of cyberspace. According to Relia, the bad non-state actors can be further sub-divided into those non-state actors who form part of radical organisations such as the ISIS and Al-Qaeda; also referred to as cyberterrorists. Then there are the cyber militia who can be defined as a group of volunteers willing and able to use cyberattacks or other forms of disruptive cyber actions on behalf of a nation state in order to achieve a political goal. Increasingly, state agencies are co-opting such militia to carry out terrorist activities in cyberspace.

In the geopolitical and geostrategic realm, non-state actors are not a novel phenomenon in Asia, and Gillane Allam highlights the rise of Da'esh. She establishes the historical linkages of the evolution of this phenomenon and elaborates the appearance of armed non-state actors, notably Da'esh, addressing whether its modus operandi inclusive of cyberspace is a threat to Asia's ambitious economic progress affirmatively. Her central argument revolves around the pressing need at the international level to wage a committed digital counter-insurgency campaign, in addition to the national and international military campaigns. Drawing in the perspective from Egypt, which is undertaking its responsibilities as Chair of the United Nations Security Council Counter Terrorism Committee for 2016-2017, Allam gives a comprehensive account of the activities of Egypt and builds a case for Asia's booming economies to be part of these endeavours, to ensure sustainable economic prosperity, stability and security.

Uchenna Jerome Orji examines the legal frameworks developed at the regional and sub-regional level in the African continent by various organisations including the Economic Community of West African States (ECOWAS), Common Market for Eastern and Southern Africa (COMESA), Southern African Development Community (SADC) and African Union (AU). He notes that, while harmonisation of laws is a laudable goal, it can only succeed if there is requisite political will. Thus, while the AU Cybersecurity Convention is closely modelled on the Budapest Convention on Cybercrime, it falters when it comes to the specific legal provisions that would facilitate prosecution of cybercriminals. The lack of political will and sensitisation on cybersecurity has meant that though the AU Convention requires only 15 out of 55 African states to sign in order to be ratified into law, not a single state has signed so far. The large size of the African continent with its 55 sovereign states and their diverse legal traditions and how they receive and implement international treaties is indeed a challenge to effective national harmonisation of regional cybersecurity measures.

## Section II: Asian Perspectives on Cybersecurity

Asia is the fastest-growing Internet region with its own unique characteristics. However, there is inadequate understanding and discussion within Asia on the opportunities and challenges confronting this region in cyberspace. As such, a particular focus of the Conference that provided the inputs for this Volume was the Asian region. In addition to perspectives on the economic opportunities and challenges, and the scope for military cooperation to secure cyberspace, it includes country perspectives from South Korea and Japan, the most advanced countries in the region.

Liam Nevill emphasises that cyberspace can be a force-multiplier in tackling the many problems of the Asian region: from lowering traditional barriers to development and providing access to health and education resources. It provides an easy way for the advanced economies to help the less developed, in the process improving connectivity as well. However, none of this would be possible without a stable and secure cyberspace. Without confidence in the security of personal data and financial information, consumers will hesitate to engage with digital commerce. Cybercrime has to be tackled at a regional level with states being provided the resources to engage in capacity building as well as sharing information. Nevill argues that harmonising regulatory and legislative measures will help in not only combating cybercrime but also enable and incentivise the formation of a regional digital economy. However, he asserts that though trade frameworks such as the Trans Pacific Partnership can help towards harmonising regulations, but may also be seen as a means to push a particular framework which might be detrimental towards developing indigenous economies.

Contemplating on cybersecurity policymaking, Candice Tran Dai notes that it has been evolving in recent years from a technology-focused issue towards a more holistic issue, encompassing economic, social, educational, legal, technical, diplomatic, and military, intelligence aspects. Two major dimensions tend to be more systematically integrated into recent cybersecurity strategy roadmaps: national sovereignty and economic policy, both aspects being often tied up into the dual concept of national independence and indigenous innovation. Governments seek to incentivise their entrepreneurs in the quest for indigenous cybersecurity capability and technology by constructing barriers and raising costs for foreign companies. This comes into conflict with broader goals of facilitating international trade by reducing barriers and also has an impact on global aims to develop norms for cybersecurity. What is called the digital revolution is fundamentally based on global networks, cross-border flows of data, and network infrastructures and content platforms largely owned by global companies. She asserts that the question, therefore, becomes: How do countries protect national security interests without inappropriately undermining the value produced by a global supply chain? And that there is a balance to be reached between what is non-negotiable from a national security point of view and what is negotiable from an international cooperation point of view.

She also considers the challenge arising out of cybersecurity having evolved into a key issue in global economic relations, and wonders to what extent it tends to become more than an objective in itself, going beyond securing cyberspace, in other words, possibly as a tool for broader objectives in the political, economic and technological realms. She concludes that there are a few hints of nascent velleities geared at nurturing what is called a cyber-industrial complex, which could possibly broadly materialise into a cyber military-industrial complex within the framework of the current digitalisation process of the military.

Nandkumar Saravade notes that cybersecurity is a global problem that requires a global solution. On the bright side, the cybersecurity challenge is emerging as a big uniting factor with unprecedented collaboration across the board – from information sharing amongst trusted parties on security threats and collaboration amongst law enforcement agencies to bring cybercriminals to justice, to devising treaties and regulations for the cyber domain. At the same time, tracking cybercriminals and bringing them to justice in sovereign countries is increasingly difficult with challenges in collection of appropriate cyber forensics data, applicability of laws and acceptance by courts. The question is whether enough is being done, as countries try to bring law enforcement agencies and the legal framework to deal with the problems of the 21st century.

For Saravade, this holds true even when it comes to adapting the business environment to new technologies such as cloud computing. While transitioning to 'Cloud' offers immense benefits especially in terms of cost, flexibility, scalability and agility, there are major challenges – non-tariff barriers in the form restrictions on the global data flows, jurisdictional issues because of location of servers in the Cloud, stringent investigation/surveillance regulations in countries that give legal rights to law enforcement agencies for accessing data, concerns over security and privacy of sensitive business and personal information hosted in the Cloud and sharing of ownership and accountability between the user and provider organisations.

Further, he adds that India lacks the necessary structures and capabilities within the country to understand complex cybersecurity issues, and the government has to build the necessary frameworks to engage in consultations with the key stakeholders and develop a position on key cybersecurity issues.

Munish Sharma and Cherian Samuel discuss the case of cybersecurity cooperation in South Asia, as much of the economic development in this region has been facilitated through the advances in information technology with the digitisation of public services and opportunities for business development in Information and Communications Technology (ICT). However, they argue, that countries in the region are yet to internalise cybersecurity as essential to their economic, political and national well-being. They note that the existing lack of capabilities and capacities to understand and remediate threats make not just individual countries, but the entire region vulnerable to threats from state and non-state actors.

Sharma and Samuel postulate that as an economic and geopolitical organisation of eight countries, the South Asian Association for Regional Cooperation (SAARC) can play a pivotal role in capacity building as well coordinating cybersecurity efforts of all the members facing non-traditional security threats, such as cyberterrorism and cybercrime, from non-state actors to both their populace and businesses. The chapter objectively examines the prospects and challenges of SAARC putting across cybersecurity as a key agenda item for cooperation, given the divergences among members over traditional security threats.

Dwelling on the role of the military with specific reference to the Asia-Pacific, Caitríona Heinl states that cybersecurity issues within the region are shaped by high national security sensitivities, exceptional levels of military modernisation and defence spending, on-going maritime and territorial disputes and increasingly malevolent non-state actors who further complicate matters. Other factors that impact on cybersecurity include culture, ideologies and perspectives which also colour the developing strategies and doctrines. She notes that there is an increased emphasis on the part of some countries to engage in non-traditional warfare through cyberspace and attack targets that disrupt information systems, decision-making processes, and cognitive perceptions on the battlefield, as well as critical infrastructure like financial, energy and transportation systems. Obviously, this would result in fundamental changes to the character of warfare.

On the flip side, Heinl reflects, while militaries might aim to be prepared to win in conflict, there should be an obligation to avoid escalation. The military is after all an important stakeholder with an interest in a safe and secure cyberspace. She suggests that having the defence community engage in military-to-military dialogues and other practical measures could complement the aims of building transparency and confidence to improve international stability through international political agreement. Military-to-military relations might even be easier to establish given common hierarchies, terminologies and structures that can transcend national differences.

Li-Chung Yuan contemplates on the role of military from the Taiwanese perspective, in regard of cross-strait relations and the national security imperatives of Taiwan. He exhibits the Taiwanese apprehensions with respect to the possibility of military confrontation, and its percolation into the cyber domain, as the Taiwan Government has been a target of cyberespionage and attacks allegedly sponsored by China. Yuan alludes that these issues have been addressed appropriately, and armed forces of Taiwan are authorised to tackle threats of, inter alia, sabotage, subversion, and especially espionage in cyberspace. The chapter further explores the role of Taiwanese military in various aspects of cyberspace, focussing on intelligence and national critical infrastructure.

Yasuaki Hashimoto elaborates on Japan's role in promoting cybersecurity policies at various strata of governance, in particular the objectives of National Security Strategy, National Cybersecurity Strategy, Basic Act on Cybersecurity and various security-related organisations. He explains the role of Japanese Ministry of Defense and Japan Self-Defense Forces in the stability of cyberspace. The chapter also takes a detailed account of cybersecurity initiatives, with a focus on the Cyber Defense Unit, its functioning, roles and responsibilities as well as the objectives. Finally, Hashimoto deducts that bilateral, multilateral, regional and global cooperation is essential for a more secured cyberspace, because cybercrime and attacks are crossing national borders.

Il Seok, Oh elaborates on the many threats faced by Korea in cyberspace and the various measures, legislative and administrative, that have been put in place by the South Korean Government. The major Acts are the Electronic Financial Transaction Act (delineating liability in fraudulent electronic transactions), Cyber Security Industry Enhancement Act (making the Ministry of Science, ICT and Future Planning the nodal agency for all cybersecurity-related activities), Network Protection and Critical Information Infrastructure Protection Act and Personal Information Protection Act. He notes that these legislative measures have gone a long way in ensuring the cybersecurity of all Koreans and could serve as a model for other countries.

Finally, a Chinese perspective of the bilateral discussion between two key players, the United States and China, rounds off this section. Comparing the American and Chinese perspectives on cybersecurity, Cuihong Cai discerns that over-interpretation of cybersecurity risks strengthens the threat cognition, which results in conflicts and control-oriented security practices. It further leads to trust deficit and weakening of rules, resulting in the self-fulfilling prophecy of heightened cyber conflicts. She confers that the cybersecurity environment is a subjective state and it is related to discourse construction as all the actors in cyberspace have the capacity to launch attacks. The chapter further elaborates the problems confronted in the global cybersecurity environment, including information inundation, information pollution, information infringement, information monopoly and cybersecurity crisis. It talks about the common characteristics as well, such as the diversity of threats, asymmetry of subjects, lagging of security technologies, absence of institutional norms, imbalance of cyber power, lack of collective security mechanism and malfunction of deterrence in cyberspace. She finally delves into the details of subjective cognition of US and China, elucidating the divergences in the understanding of core cybersecurity interests, in terms of freedom of action, commercial, social and political interests, which in turn become part of the global cybersecurity environment.

It is hoped that this volume fulfils the purpose of discussing the current issues in cybersecurity and their particular resonance to Asia. It was the institute's endeavour to bring out the volume at the earliest and we were aided in no small measure by our authors who were prompt to finalise their papers. Acknowledgements are also due to Nidhi Pant for copy-editing the papers, and Virender Negi for page-setting the manuscript. Our thanks also go to Rajan Arya and his team at Pentagon Press for bringing out the volume in record time.

# SECTION I

## International Perspectives on Cybersecurity

# 1

# Securing Cyberspace: A National Security Perspective

*Arvind Gupta*

The world is becoming increasingly turbulent. The unstoppable march of globalisation, facilitated by Information and Communications Technology (ICT), has raised many troubling questions concerning the maintenance of peace and stability. Cybersecurity is now an international security concern. It is also a top concern for most countries and figures high in their national security priorities. The focus is on managing the threats in cyberspace which affect everyone. The key question before a state is how to defend itself from the ever increasing occurrences of cyberattacks.

The year 2015 saw a number of important developments in the field of cybersecurity. President Xi's visit to the US in September 2015 will be remembered for some outspoken public comments by President Obama on US concerns over online theft of intellectual property. Aware that cyber concerns, if unresolved, could create misunderstanding and destabilise ties, the two countries agreed to bilateral cybersecurity dialogues. In President Obama's words, the two governments agreed that "neither the US nor the Chinese Government will conduct or knowingly support cyber related theft of intellectual property including trade secrets or other confidential business information for commercial advantage". President Obama, according to reports, took up strongly with President Xi the issue of cyberthreats. On his part, President Xi, declared that "China strongly opposes and combats the theft of commercial secrets and hacking attacks". The meeting took place against the backdrop of a well-publicised cyberattack in December 2014 on the Office of Personnel Management (OPM) that resulted in a major data breach, compromising the fingerprints of 5.6 million people and security clearance records of some 22 million people. Both the sides acknowledged cybersecurity as an issue between them, and that was in itself a remarkable development. During the same year (2015), China and Russia also signed a comprehensive agreement on cybersecurity.

In 2015, the United Nations Group of Governmental Experts (UNGGE) came out with its third report – an advance over the previous report. As a result of the efforts of the UNGGE, there is now a growing recognition that international law, particularly the UN Charter, applies as much as to cyberspace as to other domains. The UNGGE emphasises that principles of sovereign equality; settlement of international disputes by peaceful means; refraining from the threat or use of force against the territorial integrity or political independence of any state; respect for human rights and fundamental freedoms including the freedom of expression; and non-intervention in the internal affairs of other states are some of the principles which also apply to the ICT security. In other words, international law is technology neutral. One of the main observations of the report is that states have jurisdiction over the ICT infrastructure located within their territory.

The international law has many aspects including intervention in self-defence, economic sanctions, counter measures and so on. A debate has broken out whether intervention through cyber means in other countries' networks, under certain circumstances, is justified or not. The debate is sharp but inconclusive.

Cybersecurity issues are contentious and proving to be difficult even as the incidents of cyberattacks, cybercrime and cyberterrorism grow exponentially. Every year new types of attacks are invented and carried out. The toolkit of attackers is expanding. It is quite possible that states may be clandestinely developing arsenal of tools of cyberattack even as they discuss the need for accepted norms in cyberspace.

The challenge before states is how to defend their critical, military and civilian infrastructure from the destabilising cyberattacks. Cybercrime is on the increase. Theft of personal information and intellectual property is rampant. The distinction between state and non-state actors in cyberspace is blurring. Even as technologies of active defence are developed, the attackers are several steps ahead.

Most states are engaged in implementing strategies to defend their networks from cyberattacks; simultaneously, they are also toying with the idea of developing capabilities which would deter potential attackers. Efforts have been made to develop a theory and practice of "cyber deterrence" on the lines of nuclear deterrence.

Drawing analogies from the nuclear arms control vocabulary, it is argued that both denial and punishment are essential for deterring cyber aggression. The idea is to make it clear to the potential attacker that the cost of cyber aggression will outweigh the benefits. An effective cyber deterrence strategy will include deterrence by denial as well as penalty by punishment. Deterrence by denial will rely on strong defences. The efforts of the attacker would be rendered futile if defences and resilience i.e. the capabilities to bounce back are strong. Deterrence by punishment, on the other hand, relies on the ability to counter-attack. It is argued that if the attacker knows retaliation would be "certain, severe and immediate", it will deter him.

The question is whether cyber deterrence can work in the way similar to nuclear deterrence. Nuclear deterrence works because both sides know fairly accurately the nature, size and scope of each other's nuclear arsenal and the means of delivery. Over decades, arms control negotiations were focussed on issues such as transparency and verifiability of each other's arsenals. Detailed nuclear Confidence Building Measures (CBMs), based on verification, were developed. Attempts were made to understand each other's nuclear doctrines. In the nuclear case, actors were few – non-state actors did not possess nuclear weapons. In cyberspace, the situation is vastly different. As yet, there is no clarity even on what cyberattack means. There is no agreed definition of a cyberweapon. There are no means of verification. Multiple actors operate in cyberspace with complete anonymity.

Sceptics point out that cyber deterrence will fail because of the lack of attributability in cyberspace. In cyberspace, where anonymity is the key, it is difficult to identify precisely who the attacker is. Non-attribution is the fundamental weakness of the cyber deterrence argument. There is, however, some literature which suggests that the problem of attribution may be overcome sooner or later. Such claims are, however, unverifiable at present.

For cyber deterrence to be meaningful, a state would have to define its thresholds through appropriate signalling. It would need to indicate its cyber thresholds. Some ambiguity will no doubt be deliberate. Yet, a potential attacker must know that retaliation would be severe and unacceptable if a redline is crossed. Indicating redlines will depend upon a country's capabilities, intents and interests. Today, however, the redlines are absent. For instance, should cyberespionage, directed against military and non-military targets, be treated as an act of cyberwarfare? Is an attack on the banking networks, stock exchanges, power grids an act of war? Does cyberespionage merit a counter-attack? Should retaliation be in cyberspace or by other means? With key questions unanswered, to have a cyber deterrence on the lines of nuclear deterrence seems difficult.

The *Tallinn Manual 1.0*, originally called the *Tallinn Manual on the International Law Applicable to Cyber Warfare*, deals with conflict scenarios in cyberspace where international law would apply. While *Tallinn Manual* is not an official document, its work is sponsored by the North Atlantic Treaty Organisation (NATO) and other countries. Presently, a second version of the *Tallinn Manual*, *Tallinn Manual 2.0*, is being worked out. It deals with the application of international law to cyberspace during peacetime. A recent meeting held in The Hague on February 2-3, 2016 dealt with these issues. During discussions, attempts were made at defining a diplomatic law for cyberspace. It was suggested that attack on the computer systems of a foreign embassy should be prohibited by law. It was also professed that intervention in cyberspace may be permitted under certain circumstances.

From India's point of view, *Tallinn Manual*, while being a useful exercise, does not reflect the existing law on the subject because of the absence of state practice which is critical for the development of customary international law.

These difficulties notwithstanding, states are going ahead with the incorporation of cybersecurity into their military doctrines. Such doctrines postulate that a state, exercising its right to defend itself, could retaliate to a cyberattack by cyber or any other means. The 2015 US National Strategy says that the US could use cyber tools or other means to retaliate against cyberattacks.

The problem of cyberattacks cannot be seen in isolation. Today, cyberspace is intertwined with other domains of warfare, namely land, water, air and space. This intertwining implies that cyberattacks will not be seen merely as that. The retaliation in non-cyber form i.e. retaliation through non-cyber means including possibly military means cannot be ruled out. Cyberattacks, as means of warfare, would only enlarge the battle domain. Cyberwarfare may induce states to opt for full-spectrum deterrence.

Cyberwarfare is a contested concept. Cyberespionage, attack on critical infrastructures, etc. are routine happenings in cyberspace. So far, military means have not been used to deter attacks. Nor have economic sanctions been used because attributing a cyberattack has been so difficult. Further, many victims feel shy of reporting cyberattacks. Such incidents have not been regarded as acts of warfare thus far because no definition of cyberwarfare exists. Whether a cyberattack is seen as a component of cyberwarfare will depend upon the context of the attack. The authors of the *Tallinn Manual* have grappled for many years to come up with some acceptable definitions, but the progress has been slow.

India cannot be oblivious to these developments. Internet usage is spreading rapidly in India. Even though Internet penetration in the country is still low, nearly 400 million people are using the Internet. Digital India will take broadband Internet to every village Panchayat. With one billion SIM card subscribers, a revolution in connectivity is sweeping India. India's future progress and growth is linked with the expansion of the digital network, overcoming digital divides and ensuring that robust cybersecurity policies are adopted right from the beginning.

India has taken several steps in the recent past to strengthen its cyber defensive capabilities. To mention a few:

- A national cybersecurity policy has been announced and is being implemented.
- An elaborate national cybersecurity assurance framework is under implementation.
- The National Cybersecurity Coordinator was appointed in 2015.
- Coordination amongst various agencies has improved.
- A National Critical Information Infrastructure Protection Centre (NCIIPC) has been set up. There is a regular dialogue with the key sectors of the economy.
- Public-private partnership is being constructed. There is an active dialogue between the government and the private sector.
- A National Cyber Coordination Centre (NCCC) is being set up.
- Efforts are being made to develop cybersecurity skills in the country. New cybersecurity curricula are being introduced in the colleges.
- Cybersecurity Research and Development (R&D) policy has also been under active consideration of the government.
- The Indian Computer Emergency Response Team (CERT-In), an organisation that was set up in 2004, has done significant work in dealing with cyber incidents as well as spreading awareness.
- India is pursuing active cyber diplomacy by setting up cybersecurity dialogues with several countries and is participating in several international fora including the UN on cybersecurity.

All these synchronised and coordinated efforts are already showing results. But we cannot be complacent in the face of growing threats and evolving technologies. Due to the explosive growth of ICTs, cybersecurity scenario is likely to remain challenging. We will need to work hard on the various aspects of cybersecurity including the emerging challenges.

Like other countries, India also faces the daunting task of stopping and preventing cyberattacks on its networks. India will have to closely study the evolution of cyber deterrence idea. Building cyber deterrence capability would entail building robust networks that can be defended, encouraging comprehensive R&D in the area of cybersecurity and strengthening indigenous manufacturing of ICT products. It will also require strong cyber diplomacy to ensure that India is not at the receiving end of the emerging ICT Export Control regime under the Wassenaar Agreement. We also need to closely analyse the patterns of cyberattacks against us and build suitable response measures including the capability to conduct cyber operations if required. India would need to take note of the increasingly assertive cybersecurity doctrines that are being adopted by other countries. This will help in working out our own cybersecurity doctrines.

In conclusion, I would like to point out that there is a lack of consensus in the international community on norms of behaviour in cyberspace. We are at a stage where technology is far ahead of our thinking on cyber laws and cyber norms. The UNGGE has proved to be a useful platform to discuss these issues, but the absence of a broader representative platform where contentious issues can be hammered out and consensus arrived at is conspicuous by its absence. *Ad hoc* groups adopting *ad hoc* procedures to deliberate over *ad hoc* cybersecurity agendas will not necessarily build a consensus. The international community needs to come together to discuss how to deal with threats in cyberspace which are growing by the minute. The task may seem daunting but states should seriously reflect whether the world needs a Cyber Convention on cybersecurity. Unlike the other commons, namely the land, sea and space, wherein international law has grown immediately, cyberspace is still largely lawless. Sustained discussion by international experts is necessary to generate ideas on the way forward towards building a consensus on cybersecurity issues.

# 2

# Middle Powers and Cyber-Enabled War: The Imperative of Collective Security

*Greg Austin*

## Introduction

For a quarter of a century, middle power governments have been outclassed by more nimble small powers in coming to terms with the full import of the digital revolution transforming the world. This consistent edge of certain small powers has been captured in the annual Network Readiness Index prepared for the World Economic Forum (WEF), the 2015 version of which saw Singapore and Finland ranked yet again at the forefront (1 and 2, respectively), ahead of Sweden, the Netherlands and Norway (at 3, 4 and 5, respectively), with no G20 countries in top five.[1] This laggardness in cyber readiness among most middle powers has been particularly visible in the defence sector even though the world's only superpower, the United States, began a clear transition in its cyber-military ambitions in the mid-1990s. For example, in the case of India, in 2012, an Institute for Defence Studies and Analyses (IDSA) Task Force recommended that "India must raise a cyber command".[2] By August 2015, *The Times of India* reported that the Modi government was preparing to implement the recommendation, but the newspaper observed that "India has been quite slow to respond to the ever-expanding military challenges and threats in space and cyberspace".[3]

There has been little effort in public by middle power governments to benchmark national security needs in cyberspace in the same way as they benchmark naval, air and ground capability against strategic needs (strengths and weaknesses of potential enemies and their intentions) and balance those against national budget constraints. Most of these countries have high-profile national debates about whether they need a sovereign capability to build naval combat ships, submarines, or fighter aircraft but have been relatively silent on the type of national cyber innovation system they need for future warfare. To help stimulate such a debate, this chapter attempts to use a benchmarking approach across a combined set of political, economic, military and technical issues.

While recognising the limitations of benchmarking, the chapter sees the value of such an exercise as helping to evaluate a country's current levels of performance, build political pressure for change, identify gaps, expose other countries' pathways to success and find innovative approaches.[4]

The first main section of this chapter lays out selected aspects of what two countries of high strategic interest – China and the United States – have done or may be planning to do in the 10-20 year time frame. The second section previews trends and characteristics of cyber-enabled war, systems for attack and defence, asymmetric warfare, distributed warfare and scenario planning. In summary, the benchmarks reviewed in the chapter are derived from the sources listed in Table 1.

**Table 1: List of Benchmarks Discussed**

| |
|---|
| **Future National Defence Postures** |
| China: cyber power intent, cyber S&T intent, distributed cyberwar, militia |
| United States: prompt information dominance, cyber weapons for all, R&D innovation, military education |
| War avoidance and peace building |
| **Future 'Cyber-enabled War' Trends** |
| Future technologies of complex cyberattack and defence (multi-vector, sustained, cyber + kinetic) |
| Technologies of decision-making |
| Scenario planning |

A comprehensive study of middle power national security needs in cyberspace relying on such benchmarks would require more research, expertise and time. Therefore, the value of this chapter is more in its pointing to the need for, and potential scope of more comprehensive, public domain studies.

The discussion in the chapter is introduced by a necessary review of the boundaries of "national security in cyberspace" or "cyber-enabled war". Many policy documents are not as consistent or rigorous in differentiating keys aspects of this as they might be. The North Atlantic Treaty Organisation (NATO) Cyber

Cooperative Defence Centre of Excellence notes: "There are no common definitions for Cyber terms – they are understood to mean different things by different nations/organisations, despite prevalence in mainstream media and in national and international organisational statements."[5] In many policy documents, the term "cyber security" is too often used as a catch-all to avoid specific public elaboration of concepts like cyberwar and cyber-effect operations.

## National Security Needs in Cyberspace: Information Dominance

The term "cyberwar" is shorthand for a phenomenon that is not easily captured in a single term, much less one that may have shared meaning for people involved in national security policy around the world. The inadequacy of the word "cyber" as a prefix is illustrated quite well by the title of the book, *Cyber War Will Not Take Place*.[6] The book depends, as its main argument, on a narrow interpretation of the term, "cyberwar", as one limited to operations in cyberspace. As such, the argument is defensible but the number of countries actively preparing for what most of us call "cyberwar" is growing. They obviously believe that something like cyberwar or war in cyberspace may take place. The only way to get around this lack of terminological precision in the word "cyber" is for each publication that uses it to say how it understands the term.

There has to be a clear distinction made between "cybersecurity" on the one hand and, on the other, discussions of military and defence needs in cyberspace. The latter encompasses the former, but is very different from it and involves a much larger canvas of policy.

That said, it is worth reflecting on the concept of cybersecurity. It has at least eight "ingredients" or foundation elements, some of which are narrowly technical (but which all involve human input and institutions) and others which are simultaneously technical and deeply dependent on the character of the non-technical ecosystem. One view of these ingredients is captured in Figure 1, which describes them as vectors of attack and response, each of which can involve civil or military targets or actions.

The meaning of each of the terms describing a vector is self-evident, but "ecosystem" is worthy of recalling as we consider a national security environment. The ecosystem for military purposes must be understood to include the entire "infosphere", including attack and defence systems of potential or actual enemies and allies. This approach leads us towards cyber impacts on warfare and war planning beyond those involving traditional notions of "cybersecurity" (involving computer software and hardware).

**Figure 1: A Cybersecurity Model[7]**

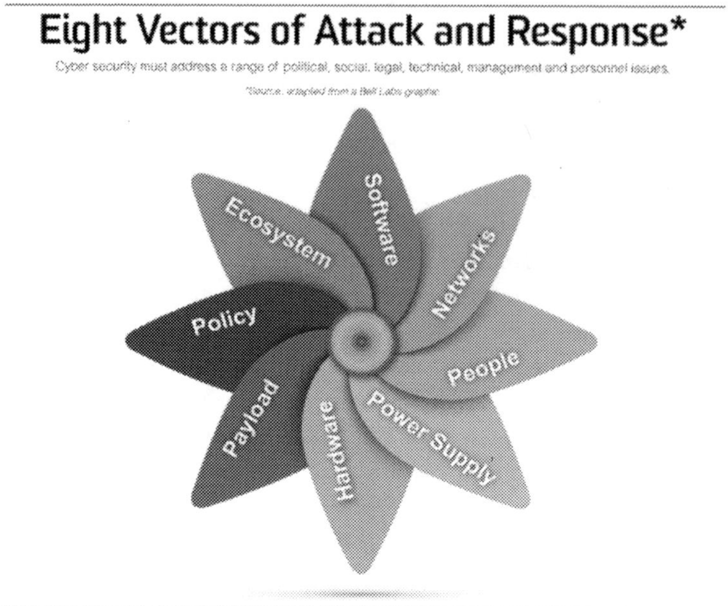


But this approach does not go far enough. It is a framework developed by engineers to address problems of protection of information and information systems largely at the enterprise or network level and in peace time. Though the eight-ingredient framework is a useful departure point for broadening our understanding of what shapes security in cyberspace in the military sphere, the framework does not do justice to wider political, economic, legal and social aspects of war fighting in the kinetic space. All military strategy and planning for any kind of war depend on the political, economic, legal and social environment as much as they do on engineering, systems management or capability-based approaches.

For this reason, we are obliged to understand the term "cyberwar" (adapting Clausewitz) to refer to the continuation of politics through cyber means with warlike intent. Cyber means must involve "machine-based computation" with or without support from kinetic military capabilities (missiles, bombs, guns). But "cyberwar" independent of the non-cyber domain, is as Rid argues, probably unimaginable. This chapter therefore see the interests of national defence planning as better served by using a concept like "cyber-enabled war", since war of any kind is an act involving the political, economic and civilian resources of states, as well as their military technological resources. We must also note that most developed countries depend on computers and IT-based communications systems for the targeting and operation of all modern missiles, bombs and guns. One of the best scholarly analyses of the impact of the information age on war may be a 2009 paper by Amit Sharma.[8] He addresses the strategic impact on war fighting goals of the information age in a way that few scholars have done, even in leading think tanks in the United States.

For this reason, among others, the use of the term "cyber-enabled war" in this chapter should not be seen as conforming in definitional terms to the meaning either of the term "information operations" or the term "cyberspace operations", as used by the US Joint Chiefs in their doctrinal publications,[9] since these US terms are intended only to convey the scope of military operations that do not by themselves constitute the totality of a state's strategic objectives and actions in any war, including in cyber-enabled war. The US Government avoids public discussion of concepts like cyberwar, even though, as is clear later in this chapter, the current administration assigns an overwhelming centrality in its military strategy to cyberspace.

A unifying element between the concept of "cyber-enabled war" and "information operations", is the concept of "information dominance" as the principal organising objective of national security policy (preparation for war) in the information era. Both the United States and China have used this concept but not always with the consistency one might expect.

In sum, the author does not see cyberspace as a separate domain[10] of military, social, economic or political life. It cuts across all domains. Cyberspace governs all economic, social, scientific, business and medical activity dependent on any sort of computerised record-keeping or more complex analysis. In military affairs, cyberspace encompasses the entire fabric of strategic command and control, weapons systems, battle space management and intelligence dissemination, on which national military security depends. Cyberspace unifies all domains of warfare, especially its political control and its political impacts.

Moreover, the US Joint Chiefs have identified three layers of policy and operational activity in cyberspace: physical, logical and the "persona", but go further by integrating these into consideration of the environments (informational, operational and political), and considerations like the relationship between information operations and cyberspace operations, and the involvement of the private sector.[11]

## International Trends in Planning for Cyber-enabled War

The national security needs of middle powers should be shaped above all by the threats or opportunities emanating from more powerful countries or non-state actors of military significance to them. This involves comprehension not only of their intent and capabilities of today but also of their likely intent and capabilities in the future. This section of the chapter looks at two cases: China and the United States. Since national security is a balance between political, economic, military and social considerations, any estimate of how the two great powers impact the security needs of other countries in cyberspace must address the full spectrum of national security: economic as well as military. The economic and social bases of national security include a country's national industry base, its scientific and technical potential, and the skills of its people. The political setting is also an essential determinant of war policy, so this section concludes with a review of the dominant trend in war policy globally: that of war avoidance in a situation of cyber arms racing.

## Cyber Military Policy in China

China is a country of immense national security interest to most middle powers, not least because of its economic weight and value as an economic partner. Chinese leaders accept the view that the cyber age is revolutionary in its impact. In February 2014, President Xi told his country and the world that the government would do everything needed for the country to become a cyber power.[12] As analysed in my book, *Cyber Policy in China*, this announcement came almost 15 years after China first committed itself to the goal of what it called informatisation: the maximum exploitation of advanced information and communications technologies to all walks of life, including military power and internal security.[13] The Xi announcement was intended by him to convey the view that China was lagging badly in cyber capability across a broad range of civil and military missions and interests and that it would henceforth work much harder to catch up.

**Cyber Power Intent:** In September 2014, Xi told the country it needed a new cyber military strategy. In December 2014, the government introduced new regulations for cybersecurity intended to help promote the rapid growth of China's domestic cybersecurity industry. In May 2015, the country issued a new Military Strategy in which the government declared for the first time in such a document the idea that "outer space and cyberspace have become new commanding heights in strategic competition among all parties".[14]

Since declaring his intent in February 2014 to do whatever was needed for China to become a cyber power, President Xi and his government have been hyperactive on all relevant fronts: political, legal, economic, organisational and diplomatic. Leadership attention to this set of issues became even more focused in May 2014 when the United States indicted five Chinese military personnel for cyber espionage involving commercial secrets of US-based corporations.[15]

Today, China is among the G20 countries with a very high level of government commitment to transform itself to exploit the information revolution. Most middle powers can learn from that level of commitment. In 2015, the WEF ranked China at 25th in the world in terms of the importance of Information and Communications Technologies (ICTs) in government vision of the future.[16] India was ranked at 71st on this criterion.

**Cyber S&T:** The scale of the China's ambition to become a world leader in the Science and Technology (S&T) base of cyber power is documented in a 2011 plan by the country's Academy of Sciences, called "Information Science and Technology in China: A Roadmap to 2050". The vision is staggeringly ambitious and complex. It sees China approaching the frontiers of science, economics and social organisation in the sphere of information technology by mid-century.

One impetus for the 2011 report was a strategy document, "Technological Revolution and China's Future: Innovation 2050", from the Academy of Sciences which served not just as an overarching mobilising document, but also marked the launch of a series of 17 subsequent sector-based roadmap reports also looking ahead to 2050. The 2009 foundation report on innovation, which had involved some 300 Academy researchers and experts for more than a year, recommended that China prepare itself for a new revolution in S&T in the coming 10 to 20 years in green energy, artificial intelligence, sustainable development, information networking systems, environmental preservation, space and ocean systems, and, most interestingly, national security and public security systems.

This all means that China as an economic and military actor in cyberspace is determined to look and feel very different in 20 years' time. For the benchmarking exercise in this current chapter, we need therefore to ask: how in the next 20 years will China change its S&T profile in cyberspace and how middle powers can benefit from that or otherwise secure their national security interests in respect of China? This benchmarking leads not just to an academic comparison of estimated static national capability at given intervals, but also provides an insight into a dynamic policy process inside China in which other countries may seek to intervene to shape China's choices to meet their own strategic interests. This has diverse aspects, not least in respect of shaping normative behaviour about cyberwar but also in respect of mutually advantageous development of both internationalised and exclusively sovereign S&T capability in both countries.

On current indications, within 20 years, China's civil economic and military capabilities in cyberspace will likely be very far ahead of most countries. A "great leap forward" by China in cyberwar S&T relative to middle powers is inevitable given China's current wealth and its scientific and industrial capability. These other countries cannot do much about that. But they must prepare now to respond to the likely impact over the longer term of Beijing's higher commitment in the past

15 years to transformation through military cyber S&T compared with their own lack of commitment in key areas of policy over the same period.

**China's Concept of Distributed Warfare:** In spite of undoubted successes in cyber espionage by the People's Liberation Army (PLA), China had until recently moved quite slowly to adjust to the opportunities and challenges presented by cyberwarfare.[17] As mentioned above, it has now made a series of new commitments and taken innovative measures to make the transition more quickly. Among these measures has been a move to joint or unified commands on the model of the US armed forces. This has been based on the strong conviction of Chinese specialists, learning from their American counterparts, that maximum exploitation of and defence against cyber assets can only be assured through inter-service operations and advanced command control systems, which in turn are integrated with space-based surveillance, intelligence and targeting capability (C4ISTAR).[18] It will take China a decade or two to bed down this transition.

Against the background of this perceived need to centralise command and control, and given China's past practices of clinging to outmoded patterns of national level command and control, including compartmented intelligence collection, it is all the more remarkable that it has in 2015 also committed to a countervailing doctrine that accepts the unique characteristic of cyberwar called "distributed warfare". This is discussed later in the chapter as a general phenomenon of high importance to any advanced country, but its application in the Chinese case is worth calling out. This is the principle that the operational combat environment of cyber-enabled war provides new opportunities for lower level formations widely dispersed to achieve strategic impacts in quite distant theatres. It also captures the consideration that the cyber environment places a premium on decapitation of superior level command authorities and even of basic communications systems in such a way that lower-level combat units may need to fight without the benefit of continuous communications and intelligence feeds.

For China, recognition of this concept of distributed authority at the same time as it is moving towards unified command centralisation is all the more remarkable. It has been captured in a turn of phrase in the 2015 military strategy: "You fight your way, I fight my way" in Section 3, "Guidelines of Active Defence": "The armed forces will adhere to the principles of flexibility, mobility and self-dependence so that 'you fight your way and I fight my way'. Integrated combat forces will be employed to prevail in system-vs-system operations featuring information dominance, precision strikes and joint operations."[19] The two sentences presented together make plain the need for self-dependence even in operations intended to achieve information dominance.

In practical terms, it will take some 5 to 10 years for China to develop its forces to any meaningful capability in this direction, but when it does achieve such a capability it will be at a scale that dwarfs that of most smaller or less wealthy countries.

One practical implication of the shift in Chinese doctrine might be that any country operating with or against Chinese forces may well face isolated warships or ground force units acting confidently but with superseded orders and/or degraded intelligence assets because they have been cut off from superior echelons. This circumstance would not be desirable in fast-moving combat regardless of whether one is fighting with or against China.

**Militia:** China has two levels of reserve forces: what might be called normal reserve forces (reasonably well trained personnel and units that can be mobilised for combat anywhere in the service of the country); and far less-trained militia units which are normally assigned to civil defence tasks in their own locality. China has been developing cyber military capabilities in some militia units. While this might be construed as related to civil defence tasks in the home province, such as protection of cyber aspects of critical infrastructure, the character of cyberwar is far different from kinetic warfare in that the latter has always been shaped by geographic proximity to one degree or another. Since this civil defence function of militia has been revived and professionalised by Chinese leaders in the past decade and since the Chinese Government has developed a massive internal surveillance and communications take-down capability based on cyber assets, China is exceptionally well placed to develop the most powerful and best-organised cyber militias in the world. It does not now have such a strong capability but it has taken steps along this path.

One added reason for China to develop cyber militia for integration into strategic and operational military tasks in wartime or in preparation for war is that it can draw on a massive pool of personnel in the civil work force who have high skills in their normal employment, in contrast to the PLA and reserve forces which will probably not have large numbers with the necessary skills on a scale that can compete with the US forces for several decades. In the 10 to 20 year time frame, China's capability in cyberwar will need to be assessed against the certain availability of a skilled workforce that no Western country could easily marshal in support of state policy short of an all-out declaration of war and general mobilisation.

Middle powers with a small highly trained cyber work force in uniform can usefully learn from the Chinese conditions that could, in a 10 to 15 year time frame, create a unique and powerful cyber militia capability.

## Cyber Military Policy in the United States

**Prompt Information Dominance:** The United States has a military strategy premised on information dominance as the foundation for what it calls "prompt global strike". This is a strategic objective in war, not just a tactical or theatre-level ambition. In conformity with this strategy, the United States is investing heavily in military uses of cyberspace and undertaking a rapid transformation of its forces. In 2015, the Pentagon issued a new Cyber Strategy[20] and the Commander of Cyber Command, Admiral Mike Rogers, issued a new planning document, titled "Beyond the Build".[21]

In US planning, "cyber effect operations" (CEO) in wartime seek to impair the confidentiality, integrity or availability of not just the machines but the data contained therein. This can include penetrating enemy intelligence systems and altering the information about one's own forces or even information about the disposition of the opposing country's forces. A Presidential Directive says that the United States will seek to apply CEO in all spheres of national activity affecting war, diplomacy and law enforcement.[22] It says that offensive CEO (OCEO) "can offer unique and unconventional capabilities to advance US national objectives around the world with little or no warning to the adversary or target and with potential effects ranging from subtle to severely damaging". The Pentagon Law of War Manual issued in June 2015 says it is lawful for a country in wartime to undertake pre-emplacement of "logic bombs" in an enemy country's networks and information systems.[23]

But there is a deeper dimension to the US concept of cyberwar beyond "information operations" or COE. It relates to the role of information and how a country's military power and strategic impact in war can be magnified by cyber means. In November 2012, the US Joint Chiefs of Staff issued a new joint training manual on "Information operations".[24] It identified the information environment as the aggregate of "individuals, organisations, and systems that collect, process, disseminate or act on information". This is a strategic-level orientation in which the United States aims above all else to disrupt the enemy's decision-making as a prelude to and adjunct for kinetic operations: the integrated employment during military operations of information capabilities "in concert with other lines of operation, to influence, disrupt, corrupt, or usurp the decision making of adversaries and potential adversaries while protecting our own". Cyberspace operations, covered in part by a separate military doctrine (Joint Publication) under that rubric, provide a sub-component to information warfare strategy.[25]

There are significant innovations in the 2015 policy statements from the Pentagon, including recognition in "Beyond the Build" that cyber defences in the Department of Defense (DoD) are weaker than the threats it faces and that military units must be able to operate with degraded systems and a lack of cyber situational awareness (including command and control, intelligence and targeting data).

The most important lesson from the 2015 "DoD Cyber Strategy" is that to be effective in cyber-enabled war a country needs to plan for it, structure its forces accordingly, train them for it and develop the foundations for public engagement in it. The strategy document makes plain that there are many foundations of cyberwar that need to be out in the open, ranging from critical infrastructure protection to industry-based research and development (R&D) and shaping a civilian cyber work force. The document makes plain that any country intent on fighting a cyber capable adversary will be more effective the more it can talk publicly about the main elements of the strategy.

By comparison, there has been little recognition among middle powers of the novel, arguably central role, of cyber-enabled warfare. There has also been no recognition of the value of public engagement in devising cyberwar polices. Of some note, for example, as of January 12, 2016, the term "cyber effect" does not appear to be found anywhere on the Australian Department of Defence website, except in a submission for the Defence White paper by this author. It is more than likely that the concept is well known in development work in the Australian Defence Force (ADF) and that the ADF has already conducted cyber effect operations of some kind.[26] On the UK Ministry of Defence website there is not a similar aversion to the term, though one finds quickly a plea on September 24, 2015 by the current Secretary for Defence in the UK, Michael Fallon, to "put cyber front and centre of our thinking".[27]

**Cyber Weapons for All:** In spite of billions of dollars spent, new forces and command entities raised, and military education and recruitment revamped, the United States recognised in 2015 how far it had yet to travel. On June 3, the Commander of US Cyber Command, Admiral Mike Rogers, observed as follows:

> Our task is to make this domain understood by other warfighters and integrated into broader military and governmental operations while providing decision-makers and operational commanders with a wider range of options while resources are constrained and threats are growing.[28]

In the short report, titled "Beyond the Build: Delivering Outcomes through Cyberspace", Adm. Rogers emphasised the need to be able to offer commanders and policymakers "cyber tools in all phases of operations" and an increase in momentum in building both "capacity and capability". One report of a large project on US decision-making for information operations found that, "the DoD does not yet understand how to measure the decision-making agility of a cyberspace operations organisation".[29] These concepts rarely receive a public airing in the debates of middle powers.

**R&D Innovation:** One key element of US national policy is its recognition of the need to "leverage the nation's ingenuity through an exceptional cyber workforce and rapid technological innovation".[30] It was expressed in just that language in the 2011 DoD strategy which was the predecessor to the 2015 strategy.

In 2011, this was held up as one of the Department's five principal strategies for cyberspace. The language was not picked up in the same way in the 2015 strategy which has a much sharper focus on operational aspects of the cyberwar problem. Yet the centrality of the civil sector underpinning of the country's cyberwar capability is visible through the 2015 document. The references from the 2011 strategy provide a more concentrated expression of the set of issues involved in innovation policy for defence purposes, and these are highly relevant to middle powers. It makes the obvious commitment to catalysing new education opportunities in a situation of high and unmet demand: "Catalyse US scientific, academic, and economic resources to build a pool of talented civilian and military personnel to operate in cyberspace." But it says that its plans in this area of skill development will be paradigm changing and will include the private sector:

- Streamline hiring practices for its cyber workforce.
- Exchange programmes to allow for "no penalty" cross-flow of cyber professionals between the public and private sectors to retain and grow innovative cyber talent.
- Adopt and scale cross-generational mentoring programmes.
- Develop reserve and national guard cyber capabilities.
- Infuse an entrepreneurial approach in cyber workforce development.
- Preserve and develop DoD's intellectual capital.
- Replicate in the DoD the dynamism of the private sector.
- Harness the power of emerging computing concepts (especially speed and incremental development rather than a single deployment of large, complex systems).
- Create opportunities for small and medium-sized businesses and entrepreneurs to move concepts rapidly from innovative idea, to pilot programme, to scaled adoption across the DoD enterprise.
- Emphasise agility, embrace new operating concepts, and foster collaboration across the scientific community.

Thus, for the United States, the national goal of ensuring military competitiveness in cyberspace depends on a "paradigm changing" approach to innovation, national education and work force development, which will then be reflected in paradigm changing approaches to military workforce development and deployment. There is almost no evidence in the public domain that any middle power has such a comprehensive view of how to make this paradigm shift.

**Military Education**: On a narrower military front, we can look at the education of junior officers and the role of their officer training academies in cyber policy development. The US Military Academy at West Point is arguably the most advanced in all aspects. Here are some highlights:

- The first undergraduate institution certified by the National Security Agency in 2001 as a "Centre of Excellence" in Information Assurance Education.
- Ranked ninth among more than 5,000 tertiary education providers in the United States in 2014 in terms of quality of education in cybersecurity.
- A Cadet Cyber Enrichment Programme offering internships in industry.
- A Cyber Leaders Development Programme providing up to 800 hours of non-academic training for each cadet.
- A community outreach programme where cadets teach local students cybersecurity.
- An Army Cyber Institute (cyberwarfare research and teaching, set up in 2014; planned for 75 members of staff by 2017, funded in excess of $ 20 million) which involves cadets in its work.
- Co-publisher of the journal, *Cyber Defense Review* (launched in February 2015).
- Cyber Research Centre (in the Electrical Engineering and Computer Science Faculty).
- Host of the first Joint Service Academy Cyber Security Summit in May 2015.

The United States Air Force Academy (USAFA) has extensive cadet-based programmes in cyber security, outer space operations and broader challenges of technological and management innovation. Its Centre of Innovation combines a range of disciplines pertinent to the broader information revolution in civil affairs or the revolution in military affairs.[31] The US Naval Academy has an undergraduate major in Cyber Operations.

## War Avoidance and Peace Building

At the beginning of this section of the chapter, I have noted that the national security needs of middle powers in cyberspace should be shaped in large part by the military intent and capabilities of other actors. But these needs must also be driven, above all else, by the goal of war avoidance. Diplomacy and politics are the main tools in war avoidance. Therefore the trend in global politics toward or away from confrontation in cyberspace or on cyberspace issues should be a major driver of national security planning for middle powers. The global trend on this front is mixed, with both increasing tensions and stepped-up efforts to reduce tensions. The seriousness of this consideration should not be underestimated.

The importance can be illustrated at one extreme end of the spectrum by developments involving the nuclear forces of Russia and the United States.[32] In June 2013, Russia and the United States agreed to set up a cyber-risk reduction centre (a hot line) staffed by technical specialists inside the existing bilateral nuclear risk reduction centre. Its purpose was to allow the two countries to exchange information on cyber incidents that might impinge on nuclear military readiness. This was an important development in the bilateral cyber military relationship.[33]

Yet in December 2014, Russia's revised military doctrine declared that the US cyber-enabled strategy of "Prompt Global Strike" is one of Russia's four main military dangers.[34] The Russian Government had been a little more circumspect in its 2010 doctrine with the statement that the only military nuclear threat it faced was "disruption of the functioning of its [Russia's] strategic nuclear forces, its systems of missile warning and control in outer space or of nuclear munitions storage facilities".[35] By October 2014, Russia had already acted on its increased concern about cyber threats, including through the deployment into its strategic missile forces of cyber defence units for the first time.[36] This link between cyber risk reduction and nuclear threats goes a long way to explaining rhetoric like a "cyber Pearl Harbor" used by former CIA Director Leon Panetta in October 2012.

The case of United States/China relations on military uses of cyberspace is also very important in terms of overall strategic stability. According to the few authoritative sources available, China's military leaders are deeply disturbed by the US policy of prompt global strike and see it as new evidence of muscle-flexing and dominating behaviour.[37] This concern is aggravated by perceptions of inadequacy in cyberwarfare capabilities relative to the United States, and a sense in China of profound weakness in the face of the information and electronic warfare power of the Americans' global alliances. The core diplomatic challenge is how to manage the asymmetry in cyber military power (which will persist for some time) without falling into a new Cold War.[38]

At the other end of the spectrum of cyberspace interaction among states are a string of cooperative measures over more than decade in multilateral and bilateral settings. These include agreements in the G20, Asia Pacific Economic Cooperation (APEC), the Association of Southeast Asian Nations (ASEAN) Regional Forum and various iterations of a Group of Governmental Experts (GGE) under the auspices of the United Nations (UN) looking at international security aspects of information and communications technology.[39] (The details of the UN GGE report are discussed later in the chapter.) By 2015, the overwhelming message of these initiatives was that global and national economic stability, as well as plain good governance, depend on constraining state-on-state cyberattacks in peacetime.[40] An equally important objective of these efforts has been to contain cyber probing and attacks in order to prevent unintended conflict escalation. The management of these issues was seen as a protracted but feasible process in an environment where all major powers not only see war amongst them as highly unlikely but also hold up this view as a major plank of policy.

The most surprising moves in 2015 were bilateral, between China and Russia and between China and the United States.

On May 8, China and Russia concluded a formal agreement with Russia not to interfere unlawfully in each other's information resources and networks.[41] (In January, China and Russia had participated in tabling a slightly revised draft of the proposed code of conduct for cyberspace initially submitted to the UN in 2011.)[42]

By signing the new bilateral agreement in May, China and Russia together appeared to have pre-empted the advisory effect of the 2015 GGE report, and its recent predecessors, to give legal effect to some of the principles proposed. The bilateral agreement also goes very close to constituting a formal military alliance in cyberspace, since it lays out a mutual obligation of assistance in the event of a wide range of cyberattacks.

The Russia/China agreement is a fulfilment of a decade of involvement by the two countries in cooperative measures on cyberspace governance, including through the Shanghai Cooperation Organisation (SCO) talks beginning in 2006. The 2015 agreement formalises, at a bilateral level, the intensifying multilateral effort building of the proposal in the UN system for a code of conduct in cyberspace. The agreement is as much about that effort as it is about strengthening each other in the face of US cyber pre-eminence. Article 1 describes malicious use of cyberspace "as a fundamental threat to international security". Article 4 commits the two countries not to undertake actions like "unlawful use or unsanctioned interference in the information resources of the other side, particularly through computer attack".

This is not a commitment to refrain from all use of military cyber assets against each other. Neither Russia nor China regards cyber espionage or preparations for war in cyberspace as "unlawful" or "unsanctioned". Of some note, Article 6.2 commits both parties to protect state secrets of the other, and references a prior bilateral treaty with that precise effect dating from May 24, 2000.

In early September 2015, in advance of a state visit by President Xi Jinping to the United States, China sent the Politburo member with responsibility for its non-military spy agencies, Meng Jianzhu,[43] to Washington for several days of official discussions to try to dampen controversies within the United States about the norms of cyber espionage.[44] This was at that time the high point in direct official contact on the subject resulting from a robust diplomatic campaign by the United States. This campaign reached a new peak in March 2013 when National Security Adviser,

Thomas Donilon, made public demands on China to abide by rules of the road prohibiting cyber espionage for commercial purposes.[45]

The Meng visit was highly productive, with the two countries agreeing not to conduct commercial espionage against each other for the benefit of their own companies and to set up a Cabinet-level working group for problem solving on cyber security issues from a law enforcement angle.[46]

Just weeks earlier, the UN had published the report of the fourth GGE.[47] With Chinese representation in the group, this report marked a new stage in intergovernmental consensus on some related issues, including most importantly the endorsement of a range of possible "voluntary, non-binding norms, rules or principles" for restraint in international cyber practices.

The 2015 GGE report proposed three potential "voluntary non-binding norms" for state behaviour in cyberspace:

- States should not attack each other's critical infrastructure for the purpose of damaging it.
- States should not target each other's cyber emergency response systems.
- States should assist in the investigation of cyberattacks and cybercrime launched from their territories when requested to do so by other states.[48]

As promising as these moves in the direction of restraint and war avoidance have been, they only begin to scratch the surface of what is needed. There is no commitment among the major military powers of the world to any idea of military sufficiency in cyberspace or the idea of the security dilemma (the concept that when a state strengthens its own military power such action can have the unintended effect of weakening its security because it prompts military rivals to increase their capabilities).

## Cyber War: Trends and Technologies

In 2009, Martin Libicki, one of the most respected scholars of cyberwarfare, concluded in a report he wrote for the United States Air Force that "strategic cyberwar is unlikely to be decisive" and that "operational cyberwar has an important niche role but only that".[49] In 2012, Thomas Rid and Peter McBurney from King's College London made an important distinction between target-specific cyberweapons that may be high value in terms of effect and those of more general application (not target-specific) that are of lower value in terms of effect.[50] They say there is a clear penalty involved in developing the high-value weapons which "increase the resources, intelligence and time required for development and deployment" and are "likely to decrease the number of targets" and the "political utility of cyber-weapons".

These assessments are sound, but they must be interpreted against the definitions of "cyberwar" or "cyberweapon" that the authors use. Libicki's definition of cyberwar is a narrow one (does not involve "real" war, that is a physical one),[51] and Rid and McBurney define a cyberweapon as "computer code that is used, or designed to be used, with the aim of threatening or causing physical, functional, or mental harm to structures, systems, or living beings".[52]

There are at least three important dimensions of the policy problem presented by cyber-enabled war that such assessments do not take into account:

- Will the cost/benefit relationship in technical development and use of cyberweapons change in the 10-20 year time frame?
- Will the political character of a cyberweapon change as countries accumulate entire cyber arsenals, rather than single cyberweapons?
- Does the political character of a cyberweapon change as countries move away from conventional military strategies to information age strategies where information dominance is judged to be the decisive capability?

My answer to all three questions would be yes. Over time, the conclusions by Libicki, Rid and McBurney are likely to be less relevant. For the purposes of this chapter, we must note the highly dynamic character of the policy field represented by cyber-enabled war as countries accumulate capability, as technological options expand, and as key governments of interest continue to move decisively toward information dominance as an over-arching military strategy.

One of the best descriptions of the trends may be a 2011 book, *America the Vulnerable*, by the former Inspector General of the US National Security Agency, Joel Brenner, who takes a distinctly non-technical approach and accords political and economic underpinnings of war and strategy a higher place than most specialists on cyberwar.[53] While not agreeing with a number of his conclusions, I would like to illustrate the preceding point by calling out his understanding of how China has reacted to US and Allied capability for information operations over the time since the first Gulf War in 1991 with a deepening and quickening attention to cyberwarfare.[54] This is laid out in different parts of the book, and is essential for understanding that the technologies and strategies of cyber-enabled warfare are not static – anything but! One essential takeaway from the Brenner book is that cyberwar as a real-life phenomenon (budgets, soldiers, politicians, industry and war-fighting) is only in its infancy and that it may be about to mature very quickly.

Another essential conclusion from the Brenner book is a very stark one that has grave national security implications for war planning by middle powers. Brenner concludes that his country "cannot defend our [its] electronic networks that control our energy supply, keep aircraft from colliding in midair, clear financial transactions, or make it possible for the President to communicate with his cabinet secretaries".[55]

For any country, as cyberwar capabilities of potential adversaries expand, those highly vulnerable aspects of cyber civil infrastructure that underpin military preparedness, including mobilisation of forces through civil airspace, also become more likely targets.

Brenner aptly titles his first chapter, "Electronically Undressed". If the United States cannot defend its critical infrastructure in cyberspace at present, and it cannot, and if the world is on the edge of a rapid expansion of cyberwarfare capabilities by the most powerful countries, this would appear to have implications which middle powers should publicly acknowledge and to which they should more consistently provide appropriate responses.

Brenner outlines a suite of policy measures, most of which are highly reasonable. While few address issues of war fighting capability or strategy, all of them represent potential contributions to national security preparedness in cyberspace. For example, he calls out the need to move toward highly secure computing ("verifiable software and firmware") by promoting public support for research in this area. The implications of this transition are spelled out at length in a 2014 paper from the EastWest Institute, which argued that governments have tolerated for too long the exposure of their security to vulnerabilities of the sort outlined by Brenner.[56] This EastWest paper called on governments to "send clear [market] signals to enable security-driven IT innovation, starting top-down with the highest security requirements in the highest value targets". As importantly, it urged governments to "cooperate internationally to realise this new paradigm quickly and to stem the evolution of high-end cyber attackers before they can inflict more damage".

Brenner's book, *America the Vulnerable,* is but one of many sources indicating the scale of the challenge in national security arising from the rapidly intensifying transformations of the information age. A policy framework that is slow, incremental and largely oblivious to the emerging trends of cyberwar (the sort of framework most middle powers have had) will fail badly. For this reason, the September 2015 document, "Beyond the Build", issued by the Commander of the US Cyber Command, Adm. Rogers, and referred to above, must stand as a summary indicator of where all countries seeking to maximise national security and cyberwar planning must head in the 10-20 year time frame.

We might note two US assessments, one official and one from an academic source:

- *US Worldwide Threat Assessment 2015*: "2014 saw, for the first time, destructive cyberattacks carried out on US soil by nation state entities ... we must be prepared for a catastrophic large scale strike – a so-called cyber Armageddon ... unpredictable instability is the new normal."[57]

- *Emerging Cyber Threats Report 2015*: "Low-intensity online nation-state conflicts become the rule, not the exception."[58]

## Special Features of Cyberwar

Leaving aside the great powers preparations for cyber war for a moment, there are other important politico-strategic aspects of war in cyberspace that have relevance regardless of the country involved and which are also likely to evolve in the next 10-20 years in ways that middle powers should take into account.

First, there is the new potential offered by cyberspace for asymmetric warfare by weak military powers (and non-state actors) against states that are clearly superior in conventional (kinetic) military terms. While this concept has been present for a long time, it is not a static phenomenon but changes with advances in technology. According to a study compiled by the US Director of National Intelligence (DNI) in 2000, of likely threats to 2015, asymmetric warfare was then the first of only three likely military threats faced by the United States.[59] The other two were strategic threats from weapons of mass destruction and regional conflict threats. The DNI report defined asymmetric conflicts as those in which "state and non-state adversaries avoid direct engagements with the US military but devise strategies, tactics, and weapons – some improved by 'sidewise' technology – to minimise US strengths and exploit perceived weaknesses". By 2015, the DNI confirmed in its annual worldwide threat assessment just this prognostication, though in slightly different words. Listing "cyber" first in its list of threats, the DNI concluded: "The likelihood of a catastrophic attack from any particular actor is remote at this time." He said that the more likely threat, rather than one that debilitates the entire US infrastructure, would be "an ongoing series of low-to-moderate level cyberattacks from a variety of sources over time, which will impose cumulative costs on US economic competitiveness and national security".[60]

Second, there is opportunity for "distributed" warfare, a capability (and arguably a practice) that will become more pervasive over time. In simple terms, distributed warfare is the translation of the national-level use of coercive power to disconnected individual units. It mirrors the same decentralisation of political authority that has been visible in the use of social media to break down the power of authoritarian regimes in places like Egypt or the Hong Kong Special Administrative Region. There are several ways of understanding this phenomenon in respect of military applications. One is to look at the role of patriotic hackers, whose potential in warfare may be likened to partisan forces capable of disrupting an enemy but which are either affiliated loosely with their home government or not connected at all, often acting against its interests or express wishes. Patriotic hacking is an important and evolving phenomenon in several regions, especially East Asia (most notably in China, South Korea and Japan).

Another dimension of distributed warfare is the contribution and role of cyber militias, people who have a civilian day job but who can be directed by the national government at short notice to participate in national security activities, including cyberwar if need be. As noted above, China has an active programme for developing cyber militia units, but also relies on its unique political system to co-opt companies and firms. The United States does not have such an explicit reliance on cyber militias, in part because it has an established network of high-tech companies who can be quickly and easily paid to feed into US national security activities if need be. According to the government, it has at least 10,000 cleared companies it can consult for advice on highly classified technical aspects of the country's intelligence needs.[61]

The above forms of distributed warfare are challenging enough to national security operations, demonstrated not least by the Snowden affair in which one of these paid employees was able to use his individual "network power" to blow open some of the most sensitive aspects of US cyberwarfare capability and preparations. The Snowden revelations on Operation Prism, which implicated nine leading US corporations in direct and large-scale involvement in US national security missions in cyberspace, produced an even more damaging outcome in that these companies reacted by distancing themselves from any political subordination to or co-optation by the national government as participants in distributed cyberwarfare capability. Microsoft, for example, has made plain its position that it treats all clients equally, including the United States and China.[62] This reverse positioning of US corporations away from integration into the distributed warfare assets of the government was evident when the US-based company, Symantec participated in the analysis of and revelations about Stuxnet, leading to the disruption of that live US intelligence operation and subsequent exposure of it.[63] One implication of this is that middle powers must continually evaluate any presumption they make about the security affiliations and dispositions (patriotic or neutral) of foreign corporations, not to mention domestic ones.

But the biggest challenge presented by distributed warfare is the fundamental change in the relations between a central command authority and its deployed units. In an era of information dominance and concrete enemy plans for decapitation of command and control, whether through cyber or kinetic means, all military units must now have ways of re-connecting with each other if key links in the central chain are broken. As mentioned above, China has responded to this much more explicitly than most countries in its recent military strategy (May 2015). It foreshadows a lessening of central command authority to foster the conditions of victory in cyberwar under the rubric of "self-dependence" for individual military units ("you fight your way and I fight my way").

One of the political consequences of distributed warfare and its asymmetric potential is that it may also break down the traditional value of military alliances, especially the provision of extended deterrence. Some countries benefit from the technical support of its intelligence alliances, especially the Five Eyes,[64] in preparing for cyberwar and conducting information operations. Even in that case however, while the UK, Canadian, Australian and New Zealand forces enjoy considerable integration into advanced command and control arrangements with the US forces for operations, there is however considerable evidence to suggest that their reliance on the United State for extended deterrence may not have as much impact in cyberspace as for kinetic operations. In NATO, the United States has agreed with its alliance partners that an attack in cyberspace can constitute an armed attack for the purpose of invoking mutual response under Article 5 of the treaty. The question however is whether cyber incursions of a warlike character or preparatory to war would in practice attract US defensive action. It is more than likely that middle powers allied to the United States would have to plan for a higher degree of self-reliance in cyberspace than in maintaining kinetic military capability because the recognition of thresholds of incursion or assault in cyberspace is far less developed and far more ambiguous than in kinetic scenarios. There is little room for doubt about intent when several bomber aircraft of one country penetrate the airspace of another without prior clearance. This would constitute a threat of armed attack. The same clarity is not yet in place for cyber incursions.

## Future Systems of Attack and Defence

Trends in technologies for cyberattack and defence have been described in many places: from government agencies, scholars, vendors, netizens and hackers. Those of significance for benchmarking national security needs range across all eight vectors of the "cyber flower" described in Figure 1, but they also include those that cut across and combine the individual vectors. These might be called "systems of systems" technologies. The scale of the challenge in forecasting technologies of attack and defence systems should not be dumbed down by any one person's understanding of security in cyberspace. The first thing that strikes a policy analyst coming to the question from a neutral, non-specialist position is the immense diversity of estimations about future technologies of attack and defence systems. There is also the consideration that novel (disruptive) cyber technologies will emerge and be deployable at short notice, in time periods as short as a matter of days in terms of warning.

From the point of view of benchmarking the national security needs of middle powers, this chapter has chosen to highlight just a few ideas of future systems that are not particularly prominent in public discussion by officials or among specialists in middle powers.

If one looks narrowly at the typical security specialist's horizon, the characterisation of threat development around complex cyberattacks is a useful place to start. In 2015, a US-based analyst, Carl Herberger, the Vice President of Security Solutions at Radware, reported that in 2013 the average cyberattack he had observed involved seven attack vectors (though some had reached over 25 attack vectors), different phases (each with several waves), with successive phases relying on methods that worked in the previous phase but adding new attack vectors.[65] This was rather well captured in a FireEye presentation in 2013 which listed four characteristics of the emerging threat landscape: coordinated persistent threat actors, dynamic polymorphic malware, multi-vector attacks and multi-phase attacks.[66]

These characterisations are very important benchmarks. But they don't take us as far as we need to look. They address only a narrow slice of the eight vectors of attack, and don't say a lot about defensive systems.

As one leading international example of future defensive systems, we might look at the topic of critical infrastructure protection and the acknowledged world leader in cyberspace defence of it, the Idaho National Laboratory (INL). The focus of this work is not military battlefield systems, but it provides many benchmarks for development of battlefield systems and for defence policy makers and military leaders who must be able to depend on certain critical infrastructure. After all, there is no victory in war without survivable critical infrastructure. That is one meaning of the word 'critical'.

We can take the case of electric power supply which is just one of eight ingredients of cybersecurity. Not only is it controlled by digital assets, but it is possibly the most ignored vector of attack and response in cyberspace. This was the subject of testimony of an Associate Director of INL, Mr Brent Stacey, on October 21, 2015, which is extracted verbatim below and which gives any country considerable cause for concern:

- The presumption that a control system is "air-gapped" is not an effective cyber security strategy. This has been demonstrated by over 600 assessments.
- Intrusion detection technology is not well developed for control system networks; the average length of time for detection of a malware intrusion is four months and typically identified by a third party.
- As the complexity and "interconnectedness" of control systems increase, the probability increases for unintended system failures of high consequence – independent of malicious intent.
- The dynamic threat is evolving faster than the cycle of measure and countermeasure, and far faster than the evolution of policy.
- The demand for trained cyber defenders with control systems knowledge vastly exceeds the supply.[67]

The sort of defensive response outlined by INL is also quite instructive. It has identified a three-tier approach rendered verbatim below:

(a) **Hygiene:** "the foundation of our nation's efforts, composed of the day-to-day measure and countermeasure battle"; "important routine tasks such as standards compliance, patching and password management"; "primarily the role of industry, with both vendors and asset owners participating".

(b) **Advanced persistent threat:** "the more sophisticated criminal and nation state persistent campaigns"; requiring "a strategic partnership with industry and government"; "these roles are still evolving"; "ICS-CERT[68] provides critical surge response capacity and issues alerts of current vulnerabilities to the government and asset owners".

(c) **High impact low frequency events:** "catastrophic and potentially cascading events that will likely require substantial time to assess, respond to, and recover from. This level is primarily the responsibility of the government."

Research at INL focuses on the two high priority tiers [(b) and (c) in the list above], aiming for a "two-four-year research-to-deployment cycle" and to "achieve transformational innovations that improve the security of our power infrastructure by reducing complexity, implementing cyber-informed design, and integrating selected digital enhancements". The laboratory "is pursuing a grand challenge to develop novel and deployable solutions to take a set of high value infrastructure assets off the table as targets". This programme assumes pervasive insecurity: It promotes "a paradigm shift in the methods used to historically develop control systems. This paradigm is predicated on the fact the traditional trust relationships in peer communications are no longer a satisfactory assumption. Instead, a resilient control system design expects a malicious actor or actions to be part of normal operation and is designed to mitigate such actions".[69]

Most middle powers have no comprehensive effort that remotely matches the approach adopted by INL, and in fact much of their governmental effort is spent on the lowest priority tier [(a) in the list above] identified by INL: the cybersecurity hygiene of operators and enterprises.

A 2012 UK analysis provides some additional insight into the processes threatening cyber resilience of another aspect of critical infrastructure, the financial services sector.[70] The study was based on consultation with industry. Interviewees identified as one of the top three technology risks the "development or emergence of new technology and poor change management in relation to new technologies".[71] A 2013 academic study on a similar subject warned against the danger of estimating risks in isolation from each other: "Estimation of CPS[72] risks by naively aggregating risks due to reliability and security failures does not capture the externalities."[73] It called out "biased security choices" that "reduce the effectiveness of security defences". Looking to future threats, it warned that CPS "are subjected to complex risks, of which very little is known despite the realisation of their significance".

## Technologies of Decision-making

High performance computing, a technology that is well established though rapidly evolving, is being seen increasingly as an essential tool of cyber defence management at the national military level, as well as a new weapon in the hands of adversaries. A 2014 paper from Sandia Laboratory lays out a future "technology of decision-making" based on high performance computing[74] that might usefully be understood by analogy as an attempt to create for cyberspace, as a global civil domain, an upscaled version of the global strategic C4ISTAR system for U.S. command of its strategic nuclear weapons, including indicators and warning.

The study took as a core operating principle the proposition that the cybersecurity terrain for national decision-making is a "continuous lifecycle with human, organisational, legal, and technical interdependencies". It identified seven high priority "wide-area problems" in the field of cyber security that have high relevance to middle powers in understanding its technologies of decision-making for cyber-enabled war. These priority problems, listed verbatim, are as follows:

(a) Disjointed response to wide-area and multi-target attack.
(b) Widely dispersed and fragmented detection and notification capabilities.
(c) Ill-defined government, commercial and academic roles and responsibilities.
(d) Divided and rigid wide-area cyber protection posture.
(e) Unresolved wide-area common and shared risks.
(f) Fragile interdependent wide-area critical access and operations.
(g) Unresolved attribution of attack and compromise.

The authors concluded by recommending areas for further research in high performance computing to support national security decision-making for cyberspace.[75]

It is unsurprising that US Government laboratories have the remit and resources to take on such challenges, and that scientists in middle powers do not have the same opportunities. Several middle powers do have well developed assets for research in and application of high performance computing, but one might reasonably assume that these have not been rigorously applied to the special demands of decision-making for cyber-enabled war at the strategic level.

If we translate the ecosystem of threat and defence implied by the mere handful of trends in technology (and response to those trends) mentioned above, we can only conclude that middle powers are staring down the barrel of almost insurmountable challenges unless they are able to develop complex responsive systems of decision-making for medium intensity war that address simultaneous multi-vector, multi-front and multi-theatre attacks in cyberspace by a determined enemy, including against civilian infrastructure and civilians involved in the war effort. And all of that before we even think about emerging technologies like quantum computing, anti-satellite weapons, mass deployment of drones as distributed airborne C4ISTAR platforms, a return to traditional HF-based communications for cyber activities, and laser-based communications.[76]

## Scenario Planning for Cyber-enabled War

There are many components to planning, funding and training a defence force for the future. One of the most important is the intelligence foundation: What are other countries doing and planning to do? What might they do in certain circumstances based on what we know? How might future technologies affect their military strategies? These are the sorts of issues canvassed above. An additional tool is that of scenario development, which is especially useful where uncertainty about the intelligence available will be high, as will be the case in cyber-enabled warfare. The value of scenario planning is widely appreciated by most military leaders, though not often exercised in respect of cyber-enabled warfare.

The merits of scenario planning, as summarised by two research scholars for a NATO-related cyber conflict conference, are worthy of mention.[77] There are classic elements, such as the elucidation of likely geopolitical environments, but these scholars also see merit in cyberspace scenarios for their ability to tease out alternative responses to future technologies and in creating a stimulus to change among policymakers and managers. They also call out, as Adm. Rogers has done, the value of providing a common 'language' and doctrinal approach to possible future trends in cyberwar. Above all, the authors highly recommend the use of scenarios as a concrete tool for reducing strategic surprise ("reduction of the impact of uncertainty through the notion of 'robustness'").

Most middle powers have been involved in scenario planning for civil cyber emergencies. Few have published details of scenarios for cyber-enabled war, but there is no shortage of demonstration scenarios. As one example, in late 2014, the United States government conducted an exercise, Cyber Flag, with a wide number of scenario elements[78] that have not been present in similar public domain announcements in most middle powers about their preparation for cyber-enabled war. These included:

- Joint force response to a regional crisis involving significant cyber military activity.
- Full spectrum military operations (with "cyber plus kinetic" combat goals).
- Alliance cyber operations with air, land and naval forces.
- Operating while being subjected to cyberattacks affecting national command and control.

This type of exercise scenario is useful but, like most scenarios, it has specific training and development purposes that need to be limited to the development stage of the forces involved in the exercise. They do not necessarily reflect the totality of the type of situation (contingency) for which military planners at the executive level of government must prepare.

For the purposes of benchmarking international best practice in scenario development or contingency planning for cyber-enabled warfare, it would be important to undertake a detailed study since none seems to exist in the unclassified domain. But for the purposes of this chapter, it may be sufficient to note that defence planning at the national level, in terms of future war, would be the "kingdom of the blind" if a country did not have an agreed vision of the likely contours of a cyber-enabled war. For the United States, one of the most cited is the case of a military confrontation with China over Taiwan.[79] This is highly credible and involves wide-ranging cyberattacks against US civil infrastructure to prevent mobilisation of US forces or delay their deployment to the Western Pacific.

An alternative way of constructing a scenario would be to take the most notable incidents of state-sponsored and criminal cyber actions that might be most relevant to a particular type of medium intensity conflict and see how they might be combined to develop a scenario of relevance to particular countries. For a middle power, the list of possible attack vectors for cyber-enabled kinetic war would be long, but we can illustrate the scope by alluding to the following potential combination:

> Estonia 2007 (a shut down of the financial and banking system) + China's kinetic anti-satellite test 2007 + Stuxnet 2010 (cyber sabotage) + release by the group Anonymous of military personnel data + cutting of undersea cable (numerous incidents) + closing down of civil satellite links (Egypt) + closing down electric grids (US operation in Yugoslavia 1999) + insertion of false data into military systems + attacks on Saudi Aramco + planting malware in civil aviation systems + opening flood gates on dams + closing down military communications.[80]

Consideration of such scenarios leads us to only one of three possible conclusions about government policy in middle powers. First, medium intensity cyber-enabled war outlined in such a scenario may be such a remote possibility that we need not plan for it. Or, second, we have not studied it sufficiently to know or to have developed a national consensus on what type of cyber-enabled war we are most likely to face. Or third, we cannot regard cyber military policy in any country as mature. This author thinks that either of the second or third possible conclusions is more logical than the first. I lean to the third, but am prepared to credit the second subject to much deeper analysis by relevant agencies, scholars and think tanks.

A government's cyber military policy cannot be considered as mature until it has:

- an open and candid conversation in public with key stakeholders about the sort of threat scenarios our armed forces and communities may face in a medium intensity cyber-enabled war;
- developed defence policies and armed forces, supported by the civil sector, that could perform credibly in those scenarios given reasonable warning time;
- articulated a diplomatic strategy to reduce the risks of such a war if it looks like emerging;
- articulated a civil defence strategy for the inevitable high impact disruption of our civil economy and communities in such a war; and
- set in place policies for development of our industry base and work force that can support all of the above to the extent that our national economy permits and limitations of alliance support dictate.

## Conclusion and Recommendations

There are many departure points for benchmarking the national security needs of middle powers for cyber-enabled war. On the one hand, there are developing capabilities in countries like China and the United States, and middle powers therefore need to respond with their own sovereign capabilities. On the other hand, there is the important consideration that the technological trends in cyberwarfare present a high complexity problem that defies past practices of defence policy formulation in most middle powers. The type of problem seems much bigger than the collectivity of institutions normally involved in framing national defence strategies for middle powers.

Just where a country needs to position itself in this highly dynamic and complex environment (the information ecosystem) is something that only the collective wisdom of its best and most expert minds, working in partnership, can answer. A prerequisite though has to be an open and public debate on the country's military, security and civil needs in cyberspace and how well its emerging capabilities match those needs. Where this has begun to happen in a few middle powers (especially among members of the European Union, in India and in Australia), the conclusion has been that the countries in question are badly lagging.

While a large slice of the national security cyber domain must remain secret, the public in any country and its key actors in policy (private sector companies, state governments, foreign suppliers, military allies, citizens, civil society groups, lawyers, judges, security agencies, university researchers and educators) need to have a clear vision, in a number of private places and in public, of where the government is headed. This vision needs to reconcile competing demands of national security in the information age with each other and with other public policy demands, such as open trade, international investment, privacy, industry regulation or industry support.

Based on this chapter, governments of middle powers and their armed forces leadership might consider articulating a comprehensive set of policies around the following benchmarks:

- A national innovation strategy that keeps the country at the forefront of international best practice in cyber technologies that can be applied in war.
- A military strategy for cyber-enabled warfare that takes account of the proven and estimated character of such an armed conflict, including public intelligence assessments of likely cyberwar threats and a top-end (but credible) scenario.
- A strategy for sovereign cyberwar capability and cyber survivability in a time of direct military confrontation with a major power.
- A capital procurement programme centred on advanced cyber-enabled war capabilities, including space-based assets and new technologies of decision-making.
- A renovation of military institutions, training and education.
- Necessary investments in niche technologies and research capabilities.
- A strategy for managing civilian-military divides and critical infrastructure protection in times of military conflict.
- A strategy for mobilising cyber-capable reservists or civilians in times of military crisis.
- A sharp distinction between the national needs for cybersecurity as largely a civil domain set of issues and the needs for cyber-enabled war fighting capability.

Above all else, a middle power needs to build a community of interest around the concept of cyber-enabled warfare with a recognised authoritative hub that can unite political, military, diplomatic, business, scientific and technical interests and expertise.

One thing is crystal clear: middle powers will not make the necessary transitions for cyber-enabled warfare at all unless they make a number of new policy commitments and substantial institutional transformations very soon.

At the same time, no middle power need stand alone in constructing a national defence strategy and diplomacy that can help secure the country's security interests in cyberspace. This consideration has been driving work by the countries most passionately engaged in the UN GGE. If national defences in most countries are unlikely ever to rise to meet the potential threat of cyber-enabled warfare, the collective security response must remain a live and fruitful avenue for policy investment. There is ample evidence in this chapter that the tension between advancing threats and multilateral restraint still favours the threats and not the defences. This is likely to remain the case as long as other developments (such as Russia's action in Crimea or a threat dynamic between China and the United States) reinforce geopolitical divides without compensating collective defence measures.

Cyberspace, in spite of its military threat potential and high sensitivity for national security, may well be a locus for collective defence measures in the civil sector that can help to compensate for persistent or increasing geopolitical tensions either around military uses of cyberspace, or for other reasons. Such a conclusion is based in part on the symbiotic relationship between cyberspace and national economic stability of even the most powerful military actors. It is also based on evident commitment, for the most part, of all great and middle powers to war avoidance.

## NOTES

1. World Economic Forum (WEF), *Global Information Technology Report 2015*, Oxford University Press, New York, p. 8, at http://www3.weforum.org/docs/WEF_Global_IT_Report_2015.pdf.
2. Institute for Defence Studies and Analyses, *India's Cyber Security Challenges*, a Task Force Report, 2012, p. 46, at http://www.idsa.in/book/IndiasCyberSecurityChallenges.
3. "Govt Gets Cracking on Three New Tri-Service Commands", *The Times of India*, August 20, 2015, at http://www.defense-aerospace.com/articles-view/release/3/166156/india-plans-3-joint-commands-for-space,-cyber-and-covert-ops.html.
4. Sigurdur Helgason, "International Benchmarking: Experiences from OECD Countries", paper presented at a conference organised by the Danish Ministry of Finance on International Benchmarking, Copenhagen, February 20-21, 1997, p. 2, at www.oecd.org/governance/budgeting/1902957.pdf.
5. See the CCDCOE website: https://ccdcoe.org/cyber-definitions.html.
6. Thomas Rid, *Cyber War Will Not Take Place*, Hurst, London, 2013.
7. Graphic adapted from a Bell labs graphic and designed by Kurt Barnett, UNSW Canberra.
8. Amit Sharma, "Cyber Wars: A Paradigm Shift from Means to End", 2009, at https://ccdcoe.org/publications/books/VirtualBattlefield.pdf.
9. US Joint Chiefs of Staff (JCS), *Cyberspace Operations*, 2013, at www.dtic.mil/doctrine/new_pubs/jp3_12R.pdf; *and* US JCS, *Information Operations*, 2012, at www.dtic.mil/doctrine/new_pubs/jp3_13.pdf.
10. The choice by the United States Government and its armed forces to refer to cyberspace as a fifth domain is understandable from an organisational point of view. It was easier politically for the government to stand up its new Cyber Command as a separate command if it was

presented as an add-on to the existing single services not taking over key parts of them. But it is important to note that language around "fifth domain" is politically loaded and organisationally driven, rather than a statement of reality. The US decision to set up a national Cyber Command announced in mid-2009, was followed by formal establishment of cyber commands in the single services (air force in August 2009, marine corps in October 2009, navy in January 2010, army in October 2010). The USAF had been in lead with its efforts to set up a new cyber command beginning in 2006 but this was subsumed into the idea of setting up a unified command. The single service cyber commands now owe their loyalty as much to the joint Cyber Command and other unified commands as to the single services.

11. See US JCS, *Cyberspace Operations*, No. 9, pp. 2-8.
12. *Xinhua*, "Xi Jinping Leads Internet Security Group", February 27, 2015, at http://news.xinhuanet.com/english/china/2014-02/27/c_133148273.htm.
13. Greg Austin, *Cyber Policy in China*, Polity Press, Cambridge UK, 2014.
14. China, Information Office of the State Council, "China's Military Strategy", May 2015, at http://news.xinhuanet.com/english/china/2015-05/26/c_134271001.htm.
15. For an extended discussion of this topic, see Greg Austin, "China's Cyber Espionage: The National Security Dimension and U.S. Diplomacy", discussion paper, 2015, at http://thediplomat.com/wp-content/uploads/2015/05/thediplomat_2015-05-21_22-14-05.pdf.
16. See country profiles in WEF, No. 1.
17. This is discussed at length in Greg Austin, No. 13, Chapter 5.
18. C4ISTAR is a US military acronym standing for "command, control, communications, computers, intelligence, surveillance, target acquisition, reconnaissance".
19. See http://www.chinadaily.com.cn/china/2015-05/26/content_20820628_3.htm.
20. United States Department of Defence, "DoD Cyber Strategy", Washington DC, 2015, at http://www.defense.gov/Portals/1/features/2015/0415_cyber-strategy/Final_2015_DoD_CYBER_STRATEGY_for_web.pdf.
21. US Cyber Command, "Beyond the Build: Delivering Outcomes through Cyberspace", U.S. Department of Defence, Washington DC, June 2015, at http://www.defense.gov/Portals/1/features/2015/0415_cyber-strategy/docs/US-Cyber-Command-Commanders-Vision.pdf.
22. United States, The White House, "Presidential Policy Directive 20: U.S. Cyber Operations Policy", 2012, at http://fas.org/irp/offdocs/ppd/ppd-20.pdf.
23. United States Department of Defence, *Law of War Manual*, Washington DC, 2015, p. 995, at http://www.defense.gov/Portals/1/Documents/pubs/Law-of-War-Manual-June-2015.pdf. The chapter on cyber operations specifically allows for pre-placement in wartime of cyberweapons (often called time-release "logic bombs"): "Cyber operations can be a form of advance force operations, which precede the main effort in an objective area in order to prepare the objective for the main assault. For example, cyber operations may include reconnaissance (e.g., mapping a network), seizure of supporting positions (e.g., securing access to key network systems or nodes), and pre-emplacement of capabilities or weapons (e.g., implanting cyber access tools or malicious code)." While this statement in the manual refers to the US view of its own actions in wartime, it would also be regarded by most states as the applicable international law in peacetime.
24. US JCS, *Information Operations,* No. 9.
25. US JCS. *Cyberspace Operations*, No. 9.
26. This can be deduced from the public references to deployment of Defence civilian cyber specialists to combat areas in Afghanistan.
27. Michael Fallon MP, Speech to Cyber Symposium 2015, Paris, September 24, 2015, at https://www.gov.uk/government/speeches/cyber-symposium-2015.

28. US Cyber Command, No. 21.
29. Steven W. Stone, "Factors Influencing Agility in Allocating Decision-Making Rights for Cyberspace Operations", 20th ICCRTS Paper 096, June 2015, p. 1, at http://static1. square space.com/static/53bad224e4b013a11d687e40/t/54da5be5e4b0e9d26e577151/1423596517506/096.pdf. Cited with the author's permission.
30. United States Department of Defence. "Department of Defense Strategy for Operating in Cyberspace", Washington DC, 2011, p. 11, at http://www.defense.gov/news/d20110714cyber.pdf.
31. See the centre's website, at http://www.usafa.edu/df/dfe/dfer/centers/coi/.
32. For an extended analysis, see Greg Austin and Pavel Sharikov, "Preemption is Victory: Aggravated Nuclear Instability of the Information Age", Working Paper, January 2015.
33. For an extended discussion of the evolution of that relationship, see Franz Stefan Gady and Greg Austin, "Russia, the United States, and Cyber Diplomacy: Opening the Doors", EastWest Institute, New York/Brussels/Moscow, September 2010, at http://www.eastwest.ngo/sites/default/files/ideas-files/USRussiaCyber_WEB.pdf.
34. Russia, "Military Doctrine of the Russian Federation" (*Voennaya doktrina Russkoi Federatsii*), issued December 2014, at http://news.kremlin.ru/media/events/files/41d527556bec8deb3530.pdf.
35. Russia, "The Military Doctrine of the Russian Federation", February 5, 2010, translation, Carnegie Endowment, http://carnegieendowment.org/files/2010russia_military_doctrine.pdf. This statement is also in its 2014 doctrine; see Ibid.
36. RIA Novosti, "Russia Strategic Missile Forces Create Cybersecurity Units: Defense Ministry", October 16, 2014, at http://sputniknews.com/military/20141016/194157367/Russian-Strategic-Missile-Forces-Create-Cybersecurity-Units.html.
37. See Greg Austin, "Managing Asymmetries in Chinese and American Cyber Power", *Georgetown Journal of International Affairs*, "International Engagement on Cyber IV", October 2014, pp. 141-151; and Greg Austin, No. 13.
38. For an extended analysis of background to US/China relations on cyberspace issues and policy recommendations, see Greg Austin, "China's Approach to International Legal Norms for Cyber Space", in Anna Maria Osula and Henry Röigas (eds), *International Cyber Norms: Legal, Policy & Industry Perspectives,* NATO Cooperative Cyber Defence Centre of Excellence, Tallinn, 2016, 171-201; and Greg Austin and Franz Gady, "Cyber Detente between the United States and China", EastWest Institute, New York/Brussels/Moscow, November 2012, http://www.eastwest.ngo/idea/cyber-detente-between-united-states-and-china.
39. An extensive overview of these can be found in Greg Austin et al., "A Measure of Restraint in Cyberspace: Reducing Risk to Civil Nuclear Assets", EastWest Institute, 2014, pp. 13-14, at http://www.eastwest.ngo/sites/default/files/A%20Measure%20of%20Restraint%20in%20Cyberspace.pdf.
40. Progress in 2015 has been summarised in Greg Austin et al., "Promoting International Cyber Norms: A New Advocacy Forum", EastWest Institute, New York, December 2015, at http://www2.ewi.info/idea/slowing-cyber-arms-race.
41. For a text of the approved agreement, see Russia, Federal Government, "Order on the Signing on an Agreement between the Russian Federation and the People's Republic of China on Cooperation in the Field of Securing International Information Security" (in Russian), Moscow, April 30, 2015, at http://pravo.gov.ru/laws/acts/34/555656451088.html.
42. See United Nations, "Letter dated 9 January 2015 from the Permanent Representatives of China, Kazakhstan, Kyrgyzstan, the Russian Federation, Tajikistan and Uzbekistan to the United Nations addressed to the Secretary-General", January 13, 2015, at https://ccdcoe.org/sites/default/files/documents/UN-150113-CodeOfConduct.pdf.

43. Of special note, Meng controls the civilian spy agencies (Ministry of State Security for external intelligence and Ministry of Public Security for Domestic Intelligence). He does not control the main signals intelligence agency of China which sits in the PLA, under the control of the Central Military Commission of the Chinese Communist Party (CCP). Meng is the Secretary of the Central Political and Legal Commission of the CCP, one of the most powerful political bodies in the country because of its role is the protection of all aspects of the "political and legal" system in the country.
44. United States, White House, "Readout of Senior Administration Officials' Meeting with Secretary of the Central Political and Legal Affairs Commission of the Communist Party of China Meng Jianzhu", September 12, 2015, at https://www.whitehouse.gov/the-press-office/2015/09/12/readout-senior-administration-officials-meeting-secretary-central.
45. Greg Austin, "Cybersecurity: The Toughest Diplomatic Challenge Is China's Weakness", *The Global Journal*, April 2, 2013, at http://theglobaljournal.net/article/view/1049/.
46. For further analysis of the agreements, see Greg Austin, "Why the China-US Cyber Agreement May Prove Destructive", *The Diplomat*, October 7, 2015, at http://thediplomat.com/2015/10/why-the-china-us-cyber-agreement-may-prove-destructive/.
47. United Nations, "Report of the Group of Governmental Experts on Developments in the Field of Information and Telecommunications in the Context of International Security" (UN GGE 2015), July 22, 2015, at http://www.un.org/ga/search/view_doc.asp?symbol=A/70/174.
48. Ibid.
49. Martin C. Libicki, *Cyberdeterrence and Cyberwar*, Rand Corporation, Santa Monica CA, 2015.
50. Thomas Rid and Peter McBurney, "Cyber Weapons", 157 (1), *RUSI Journal*, February/March 2012, pp. 6–13, 6, at https://www.rusi.org/downloads/assets/201202_Rid_and_McBurney.pdf.
51. In his chapter six, Libicki looks at the relationship between the two forms of war.
52. Thomas Rid and Peter McBurney, No. 50, p. 7.
53. Joel Brenner, *America the Vulnerable: Inside the New Threat Matrix of Digital Espionage, Crime and Warfare*, Penguin Press, New York, 2011.
54. Ibid., pp. 117-9, 121-2, 125-7, 130-1, 131-36, 137-47, 153-4.
55. Ibid., p. 245.
56. Sandro Gaycken and Greg Austin, "Resetting the System: Why Highly Secure Computing Should Be the Priority of Cybersecurity Policies", EastWest Institute, New York/Brussels/Moscow, January 2014, p. 5, at http://www.eastwest.ngo/sites/default/files/Resetting%20the%20System.pdf.
57. United States Director of National Intelligence, "Remarks as delivered by The Honorable James R. Clapper, Director of National Intelligence, Opening Statement to the Worldwide Threat Assessment Hearing, Senate Armed Services Committee", February 26, 2015, at http://www.dni.gov/index.php/newsroom/testimonies/209-congressional-testimonies-2015/1175-dni-clapper-opening-statement-on-the-worldwide-threat-assessment-before-the-senate-armed-services-committee. The Clapper statement on cyber Armageddon was prefaced with the word "though", and he contrasted that future planning need with the current reality of a "constant and expanding barrage of cyberattacks for some time". He did this to highlight the seriousness of the latter, not to suggest that the cyber Armageddon was not a focal point of US planning from a defensive point of view. Clapper also said that the United States was still trying to decide how aggressive it should be in launching counter-attacks in cyberspace, but this can only have been a reference to peacetime rather than open hostilities.
58. Georgia Tech information Security Centre and the Georgia Tech Research Institute, "Emerging Cyber Threats Report 2015", 2014, p. 13, at https://www.gtisc.gatech.edu/pdf/Threats_Report_2015.pdf.

59. United States Director of National Intelligence, "Global Trends 2015: A Dialogue about the Future with Non-Government Experts", NIC2000-2, December 2000, at http://www.dni.gov/files/documents/Global%20Trends_2015%20Report.pdf.
60. United States Director of National Intelligence, "Statement for the Record, Worldwide Cyber Threats, House Permanent Select Committee on Intelligence", James R. Clapper, Director of National Intelligence, September 10, 2015, at http://www.dni.gov/files/documents/HPSCI%2010%20Sept%20Cyber%20Hearing%20SFR.pdf.
61. United States Director of National Intelligence, "Industry Snapshot: Summary of Partner Responses to the FY 2015-2019 IC S&T Investment Landscape", Washington DC, 2015, at http://www.dni.gov/files/documents/atf/In-STeP%20-%20Industry%20Snapshot.pdf.
62. This has been made plain repeatedly by Microsoft representatives speaking at international conferences attended by the author.
63. Kim Zetter, "How Digital Detectives Deciphered Stuxnet, the Most Menacing Malware in History", *Wired*, November 7, 2011, at http://www.wired.com/2011/07/how-digital-detectives-deciphered-stuxnet/.
64. The "Five Eyes" is an intelligence alliance between the US, UK, Canada, New Zealand and Australia arising from their collaboration in the Second World War.
65. See more at http://inspiratron.org/blog/2015/05/29/the-art-of-cyber-war/#sthash.LxJiSSIc.dpuf.
66. See http://www.exclusive-networks.be/wp-content/uploads/2013/11/FireEye-breakout-session.pdf.
67. United States House of Representatives, Science Subcommittee on Energy and Science Subcommittee on Research and Technology, "Written Testimony of Mr. Brent Stacey, Associate Laboratory Director for National & Homeland Security, Idaho National Laboratory", October 21, 2015, p. 3, at http://docs.house.gov/meetings/SY/SY20/20151021/104072/HHRG-114-SY20-Wstate-StaceyB-20151021.pdf.
68. ISC-CERT is an acronym for "Industrial Control Systems Computer Emergency Response Team".
69. See the website of Idaho National laboratory, at https://inlportal.inl.gov/portal/server.pt/community/distinctive_signature__icis/315/grand_challenge.
70. United Kingdom, Financial Conduct Authority, HM Treasury and the Bank of England, "Technology and Cyber Resilience Benchmarking Report 2012", London, 2013, at http://www.bankofengland.co.uk/financialstability/fsc/Documents/technologyandcyberresilience benchmarkingreport2012.pdf'
71. The other two were network and critical system outages, and access management and control of administration privileges.
72. Cloud Platform Services.
73. Saurabh Amin, Galina A. Schwartz and Alefiya Hussain, "In Quest of Benchmarking Security Risks to Cyber-Physical Systems", *IEEE Network*, January/February 2013, 19-24, 24, http://www.eecs.berkeley.edu/~schwartz/IEEEMag2013.pdf.
74. Curtis M. Keliiaa and Jason R. Hamlet, "National Cyber Defense High Performance Computing and Analysis: Concepts, Planning and Roadmap", SANDIA Report, SAND2010-4766, September 2010, pp.7-8, at http://prod.sandia.gov/techlib/access-control.cgi/2010/104766.pdf.
75. These areas for research were: trusted connection and automated processes; informatics, statistics and anomalous behaviour; mathematics analysis and intrusion detection; complexity science and emergent behaviour; modelling and simulation; analysis and correlation algorithms; sociology and psychology.

76. Robert Koch and Mario Golling, "Blackout and Now? Network Centric Warfare in an Anti-Access Area Denial Theatre", in M. Maybaum, A-M. Osula, and L. Lindström (eds.), *2015 7th International Conference on Cyber Conflict: Architectures in Cyber Cyberspace*, NATO CCDCOE, Tallinn, 2015, pp. 169-184, 178-180, https://ccdcoe.org/cycon/2015/proceedings/12_koch_golling.pdf.
77. Plausible Futures Newsletter, "The Use of Scenarios in Long Term Defence Planning", April 2007, https://plausiblefutures.wordpress.com/2007/04/10/the-use-of-scenarios-in-long-term-defence-planning/.
78. Bill Gertz, "Cyber War Games Held", *The Washington Times*, November 12, 2014, at http://www.washingtontimes.com/news/2014/nov/12/inside-the-ring-cyber-war-games-held/?page=all.
79. David C. Gompert and Martin Libicki, "Cyber Warfare and Sino-American Crisis Instability", *Survival*, 56 (4), pp. 7-22, at https://www.iiss.org/en/publications/survival/sections/2014-4667/survival--global-politics-and-strategy-august-september-2014-838b/56-4-02-gompert-and-libicki-04fc.
80. This sort of scenario was canvassed by Mark Seinor at an international conference on "Redefining R&D priorities for Australian Cyber Security" on November 16, 2015. See https://www.unsw.adfa.edu.au/australian-centre-for-cyber-security/cyber-r-d for a list of presentations, including that by Seinor.

# 3

# THE TRIAD THEORY FOR STRATEGIC CYBERWARFARE

*Amit Sharma*

"War is thus an act of force to compel our enemy to do our will."

—Clausewitz, *On War*[1]

"One hundred victories in one hundred battles is not the most skilful. Seizing the enemy without fighting is the most skilful."

—Sun Tzu, *The Art of War*[2]

Sun Tzu, the elusive military thinker of the East believed that the best form of warfare is the one in which there is almost no application of force[3] and Clausewitz, the military theorist of the West professed that the primary aim of the war should be to make the enemy submit to your demands.[4] Cyberwarfare draws on the essence of both, as it is a warfare that is capable of making the enemy to submit to your will, with almost no application of physical force. Hence, in order to elucidate the strategic aspect of cyberwarfare, a cyber strategy involving the conduct of warfare in cyberspace should be based on the directions of these great military theorists.

There is a paramount need of shattering the misinterpretations of information warfare. To address these needs, it is imperative that the framework for strategic cyber or information warfare should have a strategy, which will highlight the strategic aspect of cyberwarfare. This chapter defines a cyberwarfare strategy, which is based on destroying the Clausewitzian Trinity in cyberspace by conducting parallel warfare[5] to gain rapid dominance in cyberspace and generating a strategic paralytic effect on the victim nation called the *triad theory of cyberwarfare*. The *triad theory of cyberwarfare* provides ways and means to conduct strategic warfare in cyberspace, which will have a catastrophic effect especially on the information-dependent Western nations. The chapter will also analyze the effects of this warfare on the population of the victim nation based on the characteristics of Ulrich Beck's "risk society".[6] Based upon the cyberwarfare strategy, the analysis also illuminates the reasons as to why previous cyberattacks such as the Titan Rain[7] and the ones conducted on Estonia[8] and Georgia[9] could not achieve the strategic effect.

The chapter also elaborates a framework of strategic cyberwarfare involving *the triad theory of cyberwarfare, its operational cyber campaign plan and formation of a 'Known' and 'Credible' cyber deterrence* to generate a scenario of *Mutually Assured Destruction (MAD) in cyberspace* thus guaranteeing a *strategic status quo*; as the primary means of achieving grand strategic objectives in the contemporary world order. Once the strategic aspect of the cyberwarfare is established, the chapter concludes by providing various recommendations for creation of cyber deterrence capabilities especially involving the notion of *Prepare, Pursue, Protect and Prevent* in cyberspace to capitalise the strategic opportunities provided by this framework of strategic cyberwarfare.

## The Triad Theory of Cyberwarfare

> "A computer-based attack on the national infrastructure could cripple the nation more quickly than a military strike."
>
> —Robin Cook, former UK Foreign Secretary[10]

> "The results of cyber-terrorism … could be worse even than those caused directly by explosion."
>
> —Jack Straw, former UK Home Secretary[11]

The kind of ramifications, which Robin Cook[12] and Jack Straw[13] illustrated, may sound dire, but when a cyberattack is conducted with a well-laid strategy, then the *cascade effect* would generate ramifications far beyond the arithmetic benefits of a direct attack. In order to formulate such a strategy, Clausewitz's Trinitarian Warfare provides suitable inroads. Clausewitz believed that his wondrous trinity held the key to victory in wars[14] and his claims are well proven even in contemporary wars. His elusive trinity is composed of three dominant tendencies. He defined these three tendencies as "the primordial violence, hatred, and enmity, which are to be regarded as a blind natural force; of the play of chance and probability within the creative spirit is free to roam; and of its element of subordination, as an instrument of policy, which makes it subject to reason alone".[15]

He abstracted these tendencies as the military or means to fight a war; the people or support for war in terms of manpower and finances; and the political instrument or government to provide leadership and direction in a war. Over centuries these tendencies have interacted with each other, although their interactions and interrelationships have changed over time. Nevertheless, until the time all these three tendencies are active and are constantly interacting with each other, a nation can withstand any attack or hostile event of any magnitude. He predicted that even if anyone of the tendencies is completely destroyed, the trinity is resilient enough and the other two tendencies will revive it back and the trinity will survive.[16]

**Figure 1: The Notion of Trinity in Terms of Strategic Cyberwarfare**[17]

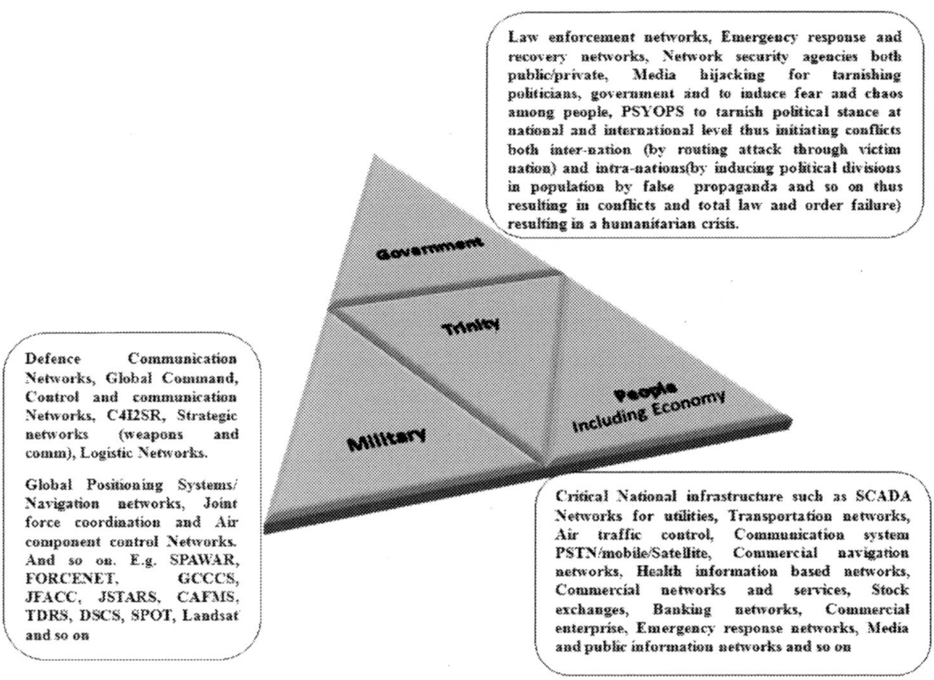


In the contemporary world, all of these three tendencies are extensively dependent on information and information assets in one form or the other (Figure 1). Denning elaborates that every aspect of modern nations, right from military systems to household essentials, everything is in one way or the other, dependent on information systems.[18] Goldman reaffirms that these information assets are not only adjunct, but have become a crucial arena for conflict.[19] As this reliance on information assets is increasing, the susceptibility of nations to strategic information warfare attack is also increasing.[20]

Under these circumstances, the three tendencies of the Clausewitzian Trinity are becoming extensively susceptible to strategic information warfare. Hence, if all the three tendencies are simultaneously attacked, or in conventional terms are subjected to parallel warfare in cyberspace, then it would generate a *cascade effect*, inducing strategic paralysis, and the victim nation would crumble as a system, resulting in chaos and mayhem. To analyse the effect as a whole, the effect on each individual tendency must be examined in terms of its susceptibility to such kind of a coordinated parallel warfare in cyberspace.

## Military: *The Means to Fight a War*

As mentioned earlier, information and information assets have become an integral component of almost every aspect of warfare. The last couple of years have seen a colossal change in the conduct of conventional warfare with information assets at its centre. The change in the conduct of militaries is often termed as information-enabled or information-enhanced warfare, which has proven its metal especially during contemporary conflicts such as the Operation Desert Storm (1991),[21] Operation Iraqi Freedom (2003)[22] and the Operation Enduring Freedom (Afghanistan, 2001),[23] acting as a potent force-multiplier capable of turning the tides of the war.

Modern military forces, especially in scenarios of global deployment and of joint/coalition warfare, are extensively dependent upon information assets for the mobilisation, coordination, and to an extent for operation. Militaries rely extensively on information assets in the form of systems like the Command, Control, Communication, Computer, Intelligence, Surveillance, Reconnaissance (C4ISR) systems; Strategic Information Dissemination Systems (SIDS); Net-Centric Warfare/Network Enabled Operations (NCW/NEO); Global Positioning System (GPS) for guidance and targeting; surveillance systems right from the theatre-level target acquisition systems involving drones, Battle Field Surveillance Radars (BFSR) and data acquisition networks, to strategic surveillance systems involving surveillance satellites, acquisition and analysis systems; communication networks involving theatre-level data links such as Link 16 to strategic level communication systems involving communication satellites and Global Command and Control System (GCCS); and so on.

These information assets have changed the way in which wars are fought and have acted as a force-multiplier, but this dependence has also enhanced the susceptibility of military systems to strategic information warfare. Most of these networks have a significant number of vulnerabilities and are susceptible to attack.[24]

There are numerous instances where military systems have been compromised. It is estimated that every year nearly 250,000 systems at the US Department of Defense are subjected to cyberattacks.[25] With the most notable being the *Titan Rain,* which involved a chain of cyber intrusions from the year 2003 to present day on US Defence establishment, with the source of attack traced back to China.[26] Although most of the commentators classify Titan Rain as a tactical espionage attempt, the author in *Chinese Zixunhua Budui: A myth or a reality,* believes that Titan Rain was not just a tactical espionage attempt, but part of long-term cyber intelligence to wage strategic cyberwarfare.[27] This strategic vulnerability is not limited to just military networks but over a period of time multiple cyberattack vectors have been developed and orchestrated on even military hardware such as drones,[28] air defence systems,[29] GPS networks,[30] satellites[31] and missiles command and control networks.[32] The primary defence cited by the militaries around the world against such types of attacks is the air-gapping of these systems from the internet, but the Stuxnet and Duqu malware[33] and air-gap jumping vulnerabilities and exploits such as LNK[34] have clearly demonstrated that such air gapping is futile as malwares with capabilities to hop between air-gapped networks are already in wild. Hence cyberattacks on military infrastructure is a viable option.

## People: *The will to Fight a War*

Similar to the tendency involving the military, the tendency involving the people is also extensively dependent upon information resources. This dependence can be seen throughout the world, but the scenario is even worse in developed Western nations where people are dependent on information assets to such an extent that they are considered an integral lifestyle component. Philippa Trevorrow argues that this reliance of modern societies on information infrastructures has resulted in the emergence of new threats and vulnerabilities.[35] Streltsov also reiterates that the dependence on information infrastructures has increased the susceptibility of economic, social and public administration infrastructures to strategic information warfare.[36]

Right from the basic utilities such as gas, electricity and water supply, through to intricate systems such as nuclear power stations and dams, are all based upon information assets and networks known as the Supervisory Control And Data Acquisition (SCADA) and Distributed Control Systems (DCS). Malwares such as the Stuxnet,[37] which targeted the Iranian nuclear reactor, and Havex,[38] targeting the industrial control systems and utilities providing companies in Europe, have clearly demonstrated the development of advanced cyberweapons, which are primarily designed to destroy critical infrastructure of nations. The recent cyberattack, which disrupted the Ukrainian power grid,[39] is a blatant instance of inevitability of the threat to the critical infrastructure.[40] Linden argues that these networks are susceptible to cyberattacks and hence pose a considerable vulnerability.[41] Forrest reaffirms this claim, categorising them as a national security threat.[42] The scenario is the same with the systems related to Air Traffic Control (ATC),[43] transportation systems, navigation systems and communication systems involving Public Switched Telephone Network (PSTN) and mobile networks.[44] Apart from these basic utilities, one of the key sector which is vulnerable to strategic information warfare is the economy, especially in the light of globalisation. Banking infrastructure and stock exchanges where money in the form of financial data depends extensively on information assets. Martin Libicki terms this sector to be a critical arena of information warfare, known as the economic information warfare.[45] Another aspect that further exacerbates the problem is the interdependence of these systems on each other. For instance, the global stock exchange networks are extensively dependent on time synchronisation for effective trading, for which they usually rely on GPS networks. Therefore, a spoofing attack on the GPS network would have indirect, but catastrophic impact on trading, possibly to an extent of a massive meltdown of the monetary exchanges, a phenomenon explained in detail by Todd Humphreys, University of Texas[46]. All these instances clearly elucidate the fact that cyberattacks on the *People* component of the Trinity will not only have a direct, but also indirect effects have with exponential impacts, in the form of unintended consequences.

There are numerous instances where these critical national infrastructures have been subjected to attacks, intentionally or unintentionally. The most notable being the shutdown of the Ohio nuclear plant due to slammer worm, a denial of service worm that crippled nearly 75,000 victims within ten minutes of its initiation;[47] the crippling of financial and political institutions in Estonia;[48] and Georgia;[49] the Titan Rain;[50] and the most recent one in July 2009 involved the crippling of systems at the New York Stock Exchange and at various government and security agencies situated across the US and South Korea.[51]

An important aspect in this context is the impact of such attacks on the population. To understand the nature of this impact, it is imperative that the state of the people be analysed first. Commentators such as Bill Durodie,[52] Frank Furedi[53] and Robert Putnam[54] suggest that the modern, especially Western, societies are becoming increasingly individualistic. People are becoming socially disconnected; politically disengaged; and are in scientific disbelief. This society, classically termed as Ulrich Beck's "risk society",[55] is continuously living in an atmosphere of fear, where "perceptions overweigh reality".[56]

A society that is in such a state and is completely dependent on information assets for almost every aspect of their life, the sudden disappearance of information resources in the event of a cyberattack would render people in a state of shock. For example, in a scenario where suddenly people find that they have lost all their money; are left with none of the basic utilities; and the safety of their near ones is at stake due to critical infrastructure failure of transportation networks and Air Traffic Control; and exacerbated by the fact that they cannot contact their near ones due to loss of communication system. Such a scenario would result the society to be thrown into chaos, resulting in mayhem and bedlam.

## Government: *The Political Direction*

In contemporary world, governments act as an instrument of political direction and hence play a critical role in steering the nation. This is achieved by exercising control and by gaining the mandate of the people. The former goal is achieved by means of effective law enforcement, whereas the latter is achieved by means of providing secure, secular and democratic environment to the people. An important aspect of this government-people linkage is the media. In this risk society media has become the sensory organ of the people and is acting as a vital connector between the government and the people. The "CNN effect" has been a potent tool in framing the perceptions and mental framework of general public and decision-makers, alike.[57]

The effect of information assets on this tendency are very much alike the previous two (Figure 2). For example, law enforcement agencies rely extensively on criminal databases and on various emergency response and recovery networks like the 911 service and police nets.[58] These networks are not only force-multipliers, but have also induced an inherent susceptibility to cyberattacks.[59] The scenario is the same with the government and political institutions especially with the evolution of concepts like e-governance,[60] and the excessive reliance of political leaders on the Internet, right from campaigning to the spreading of political manifestos and ideologies of the party or the "E-mocracy".[61] These networks and the media networks provide a perfect target for spreading disinformation and for tarnishing the image of the politicians, government and ruling elite, by means of Psychological Operations (PSYOPS) in cyberspace.[62] The PSYOPS in cyberspace have metamorphosed to a new dimension with cyberattacks providing new means to attack television networks like the hijacking of state TV network in China[63] and TV5Monde in France.[64] Their potential is further exacerbated with the advent of social media and attack vectors such as the use of social bots[65] for manipulation of public opinion in open sources, especially the social media networks, thus demonstrating the military use[66] in future conflicts.

These PSYOPS can be used to induce fear, chaos, misconceptions and division of perceptions within a country which could further result in political upheavals and mass movements, for example, the role Twitter played in the June 2009 upheavals in Iran[67] and Belarus.[68] The hijacked networks of the victim nation can be utilised for orchestrating cyberattacks on a third nation; hence they can attribute to conflicts of the victim nation at an international level. It is paramount that as part of the triad theory of cyberwarfare, this tendency should be destroyed, as the political instrument is the most dominant of the three and is significant in combing the means and efforts to achieve political ends.

**Figure 2: The Reliance of the Government on Information Assets and the Related Vulnerabilities to Strategic Information Warfare[69]**

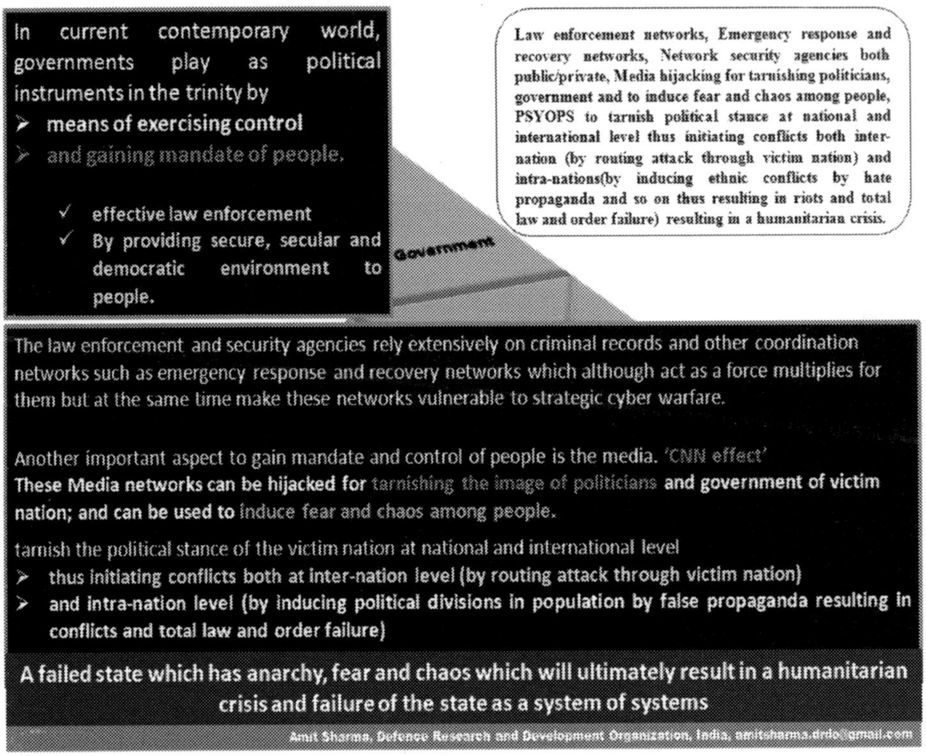


These three tendencies of the Trinity form the core of a nation, and hence to destroy a nation, it is important that all the three be destroyed. Their dependence on the cyberspace makes them a perfect target for strategic warfare in cyberspace. An important fact worth taking into account is the necessity to destroy all of the three tendencies simultaneously, as together they form a resilient system, which is capable of reviving any individual tendency target as part of the strategic information warfare.

There are numerous instances of coordinated cyberattacks, which failed to materialise the strategic effect just because they were targeting individual tendencies at a time; and since these tendencies form a resilient system hence the non-targeted tendencies usually recovered the tendency, which is targeted during the attack. The author believes that Cyberattacks such as the Titan Rain[70] and the ones launched on Estonia[71] and Georgia[72] could not achieve a strategic end primarily because they were tactical in nature and were targeting individual tendency rather than all three which resulted in their failure to achieve strategic paralytic effect on the victim nation.

**Figure 3: The Triad Theory OF Cyber Warfare Involving Cyber Trinity-based Parallel Cyber Warfare Attack to Induce Strategic Paralytic Effect on a Victim Nation[73]**

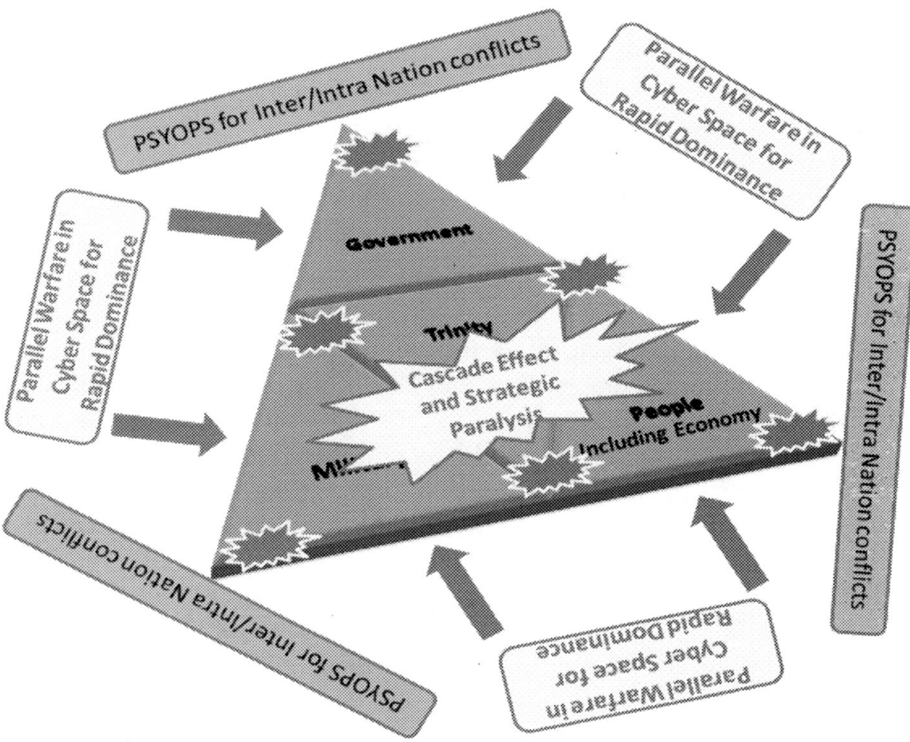


In order to achieve the strategic paralytic effect, parallel warfare[74] in cyberspace should be pursued. This parallel warfare conducted by means of rapid decisive operations to gain rapid dominance envisaged by Ullman and Wade[75] in cyberspace would not only simultaneously destroy the trinity, but also generate a *cascade effect*, resulting in chaos, anarchy and bedlam in the victim nation (Figure 3).

This form of parallel warfare based upon rapid decisive operations to gain rapid dominance by inducing strategic paralytic effect at all levels, strategic, operational and tactical, and across spectrum against the critical components of the trinity in cyberspace of the victim nation will generate the desired end result of compelling the enemy to submit to your will. The rapid dominance is essential not only for inducing the strategic paralytic effect, but also as it renders the enemy incapable of taking necessary counter-measures or initialling the process of *pulling the plug*.

The triad theory of cyberwarfare envisages the amalgam of simultaneous parallel warfare in cyberspace with the Clausewitz's Trinitarian Warfare. The excessive dependence of the three tendencies of the Trinity has made it susceptible to cyberattacks. As this dependency is increasing predominantly as a force-multiplier, the vulnerability is also increasing. An important aspect in pursuing strategic warfare in cyberspace is that all the three tendencies of the Trinity should be simultaneously destroyed, as being a resilient system the chances of revival of the Trinity and the failure of the attack to achieve a strategic effect are very high. This strategy of generating a strategic paralytic effect to compel the enemy to submit to your will not only provides for an alternative to the annihilation-based approach, but also provides a constructive conflict termination which is important to avoid protracted conflicts generated due to a flawed exit strategy, like the Iraq War.

An important aspect to underlined is the imperative need for parallel attack on all the three tendencies, as otherwise the three tendencies are resilient enough to withstand a worst case scenario, if they are individually attacked. Hence, even if an attack manages to completely destroy one of the tendencies, the other two will rejuvenate the destroyed tendency. To cater this resilient nature of the trinity, simultaneous parallel attacks should be conducted to destroy the trinity simultaneously to gain rapid dominance and strategic paralytic effect on the victim nation. The triad theory of warfare clearly establishes the fact that the contemporary cyber attacks could never have generated the strategic paralytic effect as envisaged as the consequence of strategic cyber warfare; as they were not only tactical, but were also aimed at individual tendency rather than the trinity. The author believes that the Triad theory of cyberwarfare, if not only then the primary means to achieve a strategic effect of compelling the enemy to your will as envisaged by Clausewitz.

## *The Ways and Means: The Framework of Strategic Cyberwarfare*

The analysis and operational ability of the triad theory of cyberwarfare would entail the analysis of evolutionary informationisation of mankind as a system. The last few decades have seen an information renaissance, with information becoming an integrated component of our societies to such an extent that it is been termed the 'information age society'[76] or 'network society'.[77] Commentators such as Manuel Castells[78] and Darin Barney argue that this information renaissance in the contemporary world, has resulted in the rise of new power structures and has radically and irreversibly changed the geo-political and socio-economic structure of current times to an extent that borders are slowly becoming meaningless in this cyber village.

The impact of information is not only on the civil society, but throughout the history of warfare, military experts such as Sun Tzu,[79] Carl Von Clausewitz[80] and Count von Moltke[81] have always considered information to be an important component of warfare. Over years the information and information assets have radically changed the way in which conventional wars are being fought to such an extent that experts such as George Stein refer this radical change in warfare as the "third wave".[82] This radical change of warfare has initiated a misinterpreted notion of information warfare, which ideally would have been termed as information-enabled warfare. This information-enabled warfare has led to the creation of a paradigm where information warfare acted as a tactical force-multiplier, rather than a strategic warfare which could initiate a revolution in military affairs.

Even though experts such as John Arquilla, David Ronfeldt,[83] Stephen J. Blank,[84] Norman Davis,[85] Jeffrey R. Cooper,[86] David Lonsdale,[87] George Rattary,[88] Siroli,[89] Molander[90] and Martin Libicki[91] have tried to define this relatively misinterpreted notion, but most of the attempts still highlight information warfare as a decisive, but a tactical force-multiplier. The strategy elucidated in this research endeavour tried to define the long-lost strategic aspect of information warfare, where information warfare would act as primary means to achieve strategic objectives, thus ushering a paradigm shift from information warfare acting as a force-multiplier to a strategic warfare capable of achieving political objectives. This theory is designed to befit Liddell Hart's[92] and Edward Luttwak's[93] criteria for strategic warfare; and utilises the elixir of Sun Tzu[94] and Clausewitz.[95] It aims at destroying the Clausewitzian Trinity in cyberspace.

In order to orchestrate the triad theory of cyberwarfare a cyber campaign plan based on conventional phasing,[96] but with an almost near parallel execution to achieve a simultaneous destruction of the Trinity in cyberspace is required. In line with conventional, kinetic, warfare, the success of the strategy for waging strategic cyberwarfare, or the triad theory of cyberwarfare, extensively relies on a well-laid cyber campaign plan. Hence, the cyber campaign should assess the means and orchestrate the triad theory of warfare, to achieve a desired operational end-state of strategic paralytic effect envisaged by the cyberwarfare strategy. An important aspect of campaign should be the campaign-objectives, which should be consistent with national strategy and should consist of decisive conditions required to achieve the desired end envisaged within the strategy. For the cyber campaign, the strategic end objective should be the strategic paralytic effect on the victim nation; although for achieving this objective multiple intermediate decisive conditions or scenarios should be achieved (Figure 4).

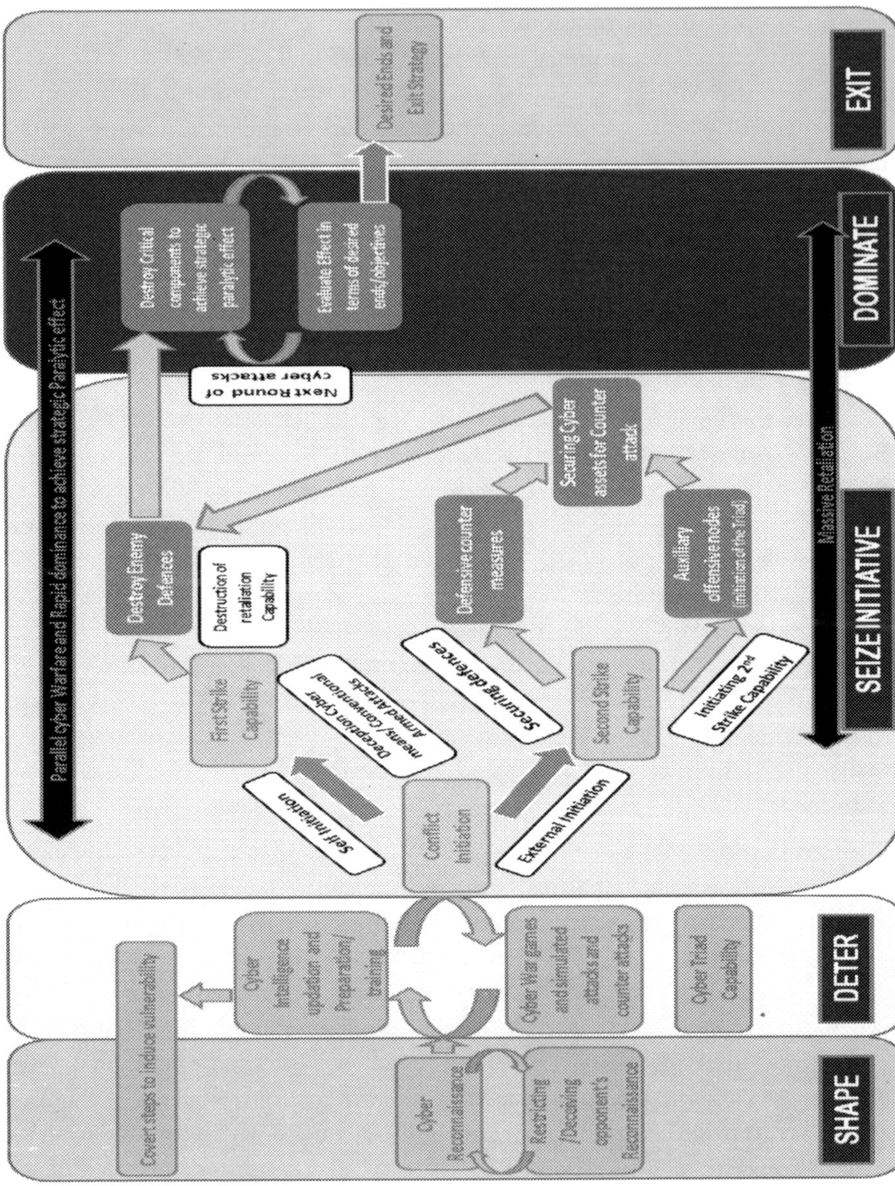


Figure 4: Cyber Campaign Plan for Orchestrating the Triad Theory of Cyberwarfare[97]

Like the traditional campaigns, the cyber campaign is also based upon phasing, but in exception to the conventional dictum of execution of campaign phases in sequential or near-sequential order, the phases in the cyberwarfare would be executed in almost near parallel order to achieve a simultaneous destruction of the trinity in cyberspace in such a way, that at any temporal instance each of the phases will have some impact at any spatial instance across the theatre. Hence, in order to orchestrate the triad theory of cyberwarfare within a given time and space, all the phases would extensively overlap and be integrated in the form of simultaneous waves.

Such cyber campaigns plan like a traditional campaign should consist of five phases, namely Shape, Deter, Seize Initiative, Dominate and Exit.[98] These phases can be further categorised as pre-conflict, conflict and post-conflict phases with Shape and Deter being the pre-conflict phases; Seize Initiative and Dominate being the conflict phases and Exit being the post-conflict phase. This cyber campaign plan should consist of aggressive *cyber reconnaissance* to identify and induce vulnerabilities in enemy cyber infrastructure through various covert and overt/cyber means. The campaign capitalises on the information gained during cyber reconnaissance and revolves around fortify-friendly cyber defence based on 'defence in depth'[99] and aims at exploiting enemy vulnerabilities by performing testing and simulated attacks, which can be initiated during an actual conflict. Such a campaign should also provide detailed guidelines for achieving a credible and known *cyber deterrence* based on *cyber triads* similar to *nuclear triads*;[100] and on *cyber countervailing strategy* based on *nuclear countervailing strategy*.[101] The plan should have specific guidelines to seize imitative and to dominate in cyberspace, in order to gain *rapid dominance* and *strategic freedom of operations*. An important aspect of the cyber campaign is its exit strategy which is essential to have a constructive conflict termination in cyberspace, thus inhibiting the chances of a *cyber insurgency*.

An important aspect for the operationalisation of the triad theory of cyberwarfare as a means for strategic cyberwarfare would be to develop a *'credible'* and *'known'* cyber deterrence (Figure 5) that guarantees an achievable second strike option, thus assuring Mutually Assured Destruction (MAD)[102] in cyberspace.

In order to achieve the 'credible' aspect of the cyber deterrence, a *cyber triad capability* should be developed similar to the *nuclear triads*.[104] This cyber triad capability (Figure 5) can consist of; regular military and civilian, offensive/ defensive cyber assets as first section of the triad; the second section can consist of a conglomerate of air-gapped networks situated across friendly nations under cooperative cyber defence initiatives, which can be activated during a conflict as a credible second strike option; and the third section of the triad can consist of a loosely connected network of *cyber militia* consisting of patriotic hackers, commercial *white hats* and cyber defence contractors. This cyber militia along with their capabilities involving their botnets[105] can be initiated in scenario of initial strikes, or in scenarios of early warning of a strike; or in worst case to initiate a protracted conflict or *cyber insurgency.*

**Figure 5: Cyber Triad-based Credible Cyber Deterrence[103]**

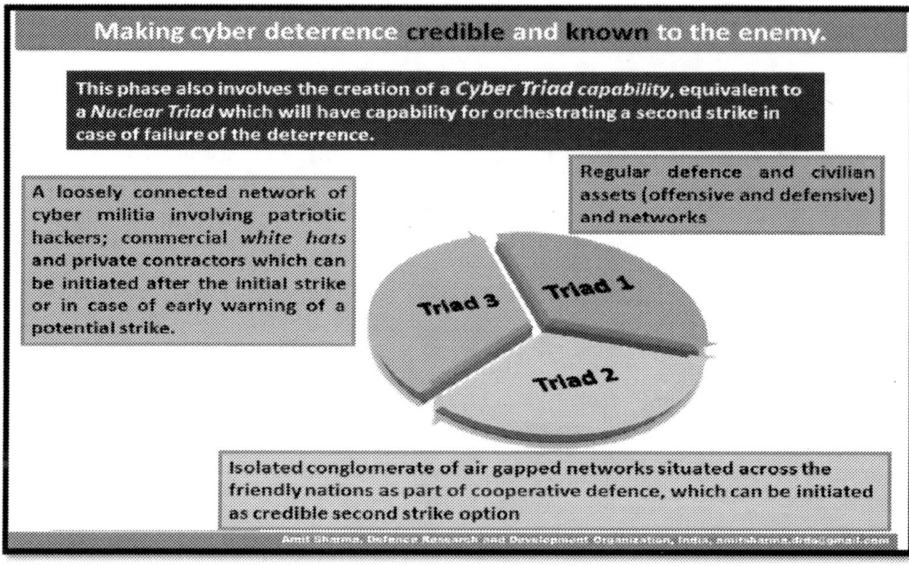


The 'known' aspect of cyber deterrence can be achieved using a *countervailing strategy* for cyberspace, similar to the *nuclear countervailing strategy*[106] used by North Atlantic Treaty Organisation (NATO) forces during the Cold War. This strategy aims at making the enemy aware of the punitive aftermath, it may face in an event of a conflict and creates an environment where the enemy is made to believe, that the implications of a first strike would be far greater than its potential gains.[107] The scenario will remain the same in a *cyber countervailing strategy.*

The cyber countervailing strategy may consist of, excessive media coverage of the friendly cyberwarfare (offensive and defence) capability; cyber war-games and simulated exercises to demonstrate both offensive and defensive cyberwarfare capability; and although it may have legal implications, the covert initiation of limited cyberattacks on potential adversary. These measures ensure the 'known' aspect of cyber deterrence capability. The author believes that the attacks on Georgia,[108] Estonia,[109] the Titan Rain[110] attacks on the US, believed to have emanated from China, and the cyberattacks on countries such as the UK, Germany and India[111] were certainly part of the cyber countervailing strategy with the sole aim of giving clear indications to potential adversaries of the losses they may face in a cyber conflict.

The cyber campaign plan provides means to orchestrate the triad theory of cyber warfare by envisaging a phased approach to achieve operational objectives, thus acting as an extension to the commander's strategy,[112] which in current context is the triad theory of cyberwarfare. The cyber campaign consisting of the pre-conflict, the conflict and the post-conflict phases may sound sequential in nature, but, as in any conflict, the phases in this campaign overlap extensively and have some effect on the entire theatre both spatially and temporally. This overlapping is especially important to achieve parallel warfare in cyberspace to induce strategic paralytic effect onto the victim nation.

This cyber campaign, as an orchestration of cyber strategy is long-term in nature involving years, especially during the pre-conflict phases, to prepare for the conflict. The campaign not only elucidates the need to shape the enemy as envisaged by Sun Tzu,[113] but also highlights the need to induce vulnerability in enemy's cyber infrastructure, which can be leveraged during a conflict. An essential component of this campaign requires the creation of a 'credible' and 'known' cyber deterrence based upon the concepts of *cyber triad* and *cyber countervailing strategy*. The campaign lays great emphasis on laying a prudent exit strategy, once the desired ends are achieved, to ensure a constructive conflict termination in cyberspace.

This campaign provides the inroads for an effective orchestration of the triad theory of warfare to achieve rapid dominance and destruction of trinity in cyberspace in offensive terms; and a deterrence mechanism to create a scenario of Mutually Assured Destruction in cyberspace, thus guaranteeing a *strategic status quo*. Nevertheless, extreme caution should be exercised while achieving cyber deterrence capability, as there can be intricate scenarios in which actions like these may be perceived as initiation of hostilities; or such an action may result in further destabilisation of the geopolitical power balances, thus initiating an arms race in cyberspace.

The triad theory of cyberwarfare*; its operational cyber campaign plan and formation of a 'Known' and 'Credible' Cyber deterrence* to generate a scenario of *Mutually Assured Destruction in cyberspace* thus guaranteeing a *strategic status quo,* heralds the genesis of strategic aspects of cyberwarfare rather than a tactical force multiplier or more appropriately *a framework for strategic cyberwarfare.* This framework not only defines the strategic and operational aspect of cyberwarfare, but in light of the formulated strategic cyberwarfare capabilities, defines a new *cyber defence* strategy. A strategy involving achieving the *deterrence by punishment* and *deterrence by denial*[114] capability to achieve a strategic status quo, based on 'known' and 'credible' cyber deterrence; along with the traditional approaches of *defence in depth;* and *legal framework of international laws.*

## Recommendations for National Security: *Prepare, Pursue, Protect and Prevent*

"*Our digital infrastructure will be treated as a strategic national asset.*"

—US President Barack Obama[115]

President Obama's statement clearly pointed out that world leaders are considering the importance and consequently the threat to information infrastructure, but an important aspect, which is clearly missing, is the recognition of the strategic aspect of information warfare. Although slowly because of rising instances of systematic compromise of digital infrastructure both in civilian and defence sectors, most notably the cyberattacks on Estonia[116] and Georgia[117] and Titan Rain,[118] decision-makers around the world are taking this 21st century threat seriously. The author believes that these situations will are likely to worsen, especially with this new framework of strategic information warfare, which is capable of inducing strategic paralytic effect upon the victim nation. In purview of such a framework, it is recommended that the nations should build their strategic cyberwarfare capability.

This strategic cyber warfare capability can be orchestrated in the form of cyber command, which is the buzzword in the cyberwarfare community, especially in light of the steps taken by the US Government.[119] Not only the US, but China is also believed to have the cyberwarfare capability, which the author believes to have the capability of orchestrating strategic cyber warfare.[120]

In light of these circumstances, nations need to understand the potency of strategic cyberwarfare and develop military doctrines for this relatively new field and keeping in mind the strategic paralytic effect rendered by the *triad theory of cyber warfare*. The author also recommends that nations should develop their cyber commands, which should aim at long-term strategic ends, rather than tactical attack. In order to achieve this aim, the cyber commands should have cyber reconnaissance capability and a campaign plan similar to the one mentioned in the framework. These commands should aim at creating cyber deterrence by developing *cyber triads* and *cyber countervailing strategy*.

Initiatives should be undertaken to recruit fresh talent as cyber warriors, who should be groomed in the tactical aspects of information security, ethical hacking, malware designing and so on along with the psychological grooming as these warriors are double-edged weapons, with sufficient information and capability to act as potent non-state actors. Steps should be taken to test the cyberwarfare capabilities in the form of cyber war-games and exercises similar to the Cyber Storm exercise, conducted by the US Department of Homeland Security in 2006 and subsequently.[121]

Nations should also modify their cyber defence strategies taking in purview of the ramifications, which an attack orchestrating this framework can bring on a nation. This strategy should be defined as a combination of "defence in depth";[122] the legal instrument;[123] and by achieving the *deterrence by punishment* and *deterrence by denial*[124] capability to achieve a strategic status quo, based on the cold war strategies of *Rational Deterrence Theory*[125] and *MAD*[126] in cyberspace.

In order to achieve these goals nations should appropriately increase their cyber defence spending and commit budget allocations for long-term development of human resources, cyberweapons, software and bugged firmware; and cyber doctrines, strategies and tactics. Moreover, cyber military alliances should be formed which should aim at developing collective cyber defence/offence capabilities and facilities like the NATO Cooperative Cyber Defence Centre of Excellence[127] etc. being important components in the collective pursuit of excellence in cyberspace at international level.

## Future: *The Pandora's Box of Strategic Cyberwarfare*

The framework of strategic cyberwarfare involving the triad theory of cyberwarfare and its operational cyber campaign plan *and formation of a 'Known' and 'Credible' Cyber deterrence* to generate a scenario of *Mutually Assured Destruction in cyberspace* thus guaranteeing a *strategic status quo* has opened a Pandora's Box of possibilities and opportunities, not only in cyberspace, but in the real world too; some of which have been recognised while others are still in haze. This relatively virgin field of strategic cyberwarfare is notorious for misinterpretations and lack of strategic thought, predominantly due to the web of secrecy and technical knowhow woven around it. This framework of strategic warfare is among the initial steps to elucidate the strategic aspect of cyberwarfare, but a lot has to be done so as to define the granularities within this framework in particular and the cyberwarfare in general, especially the technical aspects of cyberweapons which in itself is a vast area of research. Moreover, appropriate research and analysis is required for understanding the ramifications and consequences of such a warfare. Although attempts are being made in this direction by organisations such as the US Cyber Consequences Unit,[128] still more in depth research is required in this field especially in relation to consequence analysis, feasibility analysis and psychological analysis. Further, detailed research is also required to understand the psychological, legal and ethical aspects in the creation of cyber warriors/soldiers. This in itself is a mammoth task as it requires the mapping of behavioural analysis with psychological and ethical aspects of training as these warriors have to walk a thin redline between efficient cyber soldiers and cyber criminals/hackers.

The field of strategic cyberwarfare is a relatively unchartered realm of warfare with endless possibilities. This field is in a novice state and in a process of evolution.

Like the Pandora's Box, it is open now. It is up to the world leaders to decide what their approach to this warfare will be, whether they will behave like an ostrich and just pretend that there is nothing out there; or they will face it and try to gain on the opportunities that accompany this warfare.

## NOTES

1. Carl Von Clausewitz, *On War,* Howard Michael and Paret Peter (eds. and trans.), Princeton University Press, New Jersey, 1989, p. 583.
2. Sun Tzu, *The Art of War,* Samuel B. Griffith (trans.), Oxford University Press, Oxford, 1963, p, 77.
3. Ibid.
4. Carl Von Clausewitz, No. 1.
5. David Deptula, *Firing for Effect: Change in the Nature of warfare,* Aerospace Education Foundation, Arlington, 1995.
6. Ulrich Beck, *Risk Society: Towards a New Modernity*, Sage Publications, New Delhi, 1992.
7. Amit Sharma, "Chinese Zixunhua Budui: A Myth or a Reality", *Defence Science Journal,* forthcoming.
8. Centre of Excellence Defence Against Terrorism, "Overview of Cyber Attacks against Estonia", *Responses to Cyber Terrorism,* IOS Press, Ankara, 2008, pp 96-102; "'Cyber Attacks' Hit Estonia", *The Telegraph,* May 17, 2007, at http://www.telegraph.co.uk/news/worldnews/1551851/Cyber-attacks-hit-Estonia.html; Ian Traynor, " Russia Accused of Unleashing Cyberwar to Disable Estonia", *The Guardian,* May 17, 2007, at http://www.guardian.co.uk/world/2007/may/17/topstories3.russia.
9. Stephen W. Korns and Joshua Kastenberg, "Georgia's Cyber Left Hook", *Parameters,* Winter 2008, pp. 60-76.
10. Dependability Development Support Initiative (DDSI), *European Dependability Policy Environments, United Kingdom,* Rand, the Netherlands, 2002, p. 16, at http://www.ddsi.org/htdocs/Documents/CR/united_kingdom.pdf.
11. RAND Europe, *Protecting the Digital Society: A manifesto for the UK,* Information Assurance Advisory Council, March 2002, p. 10, at http://www.iaac.org.uk/Portals/0/IAACManifesto270202.PDF.
12. DDSI, No. 10.
13. RAND Europe, No. 11.
14. Carl Von Clausewitz, No. 1, p. 89.
15. Ibid.
16. Ibid.
17. Amit Sharma, "Cyber Wars: A Paradigm Shift from Means to Ends", in Christian Czosseck and Kenneth Geers (eds.), *Virtual Battlefield: Perspectives of Cyber Warfare,* IOS Press, 2009, Chapter One. [The research paper was presented at the Cyber Warfare Conference, NATO Cooperative Cyber Defence Centre of Excellence, Tallinn, Estonia, June 17, 2009.]
18. Peter J Denning, *Computers under Attack, Intruders, Worms and Viruses,* Addison Wesley, New York, 1990, pp. 6-9.
19. Matt Bishop and Emily O. Goldman, "The Strategy and Tactics of Information Warfare", in Emily O. Goldman (ed.), *National security in the Information Age,* Frank Cass Publishers, London, 2004, pp. 113-138.
20. A. Streltsov, "International Information Security: Description and Legal Aspects", *Disarmament*

*Forum ICTs and International Security 2007,* 3, pp. 5-14, at http://www.unidir.ch/bdd/fiche-article.php?ref_article=2642.

21. Shitanshu Mishra, "Network Centric Warfare in the Context of 'Operation Iraqi Freedom'", *Strategic Analysis*, 27 (4), October-December 2003, pp. 546-562; Paul Mitchell, *Network Centric Warfare and Coalition Operations: The New Military Operating System*, Routledge, London, 2009, pp. 63-70; Clay Wilson, *Network Centric Warfare: Background and Oversight Issues for Congress*, Congressional Research Service, Washington, 2004; Gregory Fontenot, E. J Degen. and David Tohn, *On point: the United States Army in Operation Iraqi Freedom*, Naval Institute Press, Annapolis, 2005, p. 415; James R. Blaker, *Transforming Military Force: the Legacy of Arthur Cebrowski and Network Centric Warfare, Greenwood Publishing Group*, Westport, *2007*; John Ferris, "Netcentric Warfare, C4ISR and Information Operations: Towards a Revolution", *Intelligence and National Security,* 19 (2), pp. 199-225.
22. Ibid.
23. Ibid.
24. Amit Sharma and Manoj Tyagi, "Bayesian Belief Network approach for Vulnerability Assessment of Strategic Information Dissemination System", Proceedings of Third International Conference on Quality, Reliability and Infocom Technology, December 2006.
25. Patrick Morgan, "Information Warfare and Domestic Threats to America Security", *Contemporary Security Policy*, 24 (1), 2003, pp. 161-189.
26. Amit Sharma, No. 7; US-China Economic and Security Review Commission, *China's Proliferation Practices and the Development of Its Cyber and Space Warfare Capabilities*, Hearing before the US-China Economic and Security Review Commission, 110th Congress, second session, May 20, 2008, at www.uscc.gov; Jason Fritz, "How China Will Use Cyber Warfare to Leapfrog in Military Competitiveness", *Culture Mandala*, 8 (1), October 2008, at http://www.international-relations.com/CM8-1/Cyberwar.pdf.
27. Amit Sharma, No. 7.
28. Kimm Hartmann and Christoph Steup, "The Vulnerability of UAVs to Cyber Attacks – An Approach to the Risk Assessment", in K. Podins, J. Stinissen and M. Maybaum (eds.), *2013 5th International Conference on Cyber Conflict 2013,* NATO CCD COE Publications, Tallinn, at https://ccdcoe.org/cycon/2013/proceedings/d3r2s2_hartmann.pdf; Alan Kim, Brandon Wampler, James Goppert, and Inseok Hwang, "Cyber Attack Vulnerabilities Analysis for Unmanned Aerial Vehicles", College of Engineering - Purdue University, at https://engineering.purdue.edu/HSL/uploads/papers/cybersecurity/cyber-attack-lit-review.pdf.
29. Stuart Hooper, "NATO Blind: Top Secret Russian Air Defence System Deployed in Syria", *21st Century Wire,* October 19, 2015, at http://21stcenturywire.com/2015/10/19/nato-blind-top-secret-russian-air-defence-system-deployed-in-syria/
30. S. Waterman, "North Korean Jamming of GPS Shows System's Weakness", *The Washington Times*, August 23, 2012, at http://www.washingtontimes.com/news/2012/aug/23/north-korean-jamming-gps-shows-systems-weakness/? page=all; L. Huang, and Q Yang, "GPS Spoofing: Low-Cost Simulator", DEF CON, August 8, 2015, at https://media.defcon.org/DEF%20CON%2023/DEF%20CON%2023%20presentations/DEFCON-23-Lin-Huang-Qing-Yang-GPS-Spoofing.pdf.
31. Leonardo Egea Nve," Playing in a Satellite Environment", Black Hat, at *http://www.blackhat.com/presentations/bh-dc-10/Nve_Leonardo/BlackHat-DC-2010-Nve-Playing-with-SAT-1.2-slides.pdf;* Jim Geovedi, Raditya Iryandi and Raoul Chiesa, "Hacking a Bird in the Sky: The Revenge of Angry Birds", Hack In The Box Security Conference (HITBSecConf), at http://conference.hackinthebox.org/hitbsecconf2011ams/materials/D2T2%20-%20Jim%20Geovedi%20and%20Raoul%20Chiesa%20-%20Hacking%20a%20 Bird%20in%20the%20Sky.pdf.

32. Conor Gaffey, "German missiles 'Hacked by Foreign Source", *Europe Newsweek*, August 7, 2015, at http://europe.newsweek.com/german-missiles-hacked-by-foreign-source-329980; Andrew Futter, "Hacking the Bomb: Nuclear Weapons", Cyber Age draft working paper for ISA Annual Conference, New Orleans, February 2015, at https://www2.le.ac.uk/departments/politics/people/afutter/copy_of_AFutterHackingtheBombISAPaper2015.pdf; Eric Schlosser, "Neglecting Our Nukes", *Politico,* September 16, 2013, at http://www.politico.com/story/2013/09/neglecting-our-nukes-96854.html; Aliya Sternstein, "Officials Worry about Vulnerability of Global Nuclear Stockpile to Cyber Attack", *Global Security Newswire,* March 14, 2013, at http://www.nti.rsvp1.com/gsn/article/officials-worry- about-vulnerability-global-nuclear-stockpile-cyberattack/?mgh=http%3A%2F%2Fwww.nti.org&mgf=1.
33. Liam O'Murchu, "Stuxnet using Three Additional Zero-Day Vulnerabilities", Symantec Official Blog, September 14, 2010, at http://www.symantec.com/connect/blogs/stuxnet-using-three-additional-zero-dayvulnerabilities; Dan Goodin, "Nation-Sponsored Malware with Stuxnet Ties Has Mystery Warhead", *Ars Technica,* 2009; "Kaspersky Looks at the Wreckage of Wiper Malware", *Infosecurity Magazine,* August 29, 2012.
34. Hal Berghel, "A Farewell to Air Gaps, Part 1", *Computer,* 48 (6), 2015, pp. 64-68, at http://www.computer.org/csdl/mags/co/2015/06/mco2015060064.pdf.
35. P. Trevorrow, S. Wright, D. C. Webb and E. F. Halpin, "Defining the Issues", in Edward F. Halpin, Philippa Trevorrow, David C. Webb and Steve Wright (eds.), *Cyberwar, Netwar and the Revolution in Military Affairs,* Palgrave Macmillan, Basingstoke, 2006, pp. 3-11.
36. A. Streltsov, No. 20.
37. Liam O'Murchu, No. 33; Dan Goodin, No. 33; *Infosecurity Magazine* No. 33.
38. US Department of Homeland Security ICS-CERT, "Alert (ICS-ALERT-14-176-02A) ICS Focused Malware – Havex", at https://ics-cert.us-cert.gov/alerts/ICS-ALERT-14-176-02A; Swati Khandelwal, "Stuxnet-like 'Havex' Malware Strikes European SCADA Systems", *The Hacker News,* June 26, 2014, at http://thehackernews.com/2014/06/stuxnet-like-havex-malware-strikes.html.
39. Riley Walters, "Russian Hackers Shut Down Ukraine's Power Grid", *Newsweek,* January 14, 2016, at http://www.newsweek.com/russian-hackers-shut-ukraine-power-grid-415751.
40. Kate Vinton, "Hacking Gets Physical: Utilities At Risk For Cyber Attacks", *Forbes Tech,* July 10, 2014, at http://www.forbes.com/sites/katevinton/2014/07/10/hacking-gets-physical-utilities-at-risk-for-cyber-attacks/#2715e4857a0b6879a8966b30.
41. Edward V. Linden, *Focus on Terrorism*, Nova Science Publishing, New York, 2003, p. 46.
42. James J. F. Forest (ed.), *Homeland Security: Critical infrastructure,* Praeger Security International, Westport, 2006, p. 386.
43. "Air Traffic Control System Vulnerable to Cyberattack", *Security*, April 10, 2015, at http://www.securitymagazine.com/articles/86298-air-traffic-control-system-vulnerable-to-cyberattack; Matthew Day, "13 Planes Vanish from Radars over Europe", *The Telegraph,* Warsaw, June 13, 2014, at http://www.telegraph.co.uk/news/worldnews/europe/austria/10898385/13-planes-vanish-from-radars-over-Europe.html.
44. Ted G. Lewis, *Critical Infrastructure Protection in Homeland Security: Defending a Networked Nation* John Wiley, New Jersey, 2006.
45. Martin C. Libicki, *Conquest in Cyberspace,* Cambridge University Press, New York, 2007, p. 17.
46. Todd Humphreys, "GPS Spoofing and the Financial Sector", The University of Texas at Austin, June 2011, at http://radionavlab.ae.utexas.edu/images/stories/files/papers/summary_financial_sector_implications.pdf.
47. Kevin Poulsen, "Slammer Worm Crashed Ohio Nuke Plant Network", *SecurityFocus,* August 19, 2003, at http://www.securityfocus.com/print/news/6767 .

48. Centre of Excellence Defence Against Terrorism, No. 8; "'Cyber attacks' hit Estonia", No. 8; Ian Traynor, No. 8.
49. Stephen W. Korns and Joshua Kastenberg, "No. 9; David J. Smith, "The Fourth Front: Russia's Cyber-attack on Georgia", Potomac Institute for Policy Studies, March 24, 2009, at http://www.potomacinstitute.org/media/mediaclips/2009/24Saati_Cyberattack_3_24_09Send.pdf.
50. Amit Sharma, No. 7; US-China Economic and Security Review Commission, No. 26; Jason Fritz, No. 26.
51. Peter Beaumont and Matthew Weaver, "MyDoom Virus Hits Key Networks in US and South Korea", *The Guardian*, July 8, 2009, at http://www.guardian.co.uk/technology/2009/jul/08/cyber-war-mydoom-virus-attack .
52. Bill Durodie, "The Limitation of Risk Management", *Tidsskriftet Politik*, 8 (1), 2004.
53. Frank Füredi, *Culture of Fear Revisited: Risk-Taking and the Morality of Low Expectation*, Continuum International Publishing Group, London, 2006.
54. Robert Putnam, *Bowling Alone: The Collapse and Revival of American Community*, Simon and Schuster, New York, 2001.
55. Ulrich Beck, No. 6.
56. Bill Durodie, No. 54; Frank Füredi, No. 55; Robert Putnam, No. 56.
57. Piers Robinson, "CNN Effect Considered" and "Developing a Theory of Media Influence", in *The CNN Effect: The Myth of News, Foreign Policy and Intervention,* Routledge, 2002, chapters one, and two.
58. James X Dempsey, "Overview of Current Criminal Justice Information Systems", Centre for Democracy and Technology, February 9, 2000, pp. 101-106, at http://www.cdt.org/publications/overviewofcjis.pdf.
59. David A. Brown, "Steps to Secure a Law Enforcement Network", GSEC Practical Assignment Version 1.3, *SANS Institute InfoSec Reading Room*, March 2002, at http://www.sans.org/reading_room/whitepapers/casestudies/steps_to_secure_a_law_enforcement_network_706.
60. Perri 6, E-*Governance: Styles of Political Judgment in the Information Age Polity*, Palgrave Macmillan, Basingstoke, 2004.
61. Barbara A. Bardes, Mack C. Shelley and Steffen W. Schmidt, *American Government and Politics Today 2008: The Essentials*, Wadsworth Publishing, Belmont, 2008.
62. Angela Maria Lungu, "WAR.com: The Internet and Psychological Operations", *Joint Force Quarterly*, Spring/Summer 2001, pp. 13-17.
63. Malcolm Moore, "Chinese Cable Television Channel Hacked", *The Telegraph,* August 2, 2014, at http://www.telegraph.co.uk/news/worldnews/asia/china/11007494/Chinese-cable-television-channel-hacked.html.
64. Joseph Menn and Leigh Thomas, "France Probes Russian Lead in TV5Monde Hacking: Sources", Reuters, San Francisco, June 10, 2015, at http://www.reuters.com/article/us-france-russia-cybercrime-idUSKBN0OQ2GG20150610.
65. Silvia Mitter and Claudia Wagner, "Understanding the Impact Of Socialbot Attacks in Online Social Networks", *WebSci'13*, May 2–4, 2013, Paris, France. at http://www.markusstrohmaier.info/documents/2013_WebSci2013_Socialbots_Impact.pdf.
66. Nick Fielding and Ian Cobain, "Revealed: US Spy Operation that Manipulates Social Media-Military's 'Sock Puppet' Software Creates Fake Online Identities to Spread Pro-American Propaganda", *The Guardian,* March 17, 2011, at http://www.theguardian.com/technology/2011/mar/17/us-spy-operation-social-networks.
67. Evgeny Morozov, "Iran Elections: A Twitter Revolution?", *The Washington Post*, June 17, 2009, at http://www.washingtonpost.com/wp-dyn/content/discussion/2009/06/17/DI2009061702232.html.

68. Fyodor Pavlyuchenko, "Belorussia in the Context of European Cyber Security", paper presented at the Cyber Warfare Conference, NATO Cooperative Cyber Defence Centre of Excellence, Tallinn, Estonia, June 17, 2009.
69. Amit Sharma, No. 17.
70. Amit Sharma, No. 7.
71. Centre of Excellence Defence Against Terrorism, No. 8; "'Cyber attacks' hit Estonia", No. 8; Ian Traynor, No. 8.
72. Stephen W. Korns and Joshua Kastenberg, No. 9.
73. Amit Sharma, No. 17.
74. David Deptula, No. 5.
75. Harlan K. Ullman and James P. Wade, *Shock and Awe: Achieving Rapid Dominance,* National Defense University, Washington, 1996, at http://www.dodccrp.org/files/Ullman_Shock.pdf.
76. Manuel Castells, *The Information Age: Economy, Society and Culture,* Blackwell, Chichester, 1999.
77. Darin Barney, *The Network Society,* Polity Press, Florida, 2004, p. 114.
78. Manuel Castells, No. 80.
79. Sun Tzu, "Weaknesses and Strengths", No. 2, Chapter VI, pp. 96-101.
80. Carl Von Clausewitz, No. 1, p. 25.
81. George Stein, "Information Warfare", *Airpower Journal,* Spring 1995, at http://www.iwar.org.uk/iwar/resources/airchronicles/stein.htm (Accessed July 11, 2009).
82. B.G. Kutais, *Internet Policies and Issues,* Nova Publishers, New York, 2002, especially, Chapter 7, "Overview Information Warfare Issues".
83. John Arquilla and David Ronfeldt, "Cyberwar is Coming!" *Comparative Strategy,* 12 (2), Spring 1993, pp. 141-165; [as also "Chapter Two", in John Arquilla and David Ronfeldt (eds.), *In Athena's Camp: Preparing for Conflict in the Information Age,* Rand Corporation, Santa Monica, 1997].
84. Stephen J. Blank, "Preparing for the Next War", *Strategic Review,* 24 (2), Spring 1996, pp. 17–25; [as also "Chapter Three", in John Arquilla and David Ronfeldt (eds.), No 87.
85. Norman Davis, "An Information-Based Revolution in Military Affairs", *Strategic Review,* 24 (1), Winter 1996, pp. 43–53; [as also "Chapter Four", in John Arquilla and David Ronfeldt (eds.), No. 87.
86. Jeffrey R. Cooper, "Another View of the Revolution in Military Affairs", paper presented at the Fifth Annual Conference on Strategy, April 1994; [as also "Chapter Five", in John Arquilla and David Ronfeldt (eds.), No. 87].
87. David J. Lonsdale, *The Nature of War in the Information Age: Clausewitzian Future,* Frank Cass, New York, 2004, pp. 134-152.
88. Gregory J. Rattray, "Understanding the Conduct of Strategic Information Warfare", *Strategic Warfare in Cyberspace,* MIT Press, London, 2001, Chapter Two, pp. 91-93.
89. Gian Piero Siroli, "Strategic Information Warfare: An Introduction", in Edward F. Halpin et al. (eds.), No. 36, pp. 32-45.
90. Roger C. Molander, Andrew S. Riddile, Peter A. Wilson, *Strategic Information Warfare Rising,* RAND, Santa Monica, 1996.
91. Martin C Libicki, *What Is Information Warfare?,* National Defense University, Washington, 1995.
92. Basil Henry Liddell Hart, *Strategy,* Meridian, New York, 1991.
93. Edward Luttwak, *Strategy: The Logic of War and Peace,* Harvard University Press, London, 1987.
94. Sun Tzu, No. 2.

95. Carl Von Clausewitz, No. 1.
96. Development, Concepts and Doctrine Centre (DCDC), *Joint Doctrine Publication JDP 5-00 Campaign Planning,* 2nd Edition, Shrivenham, December 2008,; Director for Operational Plans and Joint Force Development, *Joint Publication 5-00.1 Joint Doctrine for Campaign Planning*, HQ Department of the Army DAMO-SSP, Washington, 2002, at http://www.dtic.mil/doctrine/jel/new_pubs/jp5_00_1.pdf (Accessed July 20, 2009).
97. Amit Sharma, No. 17.
98. DCDC, No. 105.
99. Michael E. Whitman and Herbert J. Mattord, *Principles of Information Security*, Thomson Course Technology, Canada, 2007, Chapter Five, pp. 202.
100. Robert C Aldridge, *First Strike!: The Pentagon's Strategy for Nuclear War*, South End Press, Cambridge, 1983.
101. Walter Slocombe, "The Countervailing Strategy", *International Security,* 5, Spring 1981, pp. 18-27; [as also Slocombe Walter, "The Countervailing Strategy", in Steven Miller (ed.),, *Strategy and Nuclear Deterrence,* Princeton University press, New Jersey, 1984].
102. Henry D. Sokolski, "MAD in Practice" and "Moving beyond MAD", *Getting MAD: Nuclear Mutual Assured Destruction, Its Origins and Practice*, Strategic Studies Institute, Carlisle, 2004, parts two and three.
103. Amit Sharma, No. 17.
104. Robert C Aldridge, No. 109.
105. Moheeb A. Rajab, Jay Zarfoss, Fabian Monrose and Andreas Terzis, "A Multifaceted Approach to Understanding the Botnet Phenomenon", proceedings at the sixth ACM SIGCOMM 2006, pp. 41-52, at http://www.cs.jhu.edu/~fabian/papers/botnets.pdf (Accessed July 25, 2009).
106. Walter Slocombe, No. 110.
107. Ibid.
108. Stephen W. Korns and Joshua Kastenberg, No. 9.
109. Centre of Excellence Defence Against Terrorism, No. 8; "'Cyber attacks' hit Estonia", No. 8; Ian Traynor, No. 8.
110. Amit Sharma, No. 7; US-China Economic and Security Review Commission, No. 26; Jason Fritz, No. 26.
111. Amit Sharma, No. 7; US-China Economic and Security Review Commission, No. 26; Jason Fritz, No. 26.
112. Director for Operational Plans and Joint Force Development, No.105.
113. Sun Tzu, "Sun Tzu on War FIGURE 5", No. 2, p. 42.
114. Glenn Herald Snyder, *Deterrence by Denial and Punishment*, Woodrow Wilson School of Public and International Affairs, Center of International Studies, Princeton, 1959.
115. "US Launches Cyber Security Plan", BBC World News, May 29, 2009, at http://news.bbc.co.uk/2/hi/americas/8073654.stm (Accessed August 14, 2009); Remarks by US President Barack Obama on Securing the Nation's Cyber Infrastructure, May 29, 2009, at http://news.bbc.co.uk/2/shared/bsp/hi/pdfs/29_05_09_cyber.pdf (Accessed August 14, 2009).
116. Centre of Excellence Defence Against Terrorism, No. 8; "'Cyber attacks' hit Estonia", No. 8; Ian Traynor, No. 8.
117. Stephen W. Korns and Joshua Kastenberg, No. 9; David J. Smith, No. 51.
118. Amit Sharma, No. 7; US-China Economic and Security Review Commission, No. 26; Jason Fritz, No. 26.
119. The Secretary of Defence, Memorandum for Establishment of a Subordinate Unified U.S Cyber Command under U.S Strategic Command for Military Cyberspace Operations, Department of Defence, Washington, 2009, at http://docs.govinfosecurity.com/files/external/OSD-05914-09.pdf .

120. Amit Sharma, No. 7.
121. National Cyber Security Division, *Cyber Storm Exercise Report,* Department of Homeland Security, Washington, 2006, at http://www.dhs.gov/xlibrary/assets/prep_cyberstormreport_sep06.pdf.
122. Ronald L. Krutz, *Securing SCADA Systems,* Wiley, Indianapolis, 2005, pp. 130; David Chappelle, *Protect Yourself Online: How to Cover Your Assets Every Time You Log on,* Victoria, Trafford Publishing, 2003, pp. 95.
123. Michael N. Schmitt, "Wired Warfare: Computer Network Attack and Jus in Bello", *IRRC,* 84 (846), June 2002 pp. 365-395; Matthew E. Haber, *Computer Network Attack and the Laws of Armed Conflict: Searching for Moral Beacons in Twenty-First-Century Cyber warfare,* ARMY Command and General Staff College, Fort Leavenworth, 2002.
124. Glenn Herald Snyder, No. 124.
125. Christopher H. Achen and Duncan Snidal, "Rational Deterrence Theory and Comparative Case Studies", *World Politics,* 41 (2), January 1989, pp. 143-169.
126. Henry D. Sokolski, No. 111.
127. NATO, Cooperative Cyber Defence Centre of Excellence Tallinn, Estonia, at http://www.ccdcoe.org/ .
128. Author's discussion with Scott Borg, Director and Chief Economist, US Cyber Consequences Unit (US-CCU) – an independent, non-profit research institute that investigates the strategic and economic consequences of possible cyberattacks – at the Cyber Warfare Conference, NATO Cooperative Cyber Defence Centre of Excellence, Tallinn, June 17, 2009.

# 4

# WORKING OUT THE RULES OF GLOBAL CYBERSPACE GOVERNANCE

*Alexandra Kulikova*

## Introduction

The ongoing securitisation of online space has contributed to the establishment and development of the 'norms-building' discourse over the past few years. As the development of Information and Communications Technology (ICT) poses more security and safety questions, be it at the end user, business, national or international levels, the narrative is shifting from a focus on 'response to conflict' to 'conflict prevention'. This has also led to the emergence of a few norms development initiatives and platforms driven by various actors, which introduces an element of rivalry into the process. At the end of the day, the competition will most likely be led by those formations and thought leaders that have the greater trust capital and control.

The first steps in the direction of developing behaviour standards were taken over a decade ago. Russia has been advocating the necessity to define 'rules of the road in the areas of both cybersecurity and information security at the state level. The emphasis of these efforts has been not only on cyberthreats (software and hardware compromise, cyberespionage, etc.) but also on information warfare (interfering with the internal affairs of other states, propaganda, etc.). The latter has been often linked to Russia's perceived long-term efforts to assert intergovernmental control on global Internet governance. Moreover, Russia was looking to gain ground in technological development by curbing other countries' efforts to use technology for military purposes.

Nonetheless, the United Nations Group of Governmental Experts (UN GGE), a group of individual experts representing a few countries that was created back in the 1990s, received little attention from Western states until recently. Its first truly vocal success was the report of 2013, which recognised the application of international law to cyberspace. Prior to that the first draft of the International Code of Conduct for information security was submitted to the UN Secretary General in 2011 by Russia and its partners in the Shanghai Cooperation Organisation (SCO) – China, Tajikistan and Uzbekistan. This, for the first time, laid out the principles of responsible behaviour of states to secure both information and cybersecurity. In the same year (2011), a draft UN Convention on international information security was presented by then Russian Minister of Communications, Igor Shegolev, at the first round of the Global Conference on CyberSpace (GCCS) in London (also known since then as the "London process"), but it, too, found little support.

At the GCCS in The Hague in April 2015 the idea of an international treaty defining the norms of behaviour sounded premature. However, a few months later, the UN GGE report somewhat unexpectedly, for many, gave proof of the desire and ability of otherwise antagonist states to find common denominators in terms of international efforts to provide cyber/information security and agree on at least a few soft law standards of state behaviour to prevent conflict in cyberspace.

Apart from the international efforts, there are regional formats adding to the trend of norms building. For instance, in 2013 the Organisation for Security and Cooperation in Europe (OSCE) countries signed the first set of Confidence Building Measures (CBMs) to reduce the risks of conflict stemming from the use of ICTs,[1] focusing on transparency and confidence building among states to foster cybersecurity. Their implementation might be uneven, but by March 2016 the list was extended to include a second set focusing on more practical measures of cooperation.

In July 2015, the Declaration of the BRICS [Brazil-Russia-India-China-South Africa] Summit in Ufa[2] also made note of the importance to continue work on the measures to prevent conflict in cyberspace and further develop norms, standards and principles of responsible conduct.

Moreover, the bilateral dimension adds to the ecosystem of the norms building efforts in cyberspace. It feeds into larger formats, while at the same time often defines the interaction among partners working together in various dimensions. In this regard, it is especially interesting to look at the US-China-Russia triangle, which has seen three bilateral agreements including attempts at establishing mutual CBMs as well as principles on conduct in cyber domain.

Thus, such activities (particularly, through expert platforms like the EastWest Institute, The Hague Institute for Global Justice and Chatham House) related to developing norms of state behaviour in cyberspace reveal an ecosystem of norms building formats. The increasing number of places where cross-fertilisation of ideas can happen suggests, on the one hand, more opportunity for joint work and trust building; on the other hand, the interoperability of various platforms might represent a challenge and duplication of effort. Besides, the involvement of non-state actors adds another dimension and set of priorities to the agenda. Since any state efforts in the area of cybersecurity, and especially the protection of critical infrastructure (CI), depends on collaboration with the private sector, integrating business interests and profit models into agreements is crucially important. The human dimension of cyberpolitics will certainly be further monitored by civil society and academia expert groups to ensure end users' rights are protected, both online as well as offline. Finally, there is a whole non-system element of criminal actors, which can hardly be factored into even one layer of this conglomerate of various interest groups, but could serve as a source of constant disruption despite any agreement.

Furthermore, high-level political norms, even though non-binding thus far, have finally gained traction to shape the landscape for granular and lower-level norms development and implementation. The results of this work, however, will only be seen in the practical implementation. Let us look at the achievements made and challenges faced so far.

## The Routes and Rules

The 2013 UN GGE report to the UN Secretary General[3] stated that the norms of international law (including the UN Charter and the International Humanitarian Law [IHL]) applied to cyberspace, too. It emphasised the importance of CBMs to build trust among the states as well as develop capacities for better cooperation and efficiency at the technical level. In the same year, the Tallinn Manual on the International Law Applicable to Cyber Warfare,[4] developed by an informal group of experts from the North Atlantic Treaty Organisation (NATO) Cooperative Cyber Defence Centre of Excellence, attempted to describe how this principle can be operationalised in the *jus in bello* and *jus ad bellum* situations. This guide remains the only comprehensive work of its kind, except the US Department of Defense (DoD) Law of War Manual[5] released in 2015, which echoes the Tallinn Manual. The Tallinn Manual 2.0, to be finalised later in 2016,[6] will look at how peace-time international law applies to cyberspace and how legal instruments can be used to address threshold attacks.

Given the scope of the UN GGE report and the reservations about the nuances of IHL application to cyberspace, the international community took some time to arrive at a general consensus that the mere definition, specification and adaptation of the bodies of international law applying to cyber-enabled conflicts was not enough. In this respect, to avert a conflict-like situation in cyberspace, and to allow space for deliberations, rather than have confrontations, the letter of hard law is sometimes secondary to provisions of soft law.

As mentioned above, there have been a few attempts at defining such soft norms so far; we'll focus on the most salient ones.

- In September 2011, Russia, China, Tajikistan and Uzbekistan (all members of the SCO) put forward an International Code of Conduct[7] for information security highlighting the threats that the use of ICT could represent for state sovereignty, the security of information space integrity and internal state stability. In the same year, the Concept for the Convention on International Information Security[8] was presented at the high-level meeting of international security officers in Ekaterinburg and in November 2011 at the first GCCS in London. January 2015 saw a second version[9] of the Code of Conduct (also signed by two other SCO members, Kazakhstan and Kyrgyzstan) submitted in a letter to the UN Secretary General, with little substantial change; however, the notion of information weapon was removed from the text and a section on CBMs was included in Article 10. Even though the initiative hasn't gained signatories since then, the latest UN GGE report (2015) contains a reference that takes note of the Code.
- The GCCS, which was launched in London in 2011, set up a high-level ministerial platform for the discussion about the key themes in cyberspace. The London Conference initiated a broad dialogue on the opportunities and challenges that arise from an increasingly networked world.[10] The 2012 conference in Budapest was marked by discussions around cybersecurity and norms of behaviour. The Seoul Conference of 2013 produced the Seoul Framework for and Commitment to Open and Secure Cyberspace[11] against the backdrop of Snowden's revelations. The 2015 GCCS in The Hague revamped the format by introducing a multi-stakeholder composition and setting up the Global Forum on Cyber Expertise,[12] a platform for cyber knowledge accumulation, sharing and implementation facilitated by the founding states and private companies. The Hague itself has been promoted as the centre for peace and security for cyber domain, and the Netherlands might join the next UN GGE.
- The BRICS summit in Ufa in July 2015 allocated considerable space to the cybersecurity agenda. The Ufa Declaration (Article 34) declares setting up a Working Group of Experts of the BRICS states on security in the use

of ICT to "share information and best practices relating to security in the use of ICT; effective coordination against cyber-crime; the establishment of nodal points in member-states; intra-BRICS cooperation using the existing Computer Security Incident Response Teams (CSIRT); joint research and development projects; capacity building; and the development of international norms, principles and standards". Despite some differing stances, the BRICS format is the one where Russia feels it can reinstate the shared values and vision with its partners, especially China and India. For instance, the principle of sovereignty as applied to cooperation in the prevention of cyber conflict, ICT export controls and Internet governance.

- In 2013, the OSCE member-states took further the UN GGE achievements and agreed on the initial set of CBMs[13] to reduce the risk of conflict stemming from the use of ICT. The list of 11 measures enhancing transparency, information and best practices exchange facilitation among states, respective relevant agencies and Computer Emergency Response Teams (CERTs) lays the ground for building the trust essential for any deeper cooperation on conflict prevention. The extension of the initial list of CBMs[14] was unveiled in March 2016, focusing on measures assisting CI protection. The recommendations include, *inter alia*, more cooperation on the alignment of national definitions of CI as well as threat categorisation. For the implementation of CBMs, where appropriate, the multi-stakeholder approach is suggested to factor in advice from the private sector, civil society and technical community. Further, developing private-public mechanisms in accordance with the domestic law is also recommended. Finally, drawing on the UN GGE achievements, the development of the stability measures will mean embracing the normative approach to state behaviour in cyberspace.

  With all the limitations, given the caveat of differing national jurisdictions and the level of distrust among states and CERTs, the CBMs-centred collaboration still offers states a chance to fill this gap and build trust, where possible. It also gives a favourable context and offers an opportunity for the private actors engaged in strategic cybersecurity processes to interact and cooperate with their respective governments as well as counterparts in other countries.
- In summer 2015, the report[15] of the fourth UN GGE, which comprised 20 experts, proved that despite some irreconcilable stances on cyber issues, countries can agree on a minimum of voluntary norms of state behaviour in cyberspace. Since its inception in the late 90s, mostly due to the Russian diplomatic push, the UN GGE has been looking at the threats of the use of ICTs and promoting responsible state behaviour online. Eventually, with better understanding of the damage these threats were capable of, the idea of responsible conduct gained traction.

The latest UN GGE report outlines three initial norms of responsible state behaviour in cyberspace: 1) not to conduct or knowingly support ICT activity contrary to its obligations under international law that intentionally damages CI; 2) not to conduct or knowingly support activity to harm the CERTs of another state; and 3) not to knowingly allow their territory to be used for internationally wrongful acts using ICTs. The report also highlights CBMs and capacity-building goals as important elements in states' efforts to establish long-term cyberstability architecture and to raise the global level of resistance to cyberthreats.

The Group also confirmed the commitment to carry out due investigation on malicious activity before counteractions are taken; assist in investigations of cyberattacks and cybercrime launched from a country's territory; and peaceful use of ICTs as the basis for peace and security in cyberspace and beyond. Besides, the Group subscribed to the principle of state sovereignty applying to cyberspace and non-interference with states' internal affairs via the use of ICTs.

On December 23, 2015, the UN General Assembly adopted resolution 70/237, welcoming[16] the outcome of the 2014/2015 GGE, and requested the Secretary-General to set up a new GGE that would report to the General assembly in 2017, and hold its first meeting in New York in August 2016. At the time of writing, it is known that the group will be expanded by a few more members, but the list has not been revealed yet.

## Soft Law and Hard Truth

The wealth of international platforms and initiatives looking at the ways to avert a cyberwar-like situation (even though there are no definite parameters to show where the threshold is crossed) and enact the law of armed conflict is impressive. States are discussing measures of restraint, deterrence and stability, though threats often come from sides which are not parties to any of the political agreements, and lead to below-the-threshold attacks. Cyber-enabled crime and terrorism are a much bigger peril as the actors behind them cannot be put at the negotiating table and the potential damage is huge.[17]

Is it feasible to achieve cyberstability as a result of interstate negotiated commitments given the increasing capacity of various non-state actors having access to state-of-the-art technologies and the challenge of identifying source and intent of attacks? There are certain caveats here:

- The norms building process might not be the ultimate goal but rather a medium to achieve an acceptable level of mutual trust and assurance at least among the major players. Securing stability at this level could allow shifting the focus on cybercrime and the non-state players, factoring in the global cybersecurity ecosystem or persecution. If the private sector reinforces these measures, developing and abiding by the industry norms and codes of conduct, especially in the sectors deemed critical by the respective states, it would increase the synergy of the efforts made so far. For example, Microsoft has attempted to bridge the dialogue between the states and the business/technical community on mutual expectations and trust building in ensuring cyberstability.[18]

- Though the norms, principles and recommendations set out in the GGE report are voluntary, setting a high-level political benchmark is important to frame the issue and provide background for constructive work in the field at the national and regional levels. Some players also cherish a hope that over time the practice of responsible behaviour and tools to maintain it will allow for a binding agreement among the sides. The preparatory work starts at home, in national jurisdictions, in the formats of state and business cooperation to enable cyberdefence and protect CI. These form the basis for implementing high-level agreements. But the level of maturity in public-private partnerships differs from country to country, challenging the possibility for a single solution, while the experience nonetheless then feeds the discussion at higher levels.
  Even at the European Union (EU) level, the Network and Information Security (NIS)[19] Directive, looking at fostering cooperation between member states and outlining security obligations for operators of essential services (energy, transport, health and finance) and digital service providers, is expected to be challenging to implement. It is to be integrated at the nation-state level and harmonised with already existing and differing national cybersecurity and data protection regulations; and fine-tuning the public-private partnership mechanisms is equally relevant.
- The normative frameworks are set to develop over the years as states will try to achieve more granularity in the rules of the game, including on the UN GGE platform and other fora. For instance, the norm on non-attacking CI implies the specification of the types of such infrastructure, which are not universally uniform and often have private operators. In the OSCE second set of CBMs it is recommended that the states work at harmonising the CI classification. A joint effort is needed to protect CIs, particularly financial, nuclear and water facilities, their cyber interfaces as well as the 'public core of the Internet'[20] – the root server infrastructure. This could well be a part of the next UN GGE mandate. Thus, it will be critical for the states to establish a workable format of interaction with the private sector to implement the commitments they make. At the same time, granular rules will be more challenging first to develop and adopt by consensus, and then to abide by them: it is undoubtedly a difficult balance to strike.

- The ICTs develop at a speed which by far exceeds that of law- and decision-makers. In this regard, the cyber element renders international stability, peace and security a constantly moving target. This suggests that whichever normative frameworks are developed at the UN/OSCE or other top-level international fora will have to allow space for flexibility and adjust accordingly or face irrelevance. To keep up with new challenges, the parties might eventually develop expert, technical, academic formats and platforms collaboratively feeding the political level. One of such platforms could be the Dutch-Estonian initiative.[21] The United Nations Institute for Disarmament Research (UNIDIR) series of workshops[22] on International cybersecurity issues is another example of an expert platform feeding the UN GGE work.
- The low trust level among states and other actors might undermine the advances already made. The work on CBMs (by OSCE and other organisations) helps keep up an open dialogue even if the resulting agreements are non-binding. Trust at the level of technical operators (e.g. CERTs) has always been mentioned as crucial, long in the gaining and easily lost. However, even at the technical level institutional relations do not necessarily function in a smooth un-biased way. And it is indeed the responsibility of all stakeholders to work out conditions to facilitate efficient cooperation, which often comes down to informal trusted network collaborations. Such trust is both a pre-requisite and a desirable outcome of public-private partnerships needed to be built to resist malicious cyber activity. However, how such trusted channels can be reconciled with the high levels of secrecy of classified information on vulnerabilities, which hardly any side would readily share, remains to be seen.

While multilateral groupings find it hard to reach a consensus of multiple voices, bilateral agreements suggest a more viable path for constructive and implementable decisions. While sharing vision about threats and risks as well as remedies, the sides can more meaningfully operationalise the decisions made. A bilateral axis thus could be both a way to formalise unity of stances and to implement practically the higher level decisions to which both are parties.

It is thus interesting to observe how the recent Russia-China-India summit has shifted the focus from the BRICS agenda to the RIC agenda[23] on a range of issues, including security in the use of ICTs. They emphasised their "adherence to universally recognised principles of international law in the use of ICTs, in particular, the principles of political independence, territorial integrity and sovereign equality of states, respect for state sovereignty, non-intervention into the internal affairs of other states" – the values of priority shared by all the three sides. The three sides agreed on the importance of "elaboration and adoption of universal rules of responsible behaviour of states in the use of ICTs to prevent conflicts in information space". Reportedly, the states "underlined the importance of providing timely and appropriate responses to requests from one another for information and assistance concerning malicious incidents and activities in the use of ICTs and agreed to cooperate in this area". All these aspects echo the UN GGE and OSCE commitments where both Russia and China are partners, and by embracing India extend those diplomatic achievements. Thus, three out of five BRICS countries agreeing to adopt universal rules in the given context indicates a bigger proximity of stances, which eventually might lead to broader cooperation on cyberthreats.

A lot of attention is given to the US, China and Russia triangle, which stands out as the key to global cyberstability policy and cyberdiplomacy activity. The fact that both China and Russia feature in the US DoD cyber strategy[24] of 2015 as sources of major threats certainly makes this triangle ever more important. All the three have a history of bilateral agreements in the field of cyber/information security allowing for more meaningful discussions in other formats as well.

### *Russia-US, 2013*

In autumn 2013, at the Group of Eight (G8) summit in Scotland, Presidents Obama and Putin signed a series of agreements[25] on CBMs in cyberspace to address "the issues of threats to or in the use of ICTs in the context of international security". The measures included the use of the 24/7 direct communication link between the US and the Russian Nuclear Risk Reduction Centres (NRRCs) established in 1987 between the United States and the former USSR; information sharing arrangements between the US and Russian CERTs for regular technical update on cybersecurity risks to critical systems; a direct voice communications line between the US Cybersecurity Coordinator and the Russian Deputy Secretary of the Security Council to manage a crisis situation arising from an ICT emergency.

This bilateral breakthrough, a good example of how the two countries could cooperate to work out practical tools for dealing with a cyber emergency, only had time to serve the needs of Sochi Olympics but has been less operational since then due to political tensions. In April 2016 the two sides reportedly held talks[26] in Geneva led by the Deputy Head of the Security Council, Sergey Buravlev, and Special Assistant to the US President and Cybersecurity Coordinator, Michael Daniel, to discuss reinforcement[27] of bilateral CBMs and shared goals in the development norms of responsible state behaviour to confront the global security challenges enhanced with the use of ICTs.

While a genuine rapprochement is hard to expect at the time when diplomatic tensions are still high, such a meeting suggests at least two things: 1) Both sides face high damage risks in case of any unfortunate mistake as a result of miscommunication and misinterpretation of intent (as the experience of the Cold War suggests). 2) The existing tensions call for authoritative sources of information of the other side's intent. Besides, the experience of continuous cooperation in the UN GGE, and other formats, might have helped develop more points of mutual understanding.

### *Russia-China, 2015*

In May 2015, Presidents Xi Jinping and Putin signed in Moscow a cyber non-aggression agreement[28] confirming the intent to abstain from knowingly attacking each other with the use of ICTs. While reflecting the common take on the threats in cyberspace, in many instances, it draws on the UN GGE report as well as the SCO Code of conduct. The two countries agreed on joint development of standards of conduct in cyberspace; shared vision on importance of Internet governance globalisation; not attacking the respective CI facilities; cooperation on critical information infrastructure protection; information exchange and mutual assistance between law enforcement bodies to investigate cybercrime and ICT incidents; and setting up official channels for such information exchange.

So far, there is little evidence to suggest that Russia and China have engaged in intensive malicious activities against each other; therefore, this agreement appears to be a framework for more granular cooperation tools in the future. It could be a comfortable platform for operationalising the CBMs and a basis for further trust building to the extent acceptable to both sides.

### *US-China, 2015*

The issue of cybertheft of intellectual property was raised in John Kerry's speech in Seoul in May 2015 in which he voiced the principles[29] of cyber conflict prevention ("no country should conduct or support cyber-enabled theft of intellectual property, trade secrets, or other confidential business information for commercial gain"). This is the only point from that list which was not reflected in the UN GGE report the following summer.

The US makes a distinction between cyber intelligence gathering (which is neither ruled out by any body of international law, nor openly endorsed) and commercial cyberespionage. In April 2015, President Obama issued an executive order threatening sanctions against those actors who posed a threat to national security.[30] In August 2015, shortly before Xi Jinping's state visit to the US in September 2015, the US had reportedly planned to impose sanctions[31] specifically against Chinese hackers. Such cyber concerns certainly composed the agenda during this state visit, as was reflected in the White House statement,[32] released at the end of the visit. It proclaimed that the two sides had agreed to curb cybertheft: "neither country's government will conduct or knowingly support cyber-enabled theft of intellectual property with the intent of providing competitive advantages to companies or commercial sectors." Moreover, in certain aspects, the announcement echoed the Russia-China and Russia-US agreements, particularly regarding the commitment to "identify and promote appropriate norms of state behaviour in cyberspace within the international community", "cooperate using a hotline to investigate cybercrimes" and "mitigate malicious cyber activity emanating from their territory".

However, the communique isn't a formal agreement, though both China and the US have reiterated some UN GGE commitments. Given China's cyber military ambitions as part of the national security strategy and the intention to keep a close control[33] on data flows at home, the two countries can be expected to remain in dispute over this and other issues, with the private sector on both sides kept hostage.

One important note to make in this context is that such agreements do not stop the concerned states from developing cyberstrategies with both defensive and offensive elements. For instance, the US Cyber Command (USCYBERCOM) under the US Strategic Command focuses on both ensuring DoD information networks security and developing a full spectrum of military cyber capacity to be enabled when necessary.[34] To this end, the newly announced Cybersecurity National Action Plan (CNAP[35]), in February 2016, puts focus on "'paradigm changing' approach to innovation, national education and work force development, which will then be reflected in paradigm changing approaches to military workforce development and deployment". The Chinese white paper on China's military strategy,[36] published in May 2015 by the State Council Information Office of the People's Republic of China, for the first time stated that "outer space and cyberspace have become new commanding heights in strategic competition among all parties". The paper suggests China's intent to join the world major powers which "are actively adjusting their national security strategies and defence policies, and speeding up their military transformation and force restructuring". Therefore, the real restraining effect of the norms of state behaviour in cyberspace could only be tested in an impending or actual conventional conflict situation.

## Conclusion

The not-completely-new discourse of developing norms, standards and principles of state behaviour in the use of ICTs to avoid conflict in cyberspace is gaining traction as more state and institutional players are joining this effort. The diversity of such initiatives helps strengthen the dialogue among the key players and foster trust building relationships through establishing cooperation commitments.

Nevertheless, as can be seen, the bilateral-level agreements made within the UN GGE/OSCE frameworks represent the common ground; in other words, they are merely stepping stones to negotiating more pragmatic and relevant concerns. This certainly doesn't diminish the political achievements, as they have helped establish a minimal level of understanding and even develop a common language in the area of cyberthreats. On the other hand, such agreements should be used as tools for more meaningful cooperation for which not all sides would be necessarily equally ready, politically or technically. Otherwise, political norms of behaviour in cyberspace will struggle to fulfil their declared aim of cyberstability, international peace and security.

## NOTES

1. Organisation for Security and Co-operation in Europe (OSCE), "Initial Set of OSCE Confidence-Building Measures to Reduce the Risks of Conflict Stemming from the Use of Information and Communication Technologies", December 3, 2013, at http://www.osce.org/pc/109168?download=true.
2. BRICS, Documents, at http://en.brics2015.ru/documents/.
3. US Department of State, "Statement on Consensus Achieved by the UN Group of Governmental Experts on Cyber Issues", Washington DC, June 7, 2013, at http://www.state.gov/r/pa/prs/ps/2013/06/210418.htm.
4. Cooperative Cyber Defence Centre of Excellence (CCDCOE), Research, at https://ccdcoe.org/research.html.
5. Office of General Counsel Department of Defense, *Department of Defence Law of War Manual*, June 2015, at http://www.defense.gov/Portals/1/Documents/pubs/Law-of-War-Manual-June-2015.pdf.
6. CCDCOE, "Tallinn Manual 2.0 to Be Completed in 2016", October 9, 2015, at https://ccdcoe.org/tallinn-manual-20-be-completed-2016.html.
7. United Nations General Assembly, Developments in the Field of Information and Telecommunications in the Context of International Security, 6th session, Item 93 of the provisional agenda, September 14, 2011, at https://ccdcoe.org/sites/default/files/documents/UN-110912-CodeOfConduct_0.pdf.
8. Ministry of Foreign Affairs of the Russian Federation, "Convention on International Information Security (Concept)", at http://archive.mid.ru/bdomp/ns-osndoc.nsf/1e5f0de28fe77fdcc32575d900298676/7b17ead7244e2064c3257925003bcbcc!OpenDocument.
9. United Nations General Assembly, Developments in the Field of Information and Telecommunications in the Context of International Security, 69th session, Agenda Item 91, January 13, 2015, at http://archive.mid.ru/bdomp/ns-dmo.nsf/c85969b2329a429944257d5600225ebb/44257b100055f7e6c3257db4004192e4/$FILE/A%2069%20723%20En.pdf.
10. Global Conference on CyberSpace, The London Conference on Cyberspace, November 1-2, 2011, at https://www.gccs2015.com/london-1-2-november-2011.
11. "Seoul Framework for and Commitment to Open and Secure Cyberspace", at http://www.mofat.go.kr/english/visa/images/res/SeoulFramework.pdf.
12. Global Forum on Cyber Expertise (GFCE), at http://www.thegfce.com/.
13. OSCE, No. 1.
14. OSCE, "OSCE Participating States, in Landmark Decision, Agree to Expand List of Measures to Reduce Risk of Tensions Arising from Cyber Activities", Vienna, March 10, 2016, at http://www.osce.org/cio/226656.

15. United Nations General Assembly, *Report of the Group of Governmental Experts on Developments in the Field of Information and Telecommunications in the Context of International Security*, July 22, 2015, at http://www.un.org/ga/search/view_doc.asp?symbol=A/70/174&referer=http://www.un.org/disarmament/topics/informationsecurity/&Lang=E.
16. United Nations Office for Disarmament Affairs (UNODA), "Developments in the Field of Information and Telecommunications in the Context of International Security", at http://www.un.org/disarmament/topics/informationsecurity/.
17. The recent research by Microsoft and Group IB suggest that cybercrime costs Russia up to 0.25 per cent of its GDP annually (½ 203.3 billion). See https://news.microsoft.com/ru-ru/kiberprestupnost-2015-025-vvp/#sm.0001914kbcquodxc10w3t472vjy1l.
18. At the end of 2014, Microsoft released the paper, "International Cybersecurity Norms, Reducing Conflict in an Interdependent World", with its own conceptualisation of principles of state behaviour. Its six cybersecurity norms aim at "reducing the possibility that ICT products and services are used, abused or exploited by nation states as part of military operations". Some of them are echoed in the UN GGE report of 2015 (e.g. Norm 6: States should assist private sector efforts to detect, contain, respond to, and recover from events in cyberspace), others don't have consensus so far (e.g. Norm 3: States should exercise restraint in developing cyber weapons and should ensure that any which are developed are limited, precise, and not reusable).
19. European Commission, "Network and Information Security Directive: Co-legislators Agree on the First EU-wide Legislation on Cybersecurity", December 9, 2015, at https://ec.europa.eu/digital-single-market/en/news/network-and-information-security-directive-co-legislators-agree-first-eu-wide-legislation.
20. Dennis Broeders, *The Public Core of the Internet: An International Agenda for Internet Governance*, Amsterdam University Press, Amsterdam, 2015, at https://www.academia.edu/18500772/The_public_core_of_the_internet_an_international_agenda_for_ internet_governance.
21. CCDCOE, No. 4.
22. United Nations Institute for Disarmament Research (UNIDIR), International Security Cyber Issues Workshop Series, at http://www.unidir.org/programmes/emerging-security-threats/international-security-cyber-issues-workshop-series.
23. Devirupa Mitra, "Consultations on Asia, Illegal Drug Trade and Cyber Security: Key Takeaways from Russia-India-China Summit", *The Wire*, April 20, 2016, at http://thewire.in/2016/04/20/consultations-on-asia-illegal-drug-trade-and-cyber-security-key-takeaways-from-russia-india-china-summit-30443/.
24. USDoD, *The DOD Cyber Strategy*, April 2015, at http://www.defense.gov/Portals/1/features/2015/0415_cyber-strategy/Final_2015_DoD_CYBER_STRATEGY_for_web.pdf.
25. The White House, Office of the Press Secretary, "FACT SHEET: U.S.-Russian Cooperation on Information and Communications Technology Security", June 17, 2013, at https://www.whitehouse.gov/the-press-office/2013/06/17/fact-sheet-us-russian-cooperation-information-and-communications-technol.
26. http://www.kommersant.ru/doc/2975067.
27. http://www.scrf.gov.ru/news/1068.html.
28. http://government.ru/media/files/5AMAccs7mSlXgbff1Ua785WwMWcABDJw.pdf.
29. John Kerry, US Secretary of State, "An Open and Secure Internet: We Must Have Both", May 18, 2015, at http://www.state.gov/secretary/remarks/2015/05/242553.htm.
30. Cassie Spodak, "Obama Announces Executive Order on Sanctions against Hackers", CNN, Washington, April 1, 2015, at http://edition.cnn.com/2015/04/01/politics/obama-cyber-hackers-executive-order/.

31. Ellen Nakashima, "U.S. Developing Sanctions against China over Cyberthefts", *The Washington* Post, August 30, 2015, at https://www.washingtonpost.com/world/national-security/administration-developing-sanctions-against-china-over-cyberespionage/2015/08/30/9b2910aa-480b-11e5-8ab4-c73967a143d3_story.html.
32. The White House, Office of the Press Secretary, "FACT SHEET: President Xi Jinping's State Visit to the United States", September 25, 2015, at https://www.whitehouse.gov/the-press-office/2015/09/25/fact-sheet-president-xi-jinpings-state-visit-united-states.
33. Paul Mozurjan, "New Rules in China Upset Western Tech Companies", *The New York Times*, January 28, 2015, at http://www.nytimes.com/2015/01/29/technology/in-china-new-cybersecurity-rules-perturb-western-tech-companies.html?_r=0.
34. https://www.stratcom.mil/factsheets/2/Cyber_Command/.
35. The White House, Office of the Press Secretary, "FACT SHEET: Cybersecurity National Action Plan" February 9, 2016, at https://www.whitehouse.gov/the-press-office/2016/02/09/fact-sheet-cybersecurity-national-action-plan.
36. "Full Text: China's Military Strategy", *Xinhuanet*, Beijing, May 26, 2015, at http://news.xinhuanet.com/english/china/2015-05/26/c_134271001.hm.

# 5

# Defence, Deterrence, and Diplomacy: Foreign Policy Instruments to Increase Future Cybersecurity

*Sico van der Meer*

## Introduction

Predicting the future is hardly possible, but stating that cyber aggression – be it espionage, sabotage or even warfare – will be a continuing threat to international security and stability in the coming years seems a safe forecast. This chapter deals with the question of how states can cope with this forecast from a foreign policy perspective, focussing on cyber aggression conducted or sponsored by state actors.

Defence and deterrence, which could be labelled passive deterrence and active deterrence as well, are probably the most 'obvious' counter-measures to international cyber aggression that a state could implement. This chapter especially analyses why defence and deterrence look like promising policies, but in practice face some difficulties in the cyber realm.

Diplomatic efforts to create Confidence Building Measures (CBMs) and international accepted norms regarding cyberthreats could be more effective in actively addressing the core problems of international cyber aggression, but are little successful so far. The chapter argues that such multilateral diplomatic efforts are crucial for long-term cybersecurity and stability. Instead of an on-going 'cyber arms race', efforts could better be focussed on building mutual confidence and respect as well.

## Evolving Cyberthreats

Cyberthreats, also referred to as digital threats, are generally considered as an important, and still increasing risk in international security. Cyberthreats encompass a broad spectrum. Examples include digital warfare, digital terrorism, digital espionage, digital activism and digital crime. While the purpose of each type of activity differs, in each type the weaknesses within the cyber domain are exploited to harm an 'opponent'.

It is clear that the number of cyberattacks in the world is increasing sharply over the years. It is very difficult, however, to determine the exact number of attacks, as most attacks are never reported. Indeed, individuals or organisations often remain unaware that they have been attacked, since the purpose of many cyberattacks is precisely to hack into computers or computer networks while avoiding detection.

There are so many forms and types of cybersecurity breaches, and they are committed by such a variety of actors, that it is not reasonable to analyse them as a uniform group. Cyberattacks can literally range from students who hack into other people's computers for relatively harmless fun, to large-scale industrial espionage, to digital warfare waged for the purpose of disrupting an entire society. Nevertheless, within the limitations of this chapter, a cautious attempt is made to provide a general outline of the situation, especially focussing on cybersecurity from a state-level perspective.

Especially cyberespionage and cybercrime are currently conducted at large scale, all over the world. Cyber sabotage, cyberterrorism or cyberwarfare are far less common so far. One may forecast that this frequency of the types of cyber aggression will not easily change.

The continuing digitalisation of most societies is increasing the risk of more large-scale cyberattacks aimed at disrupting society. In terms of the security of individuals and society, the greater the reliance on digitisation, the greater the impact of malicious acts carried out by parties who abuse digital environments for their own ends.

Cyberespionage and cybercrime primarily cause economic damage. In addition to economic consequences, such as weakening the competitive economic position of a state, cyberespionage in particular is also a security issue in that it can be used by potential enemies, whether state or non-state actors, to learn a great deal about the national security situation and to discover potential weaknesses. Stolen information about vital infrastructure or military capabilities, for example, could be used to cause harm by digital or non-digital means.

Whereas cyberattacks on organisations, companies and individuals are by now fairly common throughout the world, there have so far been only a few cyberattacks aimed at causing large-scale disruption in society. The most well-known examples are the attacks that took place in Estonia in 2007 (attacks on the government, banks and media), the US (attacks on various banks) and South Korea (banks and media) in 2012. There are also examples of large-scale cyberattacks that were carried out for different purposes: Georgia in 2008 (by Russia to support its conventional military operation), Iran in 2010 (aimed at sabotaging the country's nuclear programme), Saudi Arabia in 2012 (attack on state oil company, Saudi Aramco, possibly to sabotage oil exports) and the US in 2014 (attack on Sony Pictures Entertainment, possibly to prevent the release of a movie about the North Korean leader, Kim Jong-Un). Although the economic damage was considerable in a number of these cases, large-scale cyberattacks on a country's truly vital infrastructure, such as power plants, water purification plants, or, of vital importance in a country like the Netherlands: flood protection and water management systems, have as yet not taken place. Unfortunately, it cannot be excluded at all that some states will experience such kind of cyberattacks in the (near) future.

Although alertness to cyberthreats has increased considerably in most states during recent years, technological developments in the cyber domain are occurring at such a rapid rate that cybersecurity measures must constantly be modernised to keep up in the fight against those who may have an intent of doing harm. In spite of increased awareness of risks among users of cyber technology, whether they be organisations or private individuals, such users remain a weak link in the chain in terms of countering cyberthreats. To give just one example: in a highly digitalised country like the Netherlands, a governmental assessment in 2014 estimated that approximately 35 per cent of all computer users have not installed antivirus software, even though installing such software is the first and most basic step in the context of cybersecurity.[1]

## A 5-10 year Forecast

Although there is currently a lack of clarity in terms of the exact number of cyber incidents, the international threat of cyberattacks – in any way – will certainly increase in the near future, mainly because of the further digitalisation of most societies, also in vital sectors. The number of devices and appliances (medical devices, household appliances and automotive devices, to mention only a few examples) that are connected to each other and to the Internet will increase exponentially worldwide to approximately 25 billion in 2020.[2] The greater the dependence on cyber technologies, the more vulnerable any society will be to cyberthreats. Because a growing number of processes are occurring in the digital domain and a growing number of devices and appliances are connected to cyber networks, the risk of these processes, devices and appliances being manipulated by unauthorised parties is increasing correspondingly.

While in many states considerable progress is being made with respect to the security of the cyber domain in terms of, for example, increasing awareness of the risks and the technological level of security of vital cyber infrastructure, other actors are also very much on the move. Many state as well as non-state actors are investing in offensive cyberwarfare capabilities; references are regularly made in this context to a cyber arms race.[3]

Because cyberattackers immediately look for other weaknesses as soon as a gap in security has been closed, they virtually always have an advantage over cyber defenders, especially because it is impossible to close every security gap in the cyber infrastructure. Cybersecurity will therefore always be a competition between attackers who are exploiting or seeking to exploit a newly discovered weakness and defenders who work to close a detected security gap as quickly as possible.

Cybercrime and cyberespionage will continue to pose the most common threats in the future. Cybercriminals are becoming more professional and their cyberattacks are becoming more sophisticated and greater in scope. Cyberespionage carried out by states as well as private organisations (industrial espionage) will likewise increase.

In addition, a major cyberterrorist attack remains a possible nightmare scenario. A great deal of damage could be caused by cyberterrorists who succeed in sabotaging, for example, energy supply systems, hospitals, chemical plants, nuclear installations, air and railway traffic control systems, flood protection and water management systems, or payment systems. Such an attack would likely lead to social unrest. In this sense, what applies to terrorism in general also applies to cyberterrorism: although the probability of an attack may be relatively low in statistical terms, the impact of such an attack would be considerable.

Actual cyberwarfare will presumably, mostly be combined with conventional warfare. It is safe to forecast that the cyber dimension of warfare will become increasingly important; even if a state has the most powerful conventional weapon systems, if an opponent is able to influence the cyber technology behind them – think of communication and command systems – they may be of less effect on the battlefield.

It is also important to bear in mind that cyber incidents in other countries can also have consequences for any state. A disruption to the American Global Positioning System (GPS), for example, could disrupt traffic in many other countries. Equally, if a cyberterrorist caused a nuclear disaster at a nuclear power plant, any radioactive fallout could also be an issue in surrounding countries, just as a cyberattack on international bank systems could disrupt payment transactions in various countries at the same time.

## Dealing with the Threat: Defence

How states could most effectively deal with international cyberthreats is a subject of ongoing discussion among researchers and policymakers. Although it is probably impossible to prevent all cybersecurity breaches, it is definitely possible to prevent many of them. Here some policy options from a foreign policy perspective will be discussed: defence, deterrence, and diplomacy.

The most obvious way to deal with cyberthreats is making such attacks more difficult for potential attackers by improving the security of cyber technology systems. One could label this as 'defence' of a state's cyber domain (although one may also label it 'passive deterrence'). One could think about technical defence measures, for example: multilayered firewalls, advanced encryption and thorough authentication methods. So-called 'honeypots' can also be used to improve security. These appear to be the kind of vulnerable areas in a system that cyberattackers are looking for, but they are in fact deliberately set traps designed to gather information about the working methods of cyberattackers. In practice, especially cybercriminals are known to avoid cyber infrastructures which are known for the use of such honeypots.[4]

Improving security increases the costs that an attacker must incur to carry out a successful cyberattack and makes it less likely that the attack will have the desired effects and gains. If cyber opponents know beforehand that the defence of a certain cyber infrastructure is well-constructed, they will less likely start a cyberattack (but instead may look for other ways to attack – or to attack another potential victim). From this perspective, defence is actually turning into passive deterrence. To achieve this, the cyber infrastructure of the potential victim must be secured in such a way as to ensure that attackers encounter barriers that considerably reduce the likelihood of their attack succeeding. Government authorities, organisations and private individuals can take a major step in such cyber defence simply by remaining very aware of the dangers of cyberattacks and ensuring that the latest security systems are always installed on their devices and networks. Networks must also continuously be monitored so that countermeasures can be taken as soon as there is any sign of an attack.

Improving cyber defence (or passive deterrence) entails fewer potential pitfalls than active deterrence or new diplomatic initiatives, as will be discussed later. This is why cyber defence is regularly regarded as the best way to deal with international cyberthreats.[5] The main problem is that cyber defence is expensive and complex and requires continuous investment; technological developments occur at such a rapid rate in the cyber domain that any stagnation means decline. In addition, it is difficult to raise full awareness on the part of all people involved; cyberattackers always exploit the weakest link in the chain that they can find and often these weakest links are human beings. In a manner of speaking, this could very well be that one inattentive employee among many other employees who downloads infected files, thereby creating an opening for the cyberattacker. As stated above, in a country like the Netherlands approximately 35 per cent of users do not even have functioning antivirus software installed on their computers. There is obviously a lot of room for improvement in terms of human awareness.

Moreover, cyberattackers always have the advantage in that they have all the time to look for weaknesses in cyber infrastructure, whereas the targeted party must respond as soon as a previously unknown weakness is exploited during a cyberattack. In other words, cyberattackers always have the element of surprise which makes defence traditionally more complicated.

## Dealing with the Threat: Deterrence

Considering that defence could also be labelled passive deterrence, here the policy option of active deterrence will be discussed. Active deterrence implies deterring potential cyberattackers by the possibility of retaliation. Retaliation of cyberattacks could be done by, for example, retaliatory measures within the cyber domain itself (a cyberattack on the attacker carried out by the party first attacked), diplomatic and/or economic sanctions, or even conventional military action against the attacker. In 2014, for example, the North Atlantic Treaty Organisation (NATO) decided that a cyberattack on one of its member states would be deemed to be an attack as defined in Article 5 of the North Atlantic Treaty, thus making it possible for the alliance to take military action against cyberattackers.[6] To a certain extent, deterrence would undoubtedly raise the threshold for cyber aggressors. A cost-benefit calculation by a potential attacker will surely be influenced by potential retaliatory measures.

Because of various specific characteristics of the cyber domain, however, it is relatively difficult to apply active deterrence as an instrument against cyberattackers.

The main obstacle to the effectiveness of such deterrence measures is the attribution problem. It is very difficult to conclusively identify the actor(s) responsible for (unclaimed) cyberattacks. Cyberweapons differ from other weapons, as the origins of cyberweapons are not clearly visible and traceable. For example, attackers can use a chain of hacked or infected computers without the owners of these computers actually being aware of any wrongdoing. Although it is technically possible to locate the source of a cyberattack by means of Internet Protocol (IP) addresses, there is always the possibility that the source identified was merely a link in the chain of the attack and that the owner was not in any way deliberately involved in the attack.

In addition, state actors can conceal their involvement by having cyberattacks carried out by non-state actors (hacker groups, for example). Conversely, non-state attackers may claim an association with a given state even if this is not actually the case. Moreover, cyberattackers can strike within a very short period of time and erase their tracks immediately after they have carried out the attack. Identifying the sources of the attack, on the other hand, is a complicated and time-consuming process. It is therefore almost impossible to take retaliatory measures during or immediately after the attack. Because it is difficult to establish the identity of the actor responsible for a cyberattack with absolute certainty, especially if the accused actor denies responsibility, there is a risk of a retaliation against an innocent party. In practice, few state actors will be willing to take this risk, something that cyberattackers are well aware of.[7]

It could be argued that indisputable and conclusive evidence is not required in some cases and that retaliatory measures can be taken if it is virtually certain that a certain state or non-state actor was involved or did not seek to stop the attackers.[8] However, leaving aside whether it is desirable to adopt this route – with the risks it entails of making false accusations – the question remains whether such an approach is actually permitted under international law. This is another area in the cyber domain where developments are still in full swing.[9]

Strong forensic capabilities in the cyber domain are crucial to identifying the party guilty of a cyberattack. A higher probability of being identified will also have a deterrent effect on potential attackers. In this regard, international cooperation, such as exchanging information about cyberweapons and cyber vulnerabilities that have been detected, is likewise essential.

In addition to the difficulty of conclusively identifying the actor responsible for a cyberattack, there are other problems associated with deterrence against such attacks as well. The credibility of deterrence and the risk of escalation are key issues. Deterrence based on the possibility of retaliation only works if the party seeking to deter communicates clearly about the retaliatory measures that may be taken in the event of a cyberattack. What acts are classified as cyberattacks that will trigger retaliation? Will retaliation take place in the cyber domain or is a conventional military strike also a possibility? If communication about possible retaliatory measures is not clear, it is unlikely that a potential attacker will take them into account and they will therefore not have a deterrent effect. After all, deterrence measures are only effective if the opponent is aware of these measures. Moreover, drawing 'red lines' in the cyber domain can also have the opposite effect to potential opponent. Cyberattackers may deliberately cross a red line to cause escalation, perhaps even while taking advantage of the attribution problem and posing as a different party. To maintain the credibility of deterrence, the party using it as an instrument must retaliate even if doing so at that specific time is not the favoured course of action. Any failure to adhere to the deterrence mechanisms communicated would dilute the deterrent effect, since potential opponents would be encouraged to think that the red lines are not all that red in practice.[10] From this perspective, deterrence may in certain circumstances even increase the risk of a vicious circle of escalating hostilities.

Another problem with deterrence based on retaliation in the cyber domain is the proportionality of the retaliatory measures. The effects of retaliation by conventional means can usually be fairly accurately assessed. The consequences of responding to a cyberattack through the cyber domain are more difficult to control, however. This is because a retaliatory cyberattack can easily have unintended consequences precisely because everything in the cyber domain is interconnected. A cyberattack on government networks, for example, may also accidentally affect networks of hospitals, water purification plants and other providers of essential services. A retaliatory attack carried out through the cyber domain may have greater effects than intended which could make the retaliating party the black sheep of the international community instead of the initial attacker.[11] The question as to when and the extent to which retaliatory measures may be taken is another problem. In the cyber domain, it is difficult to identify the boundary between acts intended to cause economic damage or disruption and obvious acts of war. There is as yet no clarity whatsoever regarding such issues.

A final key consideration is that the diversity of actors in the cyber domain makes deterrence difficult. State actors usually have interests that would be jeopardised by retaliatory action. However, non-state actors such as hacker or terrorist groups, for example, may not actually have any interests or goods of value against which a retaliatory attack could be directed, a situation which in itself undermines the credibility of retaliation. Moreover, such non-state groups, which are capable of carrying out major cyberattacks in spite of their relatively limited resources, may not always act rationally and may not even be deterred by any kind of possible retaliation.[12]

## Dealing with the Threat: Diplomacy

Diplomacy plays a role in defence and deterrence as well; think, for example, of diplomatic signalling to indicate the risk of retaliation to opponents.[13] However, diplomacy can also play an important role in increasing international cybersecurity next to defence and deterrence measures.

An important difference is that defence and deterrence are likely more effective in the short term, but diplomacy is most promising to contribute to international cybersecurity and stability in the long term. While defence and deterrence have almost direct positive effects on a state's cybersecurity, they bear the risk of continuing escalation. Ongoing investments in defence instruments may cause a 'cyber arms race' among potential opponents, and relatively minor incidents may escalate into a dangerous 'tit-for-tat' cycle of increasing seriousness because of the retaliation efforts required for effective deterrence.[14] Diplomacy may not offer any 'quick fixes' regarding cybersecurity problems, but in the long term it could offer a more secure and stable international environment in which cyber aggression becomes less likely.

Diplomacy has proven its ability to increase international security and stability regarding various other international threats, for example, the use of Weapons of Mass Destruction. The most important contribution that diplomacy has to offer to international cybersecurity are CBMs and international norms.

CBMs could enhance interstate cooperation, transparency and predictability, with the aim to reduce the risks of misperception, escalation, and conflict entailed by cyberthreats. In case of cyber aggression, CBMs could function as pressure valves, allowing a safe release of tensions before they escalate. CBMs can be both bilateral and multilateral. Various countries have agreements with other countries regarding, for example, cooperation in case of cyber aggression. An interesting regional initiative is the set of CBMs regarding cybersecurity developed by the Organisation for Security and Co-operation in Europe (OSCE).[15]

International norms established by multilateral diplomacy are to a large extent 'invisible', but very influential to international security and stability. Globally shared norms against the use of nuclear weapons, for example, make their use nearly unthinkable for many decades already. Diplomacy may contribute to establish similar norms regarding aggression in the cyber domain. Norms can provide shared understandings between states, allowing them to consider shared interests as well as finding ways to deal with diverging interests. Moreover, international norms facilitate cooperation among states through shared aims and terminology.

The diplomatic route to establish international norms regarding cybersecurity is not a short-term process. To come to broadly accepted norms, common values have to be found; states must perceive that following the norms is in their own national interest. Currently, however, many states have quite different values regarding state behaviour in cyberspace. Especially the clashing interests on the value of an open and free Internet and definitions of cybersecurity make setting international norms a difficult task.[16]

Moreover, states cannot establish norms regarding cyber issues on their own. In most states many more significant non-state players are active as well. Such non-state actors should also be incorporated in international discussions on cybersecurity norms, for example, large e-commerce firms, activists and experts. Many of them are in favour of minimum government interference in cyberspace, which may conflict with the aims of states. Although establishing international norms may thus be a difficult and time-consuming endeavour, in the end it will be worth the effort.

It should be noted that CBMs and international norms are not legally binding and thus their success completely relies on confidence between the states involved. Legally binding instruments, like treaties or conventions on state behaviour regarding cyber aggression, seem unrealistic to achieve in the current situation – not only because of a lack of shared views among states, but also because of the difficulties in verifying compliance to legally binding instruments regarding behaviour in cyberspace.

## Conclusion

In next five to 10 years, cybersecurity will be a key topic in international politics without doubt. Because of the on-going digitalisation in the world, the threats of cyber aggression in all its forms will increase as well. To deal with the risk of cyberthreats conducted or sponsored by state actors, states have several policy options available. Three of them have been discussed above: defence, deterrence and diplomacy.

While defence and deterrence policies offer good solutions in the short term, one may question whether they are able to offer international cybersecurity and stability in the long term. Both defence and deterrence policies entail a risk of an on-going cyber arms race and a cycle of escalation between potential cyber opponents.

Diplomacy may offer less results in the short term but is more promising in the long term. CBMs and international norms, which inherently must be based on mutual trust, may not always be easy to reach but in the end they could be more effective (and cheap) than a single focus on national cyber defence and deterrence strategies. In the long term, cooperation between states to establish confidence and commonly accepted norms of behaviour in cyberspace are most promising for enduring cybersecurity and stability.

## NOTES

1. National Cyber Security Centrum, "Cybersecuritybeeld Nederland: CSBN-4", 2014, p. 43, at www.ncsc.nl/actueel/Cybersecuritybeeld+Nederland/cybersecuritybeeld-nederland-4.html [in Dutch].
2. International Telecommunication Union, "Trends in Telecommunication Reform 2015", 2015, p. 4, at www.itu.int/en/publications/Documents/Trends2015-short-version_pass-e374681.pdf.
3. Michael Riley and Ashlee Vance, "Cyber Weapons: The New Arms Race", *Businessweek*, July 20, 2011.
4. TNO, KPN, National Cyber Security Centre & National Police (Netherlands), "European

Cyber Security Perspectives 2015", 2015, pp. 49-51, at www.tno.nl/en/about-tno/news/2015/3/european-cyber-security-perspectives-2nd-edition/.

5. David Elliot, "Deterring Strategic Cyberattack", *IEEE Security & Privacy*, 9 (5), 2011, pp. 38-39.
6. David E. Sanger, "NATO Set to Ratify Pledge on Joint Defense in Case of Major Cyberattack", *The New York Times*, August 31, 2014, at www.nytimes.com/2014/09/01/world/europe/nato-set-to-ratify-pledge-on-joint-defense-in-case-of-major-cyberattack.html?_r=0.
7. Emilio Iasiello, "Is Cyber Deterrence an Illusory Course of Action?", *Journal of Strategic Security*, 7 (1), 2013, p. 58; Advisory Council on International Affairs (Netherlands), "Digital Warfare", Advice No. 77, 2011, p. 13, at http://aiv-advies.nl/download/da5c7827-87f5-451a-a7fe-0aacb8d302c3.pdf.
8. Jason Healy, "Beyond Attribution: Seeking National Responsibility in Cyberspace", *Atlantic Council Issue Brief*, 2012, at http://www.atlanticcouncil.org/en/publications/issue-briefs/beyond-attribution-seeking-national-responsibility-in-cyberspace.
9. For a discussion on international law and cyberattacks, see Advisory Council on International Affairs, No. 7, pp. 19-27.
10. Martin C. Libicki, "Cyberdeterrence and Cyberwar", RAND Research Report, RAND Corporation, 2009, pp. 65-73.
11. Emilio Iasiello, No. 7, pp. 59-60.
12. Clorinda Trujillo, "The Limits of Cyberspace Deterrence", *Joint Forces Quarterly*, 75 (4), 2014, p. 49; Emilio Iasiello, No. 7, pp. 64-65.
13. Sico van der Meer, "Signalling as a Foreign Policy Instrument to Deter Cyber Aggression by State Actors", *Clingendael Policy Brief*, Netherlands Institute of International Relations 'Clingendael', 2015, at www.clingendael.nl/publication/signalling-foreign-policy-instrument-deter-cyber-aggression.
14. Sico van der Meer and Frans-Paul van der Putten, "US Deterrence against Chinese Cyber Espionage: The Danger of Proliferating Covert Cyber Operations", *Clingendael Policy Brief*, Netherlands Institute of International Relations 'Clingendael', 2015, at www.clingendael.nl/publication/danger-proliferating-covert-cyber-operations.
15. Organisation for Security and Co-operation in Europe, "OSCE Decision 1106", PC.DEC/1106, 2013, at http://www.osce.org/pc/109168?download=true.
16. Henry Farrell, "Promoting Norms for Cyberspace", *Cyber Brief*, Council on Foreign Relations, 2015, at www.cfr.org/cybersecurity/promoting-norms-cyberspace/p36358.

# 6

# Securing from Cyberthreats: Developing Defence, Deterrence and Norms

*A. Vinod Kumar*

Sometime in 2008, there was a sudden surge in attempts to infringe computers and IT systems at the Institute for Defence Studies and Analyses (IDSA), New Delhi. Initially, webmail accounts were surreptitiously created using existing user-ids (in other accounts), and malware-ridden files were sent as attachment to colleagues and other members of the strategic community, infecting a number of computers. Over the next several months, many spurious mails started doing the rounds, including of distinguished personalities, all with infected malware or phishing traps. Some of these emails had attachments which underlined specific targeting, with files on topics pertaining to our research specialisations. During the same period, there were reports of many other governmental institutions going through a similar spell.

In April 2010, the Information Warfare Monitor and Shadowserver Foundation, run by Munk Centre for International Studies in Toronto, published a report entitled, "Shadows in the Cloud: Investigating Cyber Espionage 2.0". The report revealed details of a cyberespionage network, christened the *Shadow Network*, which hacked into and stole information from personal computers in government offices from a number of countries.[1] Most of the information stolen by the hackers in this network pertained to classified and restricted documents in the Indian Ministry of Defence (MoD), as well as a many strategic organisations like the National Security Council Secretariat (NSCS), diplomatic missions, Military

Engineer Services, numerous Army brigades and Air Force stations, as well as the IDSA. The report said:

> We assess that computers at the Institute for Defence Studies and Analyses (IDSA) were compromised based on the documents exfiltrated by the attackers. During the period in which we monitored the attackers, 187 documents were exfiltrated. While many of the documents were published papers from a variety of academic sources, there were internal documents, such as an overview of the IDSA research agenda, minutes of meetings, budgets and information on a variety of speakers, visitors, and conference participants.

A *New York Times* story on this report explained the trend thus:

> ... classified assessments about security in several Indian states, and confidential embassy documents about India's relationships in West Africa, Russia and the Middle East. The intruders breached the systems of independent analysts, taking reports on several Indian missile systems ... also obtained a year's worth of the Dalai Lama's e-mail messages.[2]

The study by the Information Warfare Monitor and Shadowserver Foundation consortium on the *Shadow Network* was a follow-up to their ongoing investigation on another network, the *GhostNet,* which they claim is an earlier version of the former.[3] The study revealed that targets of *GhostNet* were 'high-value' installations in various countries, though a major focus of the attack seemed to be on the Tibetan Government-in-Exile. Both these studies traced the source of the attack to Chinese hackers, though short of implicating the Chinese Government's complicity in these campaigns. While the Chinese Government denied any links to these attacks, there was enough evidence in terms of Beijing's actions – including statements referring to 'Internet armies' and 'Information warfare units' as well as regarding its approaches to the Government-in-Exile[4] – that points to the possibility of the attackers "being directed in some manner by the agents of the Chinese State, or the obvious correlation to be drawn between the target of attacks and strategic interests of China".[5] The initial response in India to the reports of such attacks was to shrug it off as hacker activity. Though the Indian Government had supposedly prepared for such eventualities much earlier,[6] the Canadian report illustrated how the establishment woke up late to detect this clandestine activity and to discern its actual security dimensions.[7]

Events in the final years of last decade – the phenomenon of Chinese hackers spreading havoc globally, as also of the high-intensity attacks on the communication and cyber infrastructure of Estonia (2007), Georgia (2008) and Kyrgyzstan (2009), blamed on Russian hackers – changed the strategic narrative on cyberspace and the vulnerability of nations and their strategic assets to a new form of warfare-in-the-making.[8] Further, these episodes are significant in understanding the evolution of the cyber matrix in terms of the early responses in comparison to the current scenario:

First, considering that those were still early years of cyber exploitation or targeting of transnational assets, nations lacked the situational awareness or technical wherewithal to accurately determine the nature, intensity or implications of such attacks. Though governments are now more equipped and alert to this challenge, there still remains a tangible knowledge-deficit about cyber technologies or its predictability, insufficient appreciation of the threat matrix (besides alarmist notions) and the varied scenarios in which it could manifest – all adding to complexities of decision-making. As Michael Hayden, former CIA Director, remarked in 2011: "Rarely has something been so important and so talked about with less clarity and less apparent understanding ... I have sat in meetings ... unable to decide on a course of action as we lacked a clear picture on the legal and policy implications of any decision we take."[9] Things remain as fluid today as security leaderships struggle to calibrate concrete strategies in dealing with an abstract battlefield or formulating their cybersecurity models.

Second, attribution and retaliation was a problem then, as it is now. The failure or struggles to pinpoint attackers with ulterior motives, or source of attack, remains the single most potent reason why securitisation of cyberspace has attained enormous proportions. While the source of many hackers could eventually be identified, the prospect of 'false flag' and 'diversion decoys' makes cyberattacks a teething security challenge. Most states alleged to have initiated trans-border cyberattacks have hired non-state entities or have disavowed such activities in their territory, which further complicates any offensive or deterrence models that are being pursued in this domain. Further, states continue to shy away from affirmatively attributing an attack to rival state actors fearing diplomatic embarrassment or affecting relationships, as much how the offensive state could exercise 'plausible deniability'. Ample examples are how both China and Russia have escaped political scrutiny to the many instances of cyberattacks from their territory or involving their citizens in contrast to the showdowns during many geo-political crises involving these countries (though North Korea remains an exception to this trend.) Suffice it to hence postulate that transnational cyberattacks were, and continue to remain, in the beyond-your-reach realm with ample scope for impunity, despite states increasingly gearing up with retaliatory structures that are supposed to develop some form of punitive action. While most military powers are now pursuing strategies and capabilities to compete and conflict in this domain, including use of overt and covert means to repel and deter attacks, it is still a matter of imagination and conjectures on how the spectrum of cyberwarfare is set to evolve.

What would warfare in this domain entail, and how central is the vulnerability of critical national infrastructures (CNI) in this evolution? What action in this spectrum will trigger hostilities – within the domain or escalation into other means? Will nations be able to draw redlines on the nature of cyberattacks they are subjected to – in terms of the attack on CNI or the destruction caused – to engage in conflict?

## Between Security and Warfare: Discerning the Frontier

A sweeping search on the biggest cyberattacks, in terms of enormity, shows that most of the attacks since the late 1980s (including Spamhaus, Google China, Shady Rat, Stuxnet) were targeted at private entities, commercial establishments and Multinational Corporations (MNCs), with quite a few also aimed at political personalities and state institutions mainly in US, India, China and the Middle East, and the rest on CNI.[10] While these listings point to immense clandestine activity in the non-governmental domain, many industrial nations report regular attacks or efforts to breach security, disrupt control systems and steal data from their CNI and strategic assets. At a holistic level, the evolution of cyberthreats (and cyberwarfare), could be classified into two categories: The first involves what experts term as Computer Network Exploitation (CNE) – pertaining to acts of cyberespionage and information pilferage, mostly without revealing hostile intentions or complicity, and second, Computer Network Attacks (CNA) – involving concerted acts of subversion, disruption and degradation, with inherent intentions of destruction as the eventual outcome. There are many attributes to this evolution of the cyber domain and its potential emergence as the primary zone of conflict among nations:

*(a) A porous domain:* Being a seamless system that traverses borders, time and space, the cyber domain offers the most attractive frontline for hostile action with varied outcomes – of degradation, disruption, destruction as well as economic costs, along with political gains for adversaries. Besides low barriers for entry, cyberspace inherently remains a lawless and porous domain giving considerable space for camouflage and perpetuating 'invisible war'. What started as an entertaining engagement for individuals and private groups in the early days of the evolution of cyberspace, with the intention of harassing corporate entities and probing the strength of governmental systems, has now evolved into a fertile ground for states and non-state actors to pursue inimical objectives and opportunities to wriggle out even when detected. Probably, no other technology offers as much the scope for optimal exploitation, manoeuvring and asymmetric advantage in pursuing end objectives, noble or clandestine, as much as cyber tools. According to Joseph Nye, the diffusion of power is the greatest in the cyber domain where even the best of military capabilities of great powers could not offer them dominance and yet keep their vulnerabilities open for exploitation.[11]

*(b) Asymmetry redefined:* If in the Clausewitzian sense, war was "the continuation of policy by other means", cyberspace offers the platforms for "undertaking war by other means". Many a major analyses have strived to make comparisons with the traditional war theatres – land, sea, air and outer space – with the realisation that cyber will be a completely new spectrum of warfare that allows for – literal

shadowboxing bereft of direct engagement, of digital assaults from ghost-warriors, of destruction without the physical impact, of invisible combat with physical consequences, and much more. The element of asymmetry is punctuated by the fact that assaults come without warning (like guerrilla warfare), while awareness of impact and degradation might be slower, and that war could be initiated from innocuous locations with the minimal of human interface and resource deployment. Asymmetry is not just about the allure that non-state entities have for this front, but also of states that see cyberspace as the ideal frontier to target rivals without revealing identity and hostile intentions.

*(c) Proxy battle zone:* The significance of cyberspace indisputably lies in its utility as a means for proxy warfare – involving a digital battlefield where the best deceptive weapon is in use. Many of the transnational cyberattacks on global industrial majors, CNI or strategic assets have been traced to adversarial sources with latent or patent links to political actors. While China was blamed for many attacks in last decade, it may not match the near-frontal cyber assaults on Estonia, Georgia and Kyrgyzstan (and later, Ukraine), blamed on Russian groups – confirmed by circumstances of definite state backing. Another case is the reported complicity of the North Korean Government in the cyber assault on Sony. However, while Western governments vociferously point to the proxy war played out by their traditional rivals in cyberspace, they were no less complicit in the Stuxnet attack on Iranian enrichment facilities – deemed to be a US-Israeli operation.[12]

It is hence natural to draw an analogy between the many states that are supposed to sponsor proxy war through cross-border terrorism against their adversaries (notably our South Asian neighbour) and the manner in which cyberspace has been exploited by great powers across the ideological aisle to target each other's strategic interests. The deployment of non-state actors by state enterprises for transnational cyberattacks is a definite trend that defines the current contours of cyberwarfare, and explains why weak attribution and plausible deniability is a method of convenience used to optimal effect by states to avoid retribution. Such proxy actions invariably assist states in attaining notional gains of 'getting back at their rivals', without having to shed a drop of blood or incur political costs of rivalry. That there are hardly any remedies to proxy conflict, other than to retort in kind, would be reason why cyberwarfare could thrive in the coming years.

*(d) Terror of the code:* This happens to be the metaphor that could describe the current state of cyberspace. Their erudite grip on the programming code has enabled the accumulation of unprecedented power at the fingertips of a generation who could spread havoc or terror with optimal exploitation of know-how. By that standard, this domain is now ripe to perpetuate for all forms of terrorism, and probably has attained the fundamental attributes of terrorism. That one of the

objectives of attacking a CNI or strategic assets will be to create havoc and terror among the population in itself gives incentives for non-state groups as well as states to use cyber platforms for such operations at minimal costs. Thereby, any distinction between offensive characteristics of state and non-state actors seems to have blurred in this domain. For that matter, the optimal and calibrated use of cyber platforms by terrorist groups is now a *fait accompli*. In fact, terror groups have been most effective in using this domain for their full spectrum of operations – from propaganda, financing and organisation to recruitment – and without disguising their identities, unlike state actors.[13] Notwithstanding arguments about alarmist conceptions similar to nuclear terrorism, cyberterrorism is now integral to the narrative on cybersecurity and warfare, to the extent of cyberwarfare being considered by many observers as a far more potent threat to nations than terrorism.[14]

*(e) Subversion as the norm:* Cyberspace is generally seen as a lawless territory, despite nations formulating legal frameworks to penalise criminal activities in this domain. Though governments vigorously espouse the cause of stability and order in cyberspace, when it comes to the application or means of cyberwarfare, countries prefer subversion and clandestine actions as the fundamental practice. This is owing to the fact that no nation desires to be overtly seen as indulging in offensive actions in cyberspace, especially in peacetime or when a conflict situation does not prevail, or because they do not patently seek to project hostilities. A familiar pattern is the manner in which the Western narratives have sought to pinpoint the role of Chinese and Russian hackers, ostensibly backed by state enterprises, in major cyberattacks globally. Yet, the US-Israeli combine did not hesitate to use the Stuxnet worm to degrade Iran's nuclear capability – not just for a cyber offensive, but also of using a subversive means to derail a critical infrastructure project in another country. Thus, when we argue that the absence of international norms provides the leeway for subversion, the evolution in cyberspace is such that subversion and disruption are buttressed as the integral part of warfare, as states increasingly employ techno-military commands and non-state groups to target rivals. Historically, subversive tactics have been part of asymmetric battle practices mostly used against a state by guerrillas and non-state entities while in modern warfare, states have adopted covert cyber offensives as the operational norm.

*(f) Strategic rebalancing:* Cyberwarfare in general and targeting of CNI as well as economic assets in particular have attained tremendous strategic dimensions with potential for rebalancing of power among the great powers. There is a clear competition among them to attain a decisive edge in cyber capabilities – offensive, defensive and subversive. The key competitors – US, Russia and China – have all developed cyber commands, battle groups and doctrines that favour multiple battlefield applications of cyber systems – exclusively and in conjunction with other

tools of warfare. The US Department of Defence (DoD) Cyber Security Strategy 2015 talks of "displaying effective response capabilities to deter an adversary from initiating an attack; developing defensive capabilities to deny a potential attack from succeeding; and strengthening the overall resilience of US systems to withstand a potential cyberattack if it penetrates the US defenses".[15] Besides, "building forces and capabilities to conduct cyberspace operations", the strategy seeks "to provide integrated cyber capabilities to support military operations and contingency plans", to be heralded by the Cyber Mission Forces (CMF), operating under the United States Cyber Command (USCYBERCOM).[16]

In response, Russia has announced plans to increase its cyber offensive capabilities including through "development of malicious programmes which have the ability to destroy the command and control systems of enemy forces, as well as elements of critical infrastructure, including the banking system, power supply and airports of an opponent", thus clearly indicating its objectives and targets.[17] However, it is China which seems to have the most vibrant cyber offensive plans. As part of its Informationisation strategy, the People's Liberation Army (PLA) has reportedly adopted the Integrated Network Electronic Warfare (INEW) plan[18] to undertake Computer Network Operations (CNO) and information warfare tools, besides deploying militia units across the country to undertake attacks and espionage, some of which have been identified as source of many past attacks across the globe.[19]

These trends complement the assumption that the great powers want to: (a) integrate cyber operations in their military capabilities and plans, (b) optimally engage in conflicts in this domain and (c) use cyberspace to gain decisive advantage in other domains as well. The fundamental message is that cyberspace has attained a strategic dimension with operations buttressed by clear political objectives, which matter more than the nature of actors involved. Yet, there is an interesting contradiction to these patterns. Despite hectic adversarial activity in cyberspace with objectives of destruction or degradation, no military power has raised the tempo of cyber confrontations to the level of revealing intentions of conflict, hostility or mass destruction. Probably, these are still early days in the evolution of this warfare format which allows nations to initiate frontal attacks on their adversaries, and risk escalation to full-fledged wars. The inclination, instead, seems to be in favour of using cyber power as a force multiplier to other military (kinetic) capabilities in order to project power or buttress conventional military operations, as supposedly witnessed in the Georgian War of 2008. Military powers are awake to the strategy of decapitating the command and control structures and strategic assets through cyber onslaughts to make their subsequent military operations easier.

## CNI as the New Conflict Zone

The lasting significance of CNI has been reinforced in the ongoing evolution of cyber conflicts – as the pivotal frontier that makes states vulnerable to cyberattacks, as also the primary target that rivals intend to hit. Concerted attack(s) on a single or multiple CNI assets, leading to substantial damage or shut downs, could cause a destructive and demoralising impact on that nation and its society, including disrupting its way of life and creating widespread chaos, social unrest and instability. Drawing on the Estonian crisis, which many observers described as Web War I, leaderships around the world foresee catastrophic scenarios, even if by extreme imaginations, of a digital holocaust or a cyber 9/11 as the potential fall-out of cyberwar – further fuelled by fictional depictions in many Hollywood movies of CNI as the frontline. Former US Defense Secretary, Leon Panetta, warned in 2012 that the US was facing the possibility of a "cyber-Pearl Harbor" and was increasingly vulnerable to foreign hackers who could dismantle the nation's power grid, transportation system, financial networks and government.[20]

The US Department of Homeland Security, in fact, identifies 16 critical infrastructure sectors "whose assets, systems, and networks, whether physical or virtual, are considered so vital to the US that their incapacitation or destruction would have a debilitating effect on (national) security, economic security, public health or safety, or any combination thereof". In fact, most advanced societies consider sectors ranging from energy, telecommunications, water supply, transport, finance, health and other infrastructures that 'allow a nation to function' as among their critical infrastructure, though even military and national security systems form significant components of CNI.[21] While the financial sector has been a consistent target of hackers, many experts consider the energy sector (primarily the power grids) as the most vulnerable.

Various surveys have brought out an alarming picture in the CNI domain, also highlighting the high instances of attack on Western economies. A survey by the Organisation of American States of public utilities covering energy, finance, communications and security sectors finds that around 40-54 per cent of them experienced cyberattacks from attempts to shut down their systems to manipulate their equipment through controls systems, or to steal data and delete files.[22] An Enterprise Strategy Group (ESG) survey finds that 68 per cent of American CNI assets have experienced constant attacks in recent years, with incidents that could lead to disruption of critical operations and cause havoc for days on end.[23] The energy sector feels the heat with 82 per cent reporting in one survey that cyberattacks could cause serious physical damage and that all threats cannot be detected in time.[24] An Aspen Institute-Intel survey of 625 experts in US and Europe warns "the possibility of a cyberattack on critical infrastructure in the next three years that will result in the loss of life", which in turn lead to a colossal increase in the costs of cybersecurity.[25] A US Congressional Commission reportedly estimated that a large-scale and prolonged blackout could lead to 90 per cent of the US population perishing from disease, lack of food and general societal breakdown.[26] Similar conditions are conceived by many nations who fear varying levels of social instability based on their CNI dependence. India too witnessed a major power grid blackout in July 2012, affecting public utility services after the breakdown of the northern grid on two consecutive days spiralled to other grids and affected more than 300 million people.[27]

While this is the general picture of CNI vulnerabilities, the segment that holds greater significance in the narrative on cyberwarfare is the nuclear domain. Following the Stuxnet attack against Iranian centrifuge facilities, which were in their development stages, the exposure of the nuclear energy industry to threats of subversion and sabotage has come into intense focus. The Stuxnet episode was evidence that a worm could be planted by an intruder or insider in a nuclear facility to corrupt its systems and cripple or damage its operations.[28] This has led to various, and often hyperbolic, projections about the catastrophic consequences of a decapitating attack on an operating nuclear power plant – not just aggravating the paranoia on the vulnerabilities of nuclear infrastructure, but also of the apprehensions on whether nations are equipped to handle any extreme eventualities. Such projections have perpetuated, if not exaggerated, the 'radiation-scare' that has become integral to the perceptions of dangers associated with nuclear plants since the Fukushima incident.

Nuclear facilities, in fact, are as much vulnerable to cyberattacks like many other CNI segments, when intruders attempt to target their industrial mechanisms like the Supervisory Control and Data Acquisition (SCADA), which are the central process control systems that manage the automated operations of the plant and machinery through centralised computer systems. Running on industry-specific and complex architectures, these systems have levels of human-machine interface to adjust operations and function through Programmable Logic Controllers (PLC) as hardware to adjust the physical components of a plant's operations.[29] There are varying descriptions about the vulnerability of these computer systems and their hardware based on their network protocols to which they are connected or exposed. While many feel that nuclear operators are complacent about cyberattacks as most of these systems in nuclear plants are 'air-gapped' – implying that they are not connected to external networks, experts warn of the inherent danger of 'unprotected nodes or wireless access points' which might be exploited by an intruder or even an insider, which could be the pathway for total exploitation or attack. There are also concerns of operators moving towards open protocols or off-the-shelf hardware, which could cause inadvertent linking of PCS to other networks.[30]

According to a detailed study by Chatham House, "many industrial control systems are insecure by design" as cybersecurity measures were integrated later; and standard IT solutions like patching (or updating software) are difficult to implement at nuclear facilities, owing to concerns that it could break a system.[31] The report, while stating that supply chain vulnerabilities cause inherent risks to a nuclear facility, points to cultural issues between nuclear and cyber professionals and that reactive approaches, instead of proactive systems, may cause delay in awareness about an attack. The report also points to the 'insider threat' and dependency on off-the-shelf systems as among the factors pointing the worrying situation, besides lack of preparedness of most nuclear plants for a large-scale cybersecurity emergency.

While such reports underline the fact that nuclear plants, like most CNI facilities, amount to a labyrinth that could provide numerous entry/exit points to a malicious intruder, there are also sections who feel that the threat to CNI assets are only as potent from cyberspace as from any other physical or non-digital source. In other words, the threat to a highly-secure nuclear plant with 'air-gapped' process control systems and updated software and other cyber monitor systems could be similar to the physical threats like the possibility of planting explosives at an industrial plant or public infrastructure, or poisoning a centralised urban water treatment plant. Considering that an attacker has to physically intrude into a facility in order to reach a vulnerable node or breach an internal cyber warning system, the emphasis will remain on enhancing physical security as much as upgrading the cyber defences at CNI. This does not discount the continuing possibility of adversary seeking to find gaps or opportunities of 'lowering of guard'. However, the key to protecting CNI, including nuclear facilities, from cyberattacks would be the progress to a movement of high or deep securitisation, which entails the promotion of a security culture and eco-system where 'none is trusted' and 'complacency is penalised', and where every scenario of vulnerabilities has to be constantly reassessed and re-imagined to be remain on perpetual state of alert. Assuming that even the most advanced nations may not be able to construct a fool-proof security culture for CNIs, the other element will be to build defences and deterrence that denies the adversary space for exploitation or attack.

## Deterrence and Norms: Imbibing Nuclear Lessons?

Amid the debate on the evolution of cyberwarfare, a tremendous body of literature and analyses has emerged on what kind of deterrence has to be formulated by states to counter cyberattacks on their CNI and strategic assets. Needless to say, the clear inspiration for most of these analyses has been from the scholarship on nuclear deterrence. However, due to the complexities involved, scholars are faced with the fundamental challenge of developing distinct deterrence models for this domain or an exploring complete adaptation of conventional and nuclear deterrence models to determine contours of cyber deterrence. In the words of former US Secretary of Defense, William Lynn, "Traditional Cold War deterrence models of assured retaliation do not apply to cyberspace, where it is difficult and time consuming to identify an attack's perpetrator. Whereas a missile comes with a return address, a computer virus generally does not."[32]

The question of cyber deterrence could be evaluated within some larger premises on the role or application of cyber power. First, countries with substantial sway over cyber technologies could seek to pursue and project their cyber infrastructure as a currency of power, prestige and influence – in the case of great powers, using it to complement their military tools in order to enhance their strategic depth; for middle powers, using it as a strategic alternative to military strength; and in the case of deviant or pariah states (like North Korea), using it as a means for counter-balancing, including subversion, pre-emption and ensuring regime survival. Second, the utility of cyberspace for tactical as well as strategic goals could broaden with the evolution of cyber technologies. Besides its clandestine use for subversive actions against rivals (or for terrorism), states could seek to integrate cyber capabilities with military doctrines and structures to enable their application as an offensive instrument for intimation and coercion, or any purpose that offers an offensive advantage. That offence has a greater advantage in cyberspace also propels the incentives to pursue such strategies. Third, multiple manifestations of the technology has increased its strategic utility – as a means for pre-emptive strikes to precede kinetic strikes in a military campaign; the vulnerability of CNI as counter-value (civilian assets like energy, finance, transportation) and counter-force (military, nuclear, command and control assets) targets – all of which warrant the need for a broader deterrence framework to deal with the complete spectrum.

Classical deterrence models envisage some broad parameters and conditions for deterrence to be effective, some of which, in fact, may not be workable for the current framework of cyber operations. Deterrence works on the premise that the: (a) costs of attacking a state will outweigh the adversary's perceived benefits, (b) state could convey a capability and intention to retaliate to an attack by ensuring assured destruction or unacceptable damage to the adversary who initiates the conflict, (c) threat of retaliation and assured destruction should be credible in order to dissuade the adversary, and (d) adversary should be a rational actor who could be convinced on the costs and gains of initiating an attack. However, the evolution of nuclear deterrence has been such that states, when confident of offensive dominance or a defensive advantage, could seek to undertake pre-emptive strikes in order to destroy the rival's retaliatory forces or indulge in adventurism. Despite these fundamentals, nuclear deterrence had a credibility problem, aggravated by vulnerabilities kept open by nuclear powers to sustain their Mutual Assured

Destruction (MAD) equations, and enduring security dilemmas that defined their strategic competitions. Further, deterrence did not inhibit many aspirants from crossing the threshold or dissuade non-state actors from confronting the might of nuclear weapon states.[33]

Taking this broad framework into consideration, the imitation of this model in cyberspace may be fraught with incompatibilities. First, the fundamental causal for deterrence inefficacy in this domain is deemed to be attribution problem – the inability to precisely pinpoint the adversary to the extent of pursuing a decisive punitive action. By that standard, the *threat of retaliation* as the fulcrum of cyber deterrence could stand on a weak foundation. Second, the promise of *assured (or massive) destruction*, or in some terms 'unacceptable damage' could not be affirmatively conveyed or will not be considered *credible* in the light of the 'target vacuum' which the rivals could seek to exploit. Third, these are still early days to assume that a state will retaliate through kinetic options or nuclear strike to a cyberattack, unless it is of extreme proportions unseen hitherto. Fourth, the proposition that *costs could outweigh gains* may not have sufficient tenability in the cyber environment where states seek to camouflage hostile intentions and use proxies to engage in asymmetry campaigns against their rivals. Fifth, while the rationality of non-state entities cannot be credibly accounted for, adversaries may seek to exploit the irrationality of its proxies to gain maximum impact for intended cyber campaigns. Sixth, the scope or opportunity for pre-emption could exist only if the adversary is identified in advance, which, in turn could give more scope to exercise denial deterrence than retaliatory options. Finally, the scope for MAD emerges only when states reach a stage of accepting competitive or adversarial intentions in cyberspace.

### *Quest for a Cyber Deterrence Model*

Notwithstanding the limitations of applying traditional deterrence models in cyberspace, there are immense efforts to recalibrate many of these models to suit this domain, along with addressing structural issues like attribution. The US DoD Cyberspace Strategy makes a significant attempt in clearing the conceptual clutter. It says: "Because of the variety and number of state and non-state cyber actors in cyberspace and the relative availability of destructive cyber tools, an effective deterrence strategy requires a range of policies and capabilities to affect a state or non-state actors' behaviour."[34] Deterrence of cyberattacks, it postures, "will not be achieved through the articulation of cyber policies alone, but through the totality of actions, including declaratory policy, warning capabilities, defensive posture, effective response procedures, and overall resiliency of systems". The strategy is thus three pronged – response (to deter), defence (to deny) and resilience (to withstand).

Two aspects on which the US strategy attempts to bring clarity are attribution and responses. Attribution, the strategy says, is a fundamental part of cyber deterrence, and that the DoD has invested significantly in intelligence, source collection, analysis, and dissemination capabilities, to reduce the anonymity of state and non-state actor activity, identify the point of origin, and determine tactics, techniques, and procedures. As for responses (including retaliation), it says, cyberthreats 'may or may not' warrant a purely military response, but could entail non-military responses including diplomatic actions, law enforcement and economic sanctions. Syncing with this model was the decision to impose sanctions against North Korea in response to the Sony attack, which was also supposed to the first formal retaliatory action against a state for a cyberattack.[35]

However, this larger model contrasts the contentions of some analysts that the US could follow a more belligerent posture – of making states accountable for attacks originating from their territory[36] and also considering kinetic and even nuclear responses in the event of major cyberattacks on CNI or strategic interests.[37] Such postulations, in fact, come with inherent shortcomings. For instances, many attacks against the US originate from within its territory, while even some surveys have cited Chinese experts alleging US as the source of attacks against their assets. This is besides the aspect of successful attacks using 'false flags' and the entanglement of multiple national resources making attribution inaccurate and complicated. Further, the dynamics of retaliation could quite always be determined by political factors – for instance, retaliation against a friendly state or an ally whose territory is misused could be ruled out unless it is a joint-operation.

While the scope for escalation to nuclear levels might be missing in the case of a non-nuclear adversary, the current state of geo-politics may not sustain the possibility of nuclear powers pursuing an escalation ladder leading to nuclear confrontations in response to a cyber strike, unless the thresholds are quite clearly elucidated. As is clear from the US strategy, nuclear powers largely envision scenarios of mitigation of threats and covert retaliation that could be restricted to the cyberspace, and could at best spill over into political-economic actions. Events so far point to the propensity of states to use cyber power as a force multiplier, even for conventional military campaigns. Accordingly, one cannot discount the future possibility of kinetic strikes as retaliation to a severe cyberattack, though potential for escalation to other levels may depend on how the adversary state responds to the initial action. That the objective of deterrence itself would be to mitigate severe attacks or their potential escalation, the key requirement will be to have a declaratory policy that articulates the range of actions a state could consider in order for its deterrence to be credible.

Going by these trends, it could be assumed that the current shape of cyber deterrence revolves around the twin pillars of denial deterrence and retaliatory deterrence. While denial mechanisms will weigh on the durability of strong defences and resilience of the infrastructure, the validity of retaliatory postures depend on not just the declaration of intent, of certain retaliation, but its credible communication as well. Considering the targeting constraints that cyberspace carries, there will be a need for identifying 'redlines' or thresholds of tolerance to cyberattacks and retaliatory options that could address various types of attacks, unless the intention is to have a standard response model. Unlike the nuclear theatre where states use different redlines, in cyberwarfare it could be commensurate to the intended response plan. Assuming that most states could confine responses to the cyber theatre, the possibility of further escalation to other levels open up if states declare a threshold of 'unacceptable damage' to its CNI or strategic assets as reason to declare the initiation of full-fledged hostilities. The response to the North Korean attack on Sony was a case in example, which a senior White House official explained thus: "It was extremely rare for the US to attribute cyberattacks, and it was only done so because of the destructive nature of the attack, and because the White House saw it as 'crossing a threshold'."[38]

No country has so far been inclined to go to war over cyberattacks, though such eventualities cannot be ruled out if a severe attack on CNI assets result in serious national loss and source is attributed to an inimical territory. To deter such eventualities, states might have to project the inclination to go beyond in-kind responses and exercise asymmetric options at own terms. Thus, while tailored deterrence options might represent a stable deterrence posture, the shape of coming cyber conflicts will be of states initiating action on a case-to-case basis, determined by the scale of the attack and nature of actors involved.

## *A Normative Conundrum*

Adding to the deterrence narrative is a general realisation that cyberspace needs credible normative frameworks that could facilitate a common set of rules of behaviour for states, so as to instil order and stability in this domain. Like the inspiration drawn from nuclear deterrence, there are references to seek lessons from the non-proliferation system to build similar normative models for cyberspace, to ensure that it does not turn into a full-fledged conflict zone. Though many nations have promulgated domestic cyber laws to control criminal activities, governments have not shown similar enthusiasm to build international cyber norms through a convention or treaty. A major effort in this direction – the Budapest Convention on Cyber Crime 2001, initiated by the Council of Europe – has not found much traction thanks to the intricacies of geo-politics.[39]

The US has also sought to play a 'norm entrepreneurship' role by promoting the cause of a global cyber normative order, as espoused in its *International Strategy for Cyberspace 2011*. It calls for existing principles like *Upholding of Fundamental*

*Freedoms, Respect for Property, Valuing Privacy, Protection from Crime and Right of Self-Defence* to be the basis for cyber norms, along with a new set of principles including *Global Interoperability, Network Stability, Reliable Access, Multi-stakeholder Governance and Cybersecurity Due Diligence.*[40] In 2010, then Secretary of State, Hillary Clinton, called for five new internet freedoms to be adopted by the UN Human Rights Council, including the freedoms of expression, worships, from want and fear as well as to connect.[41] The dissonance on cyber norms have been such that China rejected the US campaign as a "disguised attempt to impose its values on other cultures".[42] Similarly, Russian calls for a global cyber 'arms control' agreement has been reportedly blocked by the US, terming it as a 'propaganda tool'.[43]

This political see-saw is testament to the fact that global norms on cyberspace, or against its use for warfare, may remain a mirage. In the 1960s, the commonality of interest between the two superpowers in inhibiting new nuclear weapon states enabled them to work on joint draft on a Treaty on Non-Proliferation of Nuclear Weapons (NPT). Apart from the fact that this is a different polycentric era, states are not inclined to discard the strategic gains that cyberspace endows, especially in its asymmetric utilities. Hence, while the US-led liberal security community will be targeting China and Russia for their alleged misuse of cyberspace for strategic gains, the Stuxnet episode betrays the US clamour for a normative cyber world order. Furthermore, the dialectics on cyber norms is rife with oxymoronic entities like the 'ethical hacker' – instruments that state actors engage in divergent roles. Needless to say, while states might have greater interest in applying norms against CNA activities that could cause destruction, the CNE is a frontier that all of them will want to exploit. Further, any form of cyber norms may not holistically address the temptation for subversion, which happens to be the current functional norm in cyberspace.

## Conclusion

There are significant phenomenal imprints in the manner in which cyberspace has evolved in the last decade, especially its transformation into an active field for conflict. The glaring aspect, though, is the common strands in the way governments have responded when their national assets and infrastructure were subjected to cyberattacks. Be it the attacks on Estonia's defence apparatus or the Indian MoD, their establishments have been traditionally oriented towards developing strategies to combat everything from terrorism to military campaigns to nuclear warfare. The spectre of a 'digital invasion', hence, caught them unaware, which in itself points to the major paradigmatic shift in warfare that cyberspace has heralded.

This shift points to a larger question: Is cyber the new strategic? In theory, *strategic* represents the instruments of a state that define its military, political and economic power. In practice, it also implies the use of weapon systems that could destroy or decapitate not just the war machine of a large nation, but also its instruments of national power. The manner in which cyberspace has evolved in the last decade points to the metamorphosis of its technological components into instruments of warfare that have strategic implications, even if applied in asymmetric, proxy or tactical configurations. Furthermore, cyber power has the leverage to alter, transform or influence the traditional tools of statecraft – from diplomacy and coercion to deterrence. Are we staring at a major transformation in the global order – yet again technology driven?

## NOTES

1. "Shadows in the Cloud: Investigating Cyber Espionage 2.0", Information Warfare Monitor of the Shadowserver Foundation, Munk Centre for International Studies, Toronto, April 5, 2010, at http://www.infowar-monitor.net/2010/04/shadows-in-the-cloud-an-investigation-into-cyber-espionage-2-0/ (Accessed January 2016).
2. John Markoff and David Barboza, "Researchers Trace Data Theft to Intruders in China", *The New York Times,* April 5, 2010.
3. "Tracking GhostNet: Investigating a Cyber Espionage Network", Information Warfare Monitor, March 29, 2009, at http://www.infowar-monitor.net/2009/09/tracking-ghostnet-investigating-a-cyber-espionage-network/ (Accessed January 2016).
4. One instance is of a foreign diplomat being asked to withdraw from an event for which the Dalai Lama's office had sent an email invitation. See Malcom Moore, "China's Global Cyber-Espionage Network Ghostnet Penetrates 103 Countries", *The Telegraph*, March 29, 2009, at http://www.telegraph.co.uk/news/worldnews/asia/china/5071124/Chinas-global-cyber-espionage-network-GhostNet-penetrates-103-countries.html (Accessed January 2016).
5. "Shadows in the Cloud ...", No. 1.
6. Though the Indian Computer Emergency Response Team (CERT-In) was formed in 2004 to provide security for India's communication and information infrastructure, it took the National Cyber Security Policy of 2013 to designate the agency as the nodal agency to co-ordinate and for crisis management of cybersecurity-related matters, along with other measures like the formation of a round-the-clock National Critical Information Infrastructure Protection Centre (NCIIPC). See Notification of the National Cyber Security Policy-2013, July 2, 2013, at http://deity.gov.in/sites/upload_files/dit/files/National%20Cyber%20Security%20Policy%20(1).pdf (Accessed January 2016).
7. The NDA government announced in April 2015 the setting up of a National Cyber Coordination Centre (NCCC) to coordinate between intelligence and cyber response agencies such as the Intelligence Bureau (IB) and the CERT-IN to ensure a more robust defence of critical Indian computer systems.
8. Andrzej Kozlowski, "Comparative Analysis of Cyberattacks on Estonia, Georgia and Kyrgyzstan", *European Scientific Journal,* 3, special edition, February 2014.
9. Michael V. Hayden, "The Future of Things Cyber", *Strategic Studies Quarterly*, 5 (1), Spring 2011; quoted in Joseph S. Nye Jr., "Nuclear Lessons for Cyber Security", *Strategic Studies Quarterly*, 5 (3), Winter 2011.
10. See the listings at http://www.arnnet.com.au/slideshow/341113/top-10-most-notorious-cyber-attacks-history/ and http://list25.com/25-biggest-cyber-attacks-in-history/ (Accessed January 2016).

11. Joseph S. Nye Jr., No. 9.
12. Ellen Nakashima and Joby Warrick, "Stuxnet Was Work of U.S. and Israeli Experts, Officials Say", *The Washington Post*, June 2, 2012.
13. Nicole Arce, "Cyber Attack Bigger Threat Than ISIS, Says U.S. Spy Chief", *Tech Times*, February 27, 2015, at http://www.techtimes.com/articles/35965/20150227/cyber-attack-bigger-threat-than-isis-says-u-s-spy-chief.htm (Accessed February 2016).
14. Spender Ackerman, "Cyber-attacks Eclipsing Terrorism as Gravest Domestic Threat – FBI", *The Guardian*, November 14, 2013.
15. Henry Farrell, "What's New in the U.S. Cyber Strategy," *The Washington Post*, April 24, 2015.
16. DOD Cyber Strategy, April 2015, at http://www.defense.gov/Portals/1/features/2015/0415_cyber-strategy/Final_2015_DoD_CYBER_STRATEGY_for_web.pdf (Accessed February 2016).
17. Eugene Gerden, "Russia to Spend $250m Strengthening Cyber-Offensive Capabilities", *SC Magazine*, February 4, 2016, at http://www.scmagazine.com/russia-to-spend-250m-strengthening-cyber-offensive-capabilities/article/471196/ (Accessed February 2016).
18. Deepak Sharma, "China's Cyber Warfare Capability and India's Concerns", *Journal of Defence Studies*, 5 (2), April 2011.
19. George H. Wittman, "China's Cyber Militia", *The American Spectator*, October 21, 2011, at http://spectator.org/articles/36718/chinas-cyber-militia (Accessed February 2016).
20. Elisabeth Bumiller and Thom Shanker, "Panetta Warns of Dire Threat of Cyberattack on U.S.", *The New York Times*, October 11, 2012.
21. Peter Burnett, "The Vital Role of Critical Information Infrastructure Protection (CIIP) in Cybersecurity", *Report on Cybersecurity and Critical Infrastructure in the Americas* (Trend Micro 2015), at http://www.trendmicro.com/cloud-content/us/pdfs/security-intelligence/reports/critical-infrastructures-west-hemisphere.pdf (Accessed February 2016).
22. Ibid. Also see: Warwick Ashford, "Critical Infrastructure Commonly Hit by Destructive Cyber-Attacks, Survey Reveals", *Computer Weekly*, April 7, 2015, at http://www.computerweekly.com/news/4500243886/Critical-infrastructure-commonly-hit-by-destructive-cyber-attacks-survey-reveals (Accessed February 2016).
23. Jon Oltsik, "Networking Nuggets and Security Snippets", *Network World*, September 29, 2015, at http://www.networkworld.com/article/2987290/security/u-s-critical-infrastructure-under-cyber-attack.html (Accessed February 2016).
24. "Tripwire 2016 Energy Survey: Physical Damage", at http://www.tripwire.com/company/research/tripwire-2016-energy-survey-physical-damage/ (Accessed February 2016).
25. Eduard Kovacs, "Attacks on Critical Infrastructure Organizations Resulted in Physical Damage: Survey", *Security Week*, July 20, 2015, at http://www.securityweek.com/attacks-critical-infrastructure-organizations-resulted-physical-damage-survey (Accessed February 2016).
26. Michael Krancer, "The Biggest Cybersecurity Threat: The Energy Sector", *Forbes*, November 4, 2015, at http://www.forbes.com/sites/michaelkrancer/2015/11/04/the-biggest-cybersecurity-threat-the-energy-sector/#29da30bf60ba (Accessed February 2016).
27. Naina Khedekar, "Northern Grid Power Failure: What Went Wrong?", *Tech Two*, August 2, 2012, http://tech.firstpost.com/news-analysis/northern-grid-power-failure-what-went-wrong-32046.html (Accessed February 2016).
28. "...the Stuxnet attack tried to cause centrifuge rotors to spin too fast and at speeds that would cause them to break. The 'original' payload used a different tactic. It attempted to over-pressurise Natanz's centrifuges by sabotaging the system meant to keep the cascades of centrifuges safe." Ralph Langner claims that there were two variants of Stuxnet – one to hit the rotor speed and the other to hit the cascades of the centrifuges. See "Stuxnet's Secret

Twin", *Foreign Policy,* November 19, 2013, at http://foreignpolicy.com/2013/11/19/stuxnets-secret-twin/ (Accessed January 2016).

29. Brent Kesler, "The Vulnerability of Nuclear Facilities to Cyber Attack" *Strategic Insights*, 10 (1), Spring 2011.
30. Ibid.
31. Caroline Baylon, Roger Brunt and David Livingstone, "Cyber Security at Civil Nuclear Facilities Understanding the Risks", Chatham House Report, 2015, at https://www.chathamhouse.org/sites/files/chathamhouse/field/field_document/20151005CyberSecurityNuclearBaylonBruntLivingstone.pdf (Accessed January 2016).
32. William J. Lynn III, "Defending a New Domain: The Pentagon's Cyber Strategy", *Foreign Affairs*, September/October 2010.
33. See Richard L. Kugler, "Deterrence of Cyber Attacks", in Franklin D. Kramer, Stuart H. Starr and Larry K. Wentz (eds.), *Cyberpower and National Security*, Centre for Technology and National Security, National Defense University, Washington D.C., 2009.
34. See DoD Cyber Strategy, No. 16.
35. Jim Acosta and Kevin Liptak, "U.S. Slaps New Sanctions on North Korea after Sony Hack", CNN, January 3, 2015.
36. Patrick M. Morgan, "Applicability of Traditional Deterrence Concepts and Theory to the Cyber Realm", paper presented at a workshop on deterring cyber-attacks, Washington D.C., 2010, quoted in Christopher Haley, "A Theory of Cyber Deterrence", *Georgetown Journal of International Affairs*, February 6, 2013, at http://journal.georgetown.edu/a-theory-of-cyber-deterrence-christopher-haley/ (Accessed January 2016).
37. Annegret Bendiek and Tobias Metzger formulate an escalation model which includes a linear progress from political and economic sanctions to low-level cyber responses, followed by kinetic strikes and high-level cyberattacks, culminating in nuclear action. See "Deterrence Theory in the Cyber-century", SWP Working Paper, German Institute for International and Security Affairs, Berlin, May 2015, at https://www.swp-berlin.org/fileadmin/contents/products/arbeitspapiere/Bendiek-Metzger_WP-Cyberdeterrence.pdf (Accessed January 2016).
38. "Sony Cyber-attack: North Korea Faces New US Sanctions", BBC, January 3, 2015.
39. See Convention on Cybercrime, Budapest, November 23, 2001, at http://www.europarl.europa.eu/meetdocs/2014_2019/documents/libe/dv/7_conv_budapest_/7_conv_budapest_en.pdf (Accessed January 2016).
40. "International Strategy for Cyberspace: Prosperity, Security, and Openness in a Networked World", White House, May 2011, at https://www.whitehouse.gov/sites/default/files/rss_viewer/international_strategy_for_cyberspace.pdf (Accessed January 2016).
41. Hillary Clinton, "Remarks on Internet Freedom", *Newseum*, Washington D.C., January 21, 2010, at http://www.state.gov/secretary/20092013clinton/rm/2010/01/135519.htm (Accessed January 2016).
42. Chinese reaction based on the editorial: "The Real Stake in 'Free Flow of Information'", *Global Times*, January 22, 2010.
43. See detailed analyses on these politics in: Tim Stevens, "A Cyberwar of Ideas? Deterrence and Norms in Cyberspace", *Contemporary Security Policy*, 33 (1), April 2012.

# 7

# Role of Military in Cybersecurity

*Liina Areng*

## Cyberspace as a Domain of Warfare

Cyber operations have become an indispensable element of modern conflict. Cyber means are planned and used to provide a "military advantage" – as an effective force multiplier, an enhancement for traditional means or as a stand-alone capability. Modern military doctrine is increasingly relying on being able to use cyberattacks to paralyse computer networks and, by doing so, inflict kinetic damage on the infrastructure these information systems control, including civilian critical infrastructure. According to the National Security Agency (NSA) predictions, "The next major conflict will start in cyberspace."[1] To that end, a growing number of nations are undertaking massive efforts to digitally arm themselves for cyberwarfare, establishing cyber forces. The possibility that other states are developing cyberattack capabilities means this could be the beginning of clandestine cyber aggressions. In a time of declining defence budgets, growth in resources allocated to cybersecurity leads to believe that the investment is not going only to cyber-defence initiatives but also cyber-offensives. Although most nations maintain secrecy over their cyberwarfare capabilities and doctrine, some have openly stated that they are investing large amounts of money on cyber intelligence and counterattack capabilities.

The speed at which cyberattack capabilities are proliferating is remarkable. A decade ago just a few countries were believed to possess cyberweapons. A recent study conducted by the *Wall Street Journal* claims that over 60 countries have or are developing cyber espionage or attack tools, of which 29 countries have formal military or intelligence teams with a focus that includes cyberattack capabilities.[2] Yet, there is a lot of uncertainty about the intentions as well as real capabilities being developed. Cyberweapons have never been used in large-scale war fighting between equal opponents, and the incidents to date have not caused serious long-term disruptions. Until that occurs, it is very difficult to assess real capabilities different cyber powers possess.

As modern battlefields are increasingly shifting to cyberspace, the actors and their capabilities have changed accordingly. Digital technology can give substantial asymmetric advantage to states that are considered weaker in terms of traditional combat power. Compared to conventional security domains, the cyber domain has relatively low barriers of entry.[3] Therefore, cyber capabilities are being increasingly used to gain leverage in international security, challenging the traditional military dominance and doctrine of traditional great powers. The new toolbox of "cyber power" permits not just conventionally weaker states but also non-state actors to exercise significant power against larger opponents though a variety of widely available and inexpensive cyber tools. Malicious code and expertise can be purchased or hired to cause anything from data leaks to advanced cyberespionage and cyber-generated physical attacks.

An idealist would argue that the reason why countries invest in the development of cyberweapons might be their seeming "bloodlessness" – computers attacking computers and not humans. It might be the case for public opinion that demands precision attacks involving no civilian casualties and even no own troops killed. The UK Defence Minister, Philip Hammond, has labelled the Internet as the "'new frontier' for the military because the public was no longer prepared to accept British soldiers being killed on the front line".[4] It may seem wiser to shut down enemy's airfield rather than to blow it up. But given the "once-only" nature of cyberweapons – they are only useful as long as the vulnerability in the target system remains in place and unpatched – it is probable that instead of sophisticated cyberattacks, countries will continue to prefer kinetic alternatives to achieve simple, conventional military objectives, involving less resources and effort.

No matter how ethical cyberweapons may seem at first glance, their use may certainly involve considerable damage in physical space – destruction of infrastructure, creating chaos and causing casualties. Former US Defense Secretary, Leon E. Panetta, warned in 2012 that the "next Pearl Harbor we confront could very well be a cyberattack that cripples our power systems, our grid, our security systems, our financial systems". The 2010 Stuxnet attack on a Iranian nuclear enrichment facility, as the first known example of a truly sophisticated operation against national critical infrastructure, has been labelled as a strategic game changer. Cyber doomsayers had proven right. Civilian critical infrastructure is a strategic target and cyberweapons exclusively designed to attack this target exist. Stuxnet also demonstrated that the development of a successful cyberattack against national critical assets takes substantial resources and brainpower, and adding to that the "once-only" factor, is therefore not likely to be deployed carelessly.

On the other hand, the recent attack against Ukraine energy distribution companies that lead to a massive power outage affecting hundreds of thousands of Ukrainians in December 2015 marks another strategic shift in cyber campaigns – inflicting harm and achieving high visibility by significantly less resource-demanding attack vectors. The attackers reportedly re-used a decade-old Trojan BlackEnergy to gain access to the company's network, enhancing the attack by adding special malware components to be able to shut down critical systems and wiping their data.

When Stuxnet made us believe that attacking national critical infrastructure or military defence systems requires substantial resources to conduct pre-operational intelligence, experienced strategists and operation planners, the attack against Ukraine's critical infrastructure shows something else. It indicates how future complex cyber attacks may look like. Power is a critical vulnerability for any organisation, and the core vulnerability for any nation, creating cascading damage in other critical sectors. Therefore, possessing a capability to cause a major blackout must be a desired weapon in any opponent's arsenal.

In modern warfare, opponents can distract, disrupt and demoralise a nation by skilful use of cyber tools, exploiting timing, surprise, and an adversary's specific vulnerabilities. These vulnerabilities are not restricted to military targets; the ability to attack civilian targets such as public utilities or financial sector can be far more dangerous, and, subsequently, more effective, at discouraging and deterring potential adversaries because of its immediate social and political effects. To immobilise a nation, to render it incapable of defending itself, attackers no longer need kinetic weapons.

## The Essence of Cyber Deterrence

Modern militaries depend extensively on information technology for command and control, surveillance, logistics, navigation and targeting. This raises their efficiency and combat power, but such dependencies potentially introduce increasingly complex vulnerabilities. The ability to gather, analyse and exploit information has always been crucial to national security, and digital development has made it profoundly easier. The Internet era offers intelligence agencies and military the opportunity to enhance and expand their espionage and information warfare operations. Modern warfare is inherently a fight over the access and control of information, since the better informed is likely to win the battle.

There are numerous ways to deny access to, steal, delete or alter information without getting the glimpse of the intruder, or getting it too late. In 2014, the mean time an intruder remained in the victim's system undetected was 205 days.[5] This leaves plenty of time to prepare the battlefield – infiltrate the systems, find weak spots and pre-position to take over and destroy the information systems and networks during combat.

The possibility of using different and unpredictable strategic combinations of cyber tools affecting military, political, economic and social targets makes the opponents in asymmetric warfare more equal. The asymmetry is also created by the imbalance of attack space – larger, technologically dependent nations possess a larger network space with a greater number of weak spots vulnerable to attacks, while the smaller nation has a smaller network surface to protect. "Mass" is no longer a decisive factor in the military strategic and operational equations. Even a lone cyber warrior can wreak havoc in an opponent's networks, making information technology a powerful tool for a small but sophisticated actor that possesses sufficient skill and cunning.

In conventional force, capabilities can be tracked by observing procurement, military exercises and active intelligence. Nations developing cyberattack capabilities can do it in a much more subtle way, as there are no specific facilities or visible infrastructure that could indicate the development of such capabilities. It could be happening in a plain ordinary office building or the bedroom of a teenage ponytail. The lack of easily traceable characteristics of an existence of significant cyber force makes determining the source of attacks also very difficult. However, no cyber campaign happens in isolation. It usually involves some sort of geopolitical tensions or state's own internal political challenges. Therefore, the critical question for the policy, military and intelligence community is: "Who has the intention and capability to attack us and why?"

The problems with attribution make deterrence extraordinarily difficult. An attack that looks like coming from a nation-state might not actually be one, because non-state actors have access to the same tools and techniques. And vice versa – states use affiliated or semi-independent hackers and criminals in attack campaigns to conceal state activities and confuse attribution. That, in turn also confuses response – whether it should be law enforcement agencies or military that should get involved. There are no internationally agreed definitions regarding cyber defence and cyberattack or norms regulating state behaviour in cyberspace. Cyber operations are conducted in the regulatory space resembling Wild West where the distinction between legitimate and illegal activity is blurred.

Because of the lack of visibility of cyber capabilities and un-detectability of their build-up, cyber domain is not a domain where classical arms control would have any effects. It would be extremely difficult to monitor and verify cyber weapons. Even the International Institute for Strategic Studies (IISS)'s publication, *Military Balance* that provides annual assessment of global military capabilities says under the section of national cyber forces that these "capabilities are not assessed quantitatively". Therefore, instead of counting computers and network speeds, intelligence agencies should rather focus on spotting talent and specific skillsets in opponent's military. This is a much more difficult task than assessing adversary's quantitative capabilities.

There seems to be one resemblance with the Cold War era balance of power logic, which is the concept of Mutually Assured Destruction (MAD) that still applies. Despite excessive build-up of cyber offence capabilities, we have not yet seen an excessive use of these weapons. Modern War era MAD holds on uncertainty and distrust. The deterrence of the 21st century can be described as "mutually assured doubt" represented by the strategic ambiguity of intent and capabilities. In essence, cyberattacks in their "bloodless" nature could be a perfect tool to avoid military retaliation when the perpetrator is prudently operating just below the threshold of "armed attack". Yet, because of this political doubt and ambiguity, coupled with uncertainty of the possible technological side effects, actors' own inherent vulnerabilities against the hard-to-control spread of malicious code, there seems to be a lot of hesitation to engage these capabilities. Compared with nuclear weapons that effectively guaranteed MAD half a century ago, cyberweapons can be much more unpredictable. Once launched, it is almost impossible to control their proliferation. Moreover, their explosion perimeter is going to be much wider and it is going to be more difficult to forecast the exact spread and consequences. And since there are so many question marks about other actors' capabilities, the fear of inability to defend against counterattacks also plays a deterrent factor.

## Whole-of-nation Cyber Defence Supported by "Collective Brain"

The evolution of cyberwarfare is connected with the social and political evolution of our societies that encompass a growing number of benefits and vulnerabilities of the digital era. Some modern states have become so dependent on digital solutions and services that a larger failure of the technology supporting these functionalities would threaten the very existence of the state. Therefore, governments should dedicate resources to prepare for civilian emergencies, and ensure the resilience of society in a similar way to how they conduct national defence planning. A cyberattack against national critical infrastructure could have a cascading effect to economy, society and government in ways difficult to understand, model or predict. The failures in Information and Communications Technology (ICT) can have serious national security implications, yet the response to cyber threats cannot be conceived purely in terms of classical warfare.

Already in 1970, Marshall McLuhan predicted that "World War III is a guerrilla information war with no division between military and civilian participation",[6] and nowadays increasingly there is no division between national and international space, meaning that malicious non-state actors, such as terrorist organisations can effectively use decentralised, distributed command and control systems. The lack of central "brain" makes it very difficult to take down these structures by conventional forces. Arquilla and Ronfeldt called this conflict waged by networks of non-state actors a "netwar", in which "numerous dispersed small groups using the latest communications technologies could act conjointly across great distances".[7] Al Qaeda and other terrorist organisations, especially the Islamic State of Iraq and Syria (ISIS), have become quite sophisticated in their use of the Internet, distributing militant Islamist propaganda, recruiting youth for jihad operations, organising and coordinating attack campaigns. The recent Islamic terrorist attacks in Paris have revealed that ISIS is capable of using encrypted messaging apps to communicate.

Audry Curth Cronin argued in 2006 that digital technology has significantly changed the way mass mobilisation works in modern conflicts. Personal computers, mobile phones and other "individually accessible, ordinary networked communications ... are altering the nature of human social interaction, thus also affecting the shape and outcome of domestic and international conflict".[8] Governments and militaries will have to fight against blogs, tweets and trolls both nationally and internationally.

Information systems are becoming increasingly complex and more and more difficult to defend. Cyberattacks also evolve in complexity, and responding to these challenges needs additional resources and skills. For defence establishments, it can be very hard to attract and retain IT talent. Modern military forces need experts who would be able to navigate the cyber battlefield, giving them tools and authority to operate. Rigid, hierarchical command structures will not work well in cyber operations where speed, creativity and agility are the essential requirements to lead a successful mission.

In cyber conflicts, since an adversary's preparatory phase is either very short or hard to detect, the attacks may come with no forewarning, which for unprepared or significantly underprepared organisations means that crisis can escalate very fast. If the victim state has the ability and will to retaliate, a single incident could rapidly grow and escalate to full-scale warfare involving not only the initial aggressor and the victim, but also potentially a crowd of sympathisers within and outside of the parties of conflict. Cyber crisis is thus a complex issue for conflict de-escalation, as it requires rapid containment efforts and the coordinated cooperation and involvement of a wide array of stakeholders, including military and intelligence agencies, the private sector, and civil society, each of which has a different organisational culture and different interests.

In many nations, civilian organisations such as the cybercrime teams in Interior ministries or civilian cybersecurity agencies/national Computer Security Incident Response Teams (CSIRTs) have the most cyber resources and the national role to coordinate cybersecurity. States must develop an internal coordination mechanism to connect military and civilian ministries and agencies responsible for cybersecurity. Any information about attacks and anticipated attacks has to be shared at network speed between the Internet Service Providers (ISPs), government intelligence agencies, law enforcement, military and civilian CSIRTs, and the private sector. Many private sector organisations, such as banks, have significant capabilities to fight cybercrime, respond to incidents, and foster cooperation with other nations. Nearly every cyber conflict in history has been decisively resolved not by militaries but by the private sector. This new domain of warfare requires agility, innovative, out-of-the-box thinking, which can be hard for military organisations that are used to strictly abiding by the rules. Militaries have a lot to learn from private sector, e.g. how to conduct risk assessment – companies have been doing it for decades, also for IT risk assessments, but it is very new to governments and militaries.

The proper response to hybrid/asymmetric/non-linear/full-spectrum warfare is clearly full-spectrum defence. To operate in hard-to-predict, dynamic and quickly escalating cyber conflict battlefields the defence should be dynamic, and militaries need a serious cultural transformation to become more dynamic and flexible. Preparing for the "seen" battles is not sufficient to react to the Stuxnet-like black-swan events, which really make the paradigm shift in the security landscape. Cyberspace is only in parts controlled by governments and other state actors, as the majority of the infrastructure belongs to the private sector. Military planning should therefore be well synchronised with the preparation for civil emergencies. Only that way it can be ensured that the critical players in national defence – militaries, intelligence, governments, critical infrastructure operators and other key players of the private sector are sufficiently informed, prepared and resilient to handle large-scale cyberattacks. Broad-based national cybersecurity requires that the contingency plans of critical infrastructure operators are coordinated with national defence threat scenarios.

Pure military planning and countermeasures cannot play a critical role in cybersecurity because of the civilian nature of cyberspace and the predominantly non-military nature of the most probable attacker. Much of the cyber expertise and resources required to defend information infrastructure are located outside of the military establishments. One effective counterforce to non-military groups would be an organised but agile, community-based "collective brain" of patriotically motivated IT experts.

One of the most important lessons learnt from the 2007 attacks for Estonia was the importance of trust-based information-sharing and collaboration networks between government, military and private sectors. In the aftermath of the attacks, Estonian government decided to formalise this collaboration and began to develop a voluntary unit of cyber experts under the Defence League, a militarily-organised voluntary national defence organisation dating back to 1918. The main aim of this unit is to attract patriotically motivated IT security talent, mostly experts employed by banks, software companies, ISPs and the public sector, to prepare for and help the government to respond in large-scale cyber crises. The unit offers a "trust circle" to exchange information and best practices, to train, and to experiment. Another important element of the Cyber Defence League's activity is exercises and awareness raising. As part of military reserve, the unit members can also participate in military cyber defence, including operations and missions abroad.

Formation of such a voluntary unit allows the government to use highly skilled civilian cybersecurity expertise in various roles, including cyber crisis response and military cyber defence in a relatively agile and cost-effective setting. Coupled with the effective mitigation of the cyberattacks in 2007, the existence of the Estonian Cyber Defence League has been one of the most important elements of Estonian cyber deterrence.

"Little green men" have lately penetrated the conventional battlefields making "hybrid conflict" a new buzzword for politicians and defence planners. In cyber domain, asymmetry is an integral part of the threat picture and "little green men" have been present in most recent politically motivated cyberattack campaigns in the form of patriotically motivated state-influenced or state-controlled malicious actors. There is very limited experience or best practice, not to mention a clear doctrine on how to defend against these "little green men" in physical or virtual battlefields. Developing a "whole-of-ecosystem" threat picture and response plan is probably the most viable approach currently available.

## International Cooperation

It is clear that, in order to achieve good national cybersecurity, domestic efforts need to be complemented by strong international cooperation. The critical interdependencies and mutual vulnerabilities in cyberspace should encourage cooperation between nations in times of crisis. However, despite the overall interest in discussing cybersecurity problems, the more specific debate on international cyber crisis management remains largely undeveloped. Countries also differ in the general approach to cyber (e.g. civilian, military, or intelligence-led) and to national cooperation environments (completely voluntary or completely mandated); this creates imbalance and complexity in finding the right counterparts and creating comparable crisis management procedures. Cooperation in cyberspace comes down to trust more than anything, and trust is hard to build. Information sharing and early warning are essential components of crisis prevention and management, but usually require a long history of working together.

The chance of agreement on cooperative measures such as close-to-real-time information sharing and mutual assistance is definitely higher in regions with a long and highly developed tradition of cooperative arrangements, such as the Nordic cooperation between the Scandinavian countries or the Five Eyes Agreement on Security Partnership between the US, UK, Canada, Australia and New Zealand, representing a common linguistic region. A major problem in reaching a cooperative regional solution is likely to be the lack of trust due to past challenges or political or economic competition. And because trust is a fundamental factor in cyber cooperation, cyberthreats to national security are more likely to be addressed at a national level or, at best, through bilateral agreements between states.

The North Atlantic Treaty Organisation (NATO) has declared that cyberattacks would be covered by the collective self-defence clause—"Article 5" of the treaty—which would apply if any of the 28 member nations were attacked. The organisation has also collectively stated that response to cyberattacks would not necessarily have to be confined to cyberspace and could involve physical retaliation with conventional military forces. There is no automatism in NATO's responses to Allies' requests for assistance pursuant to Article 5; any collective action will be decided as a result of consultation. NATO's response to non-Article 5 crises that involve a cyber component is not yet well defined and regulated, because there appears to be no consensus as to what degree national responsibilities for cyber defence and security should be transferred to NATO. While cyber defence of the Allies is and will likely remain a national responsibility, it is clear that in times of crisis affecting Allies, NATO's cyber defence does not stop at the defence of its own networks but involves some sort of coordinated response, depending on circumstances. The ambiguity of crisis response and coordination mechanism might serve as a deterrent, however, the lack of determination may also weaken NATO's image and credibility as a strategic actor in cyber defence.

## Conclusion

There is still a lot to understand about the escalation patterns and consequences of conflict in cyber domain, particularly about cyberattacks that cause physical effects. The speed of cyberattack tools' proliferation increases the risk of governments getting hold of perilous weapons but failing to understand the strategic implications of their use, let alone their best operational employment and possible side effects. Therefore, uncontrolled cyber arms race can have dangerous consequences.

Although "cyber" is a politically 'hot' issue and IT security companies are routinely contributing to sensationalist headlines, many experts agree that strategic cyberwarfare remains unlikely to happen in the foreseeable future. While cyberspace definitely continues to play a substantial role in future military conflicts, cyberwarfare will not likely become a dominant feature igniting new conflicts or melting the frozen ones; however we will definitely continue to see operations in the "grey zone" of cyber conflict as show of force, including attacks against national critical infrastructure. Undoubtedly, cybercrime and cyberespionage continue to flourish as the main battlefield of hostile cyber activity. Cyberterrorism is probably becoming a larger cyber risk deserving full attention of the governments and international organisations.

Militaries should continue to be prepared for worst-case scenarios, however, strategically and resource-wise, states should rather focus on lower impact but higher probability events – economic and political espionage and terrorist attacks that can seriously affect the normal functioning of the state and society. As not only most of the infrastructure threatened by such malicious events is in the hands of private actors, but governments are also not privy to the critical knowledge required to defend these networks, it is necessary that states maintain very good cooperation with the private sector in preparing for cyber crisis scenarios. Militaries should focus on the defence of their networks and infrastructure. Beyond that, their role in cybersecurity is rather limited even if governments acknowledge it to be a serious threat to national security.

Creating a credible cyber capability is less about technology than finding the right people and skill sets, which can be difficult for militaries. Therefore, it also merits searching innovative cooperation arrangements with the civilian and private sectors to increase military's own cyber defence capability as well as participate in building a "whole-of-nation collective brain" able to respond to the full magnitude and latitude of different cyberthreat scenarios facing the nation.

Although in modern conflicts filled with "little green men", it is difficult to draw the line between a military and non-military affair, defending civilian infrastructure cannot be a military mandate. Protecting the modern, digital way of life should be a joint and shared civil-military "whole of nation" effort.

## NOTES

1. Jacob Appelbaum, Aaron Gibson, Claudio Guarnieri, Andy Müller-Maguhn, Laura Poitras, Marcel Rosenbach, Leif Ryge, Hilmar Schmundt and Michael Sontheimer, "The Digital Arms Race: NSA Preps America for Future Battle", *Spiegel Online International*, January 17, 2015, at http://www.spiegel.de/international/world/new-snowden-docs-indicate-scope-of-nsa-preparations-for-cyber-battle-a-1013409.html.
2. Jennifer Valentino-DeVries, Lam Thuy Vo and Danny Yadron, "Cataloging the World's

Cyberforces", *The Wall Street Journal*, December 28, 2015, at http://graphics.wsj.com/world-catalogue-cyberwar-tools/.

3. Joseph S. Nye, "Cyber Power", Belfer Center for Science and International Affairs, Harvard Kennedy School, 2010, p. 4, at http://belfercenter.ksg.harvard.edu/publication/20162/cyber_power.html.
4. Tom Mctague, "'Call of Duty' Will Become REAL: Ministry Of Defence Reveals How Future Wars Will be Fought Entirely in Cyberspace", *Daily Mail Online*, May 8, 2014, at http://www.dailymail.co.uk/news/article-2623523/Future-wars-fought-cyberspace-stop-soldiers-killed-line-Defence-Secretary-Philip-Hammond-claims.html.
5. Mandiant, *M-Trends 2015: A View from the Front Lines*, FireEye Corporation, at https://www.fireeye.com/current-threats/annual-threat-report.html
6. Marshall McLuhan, *Culture is Our Business*, Wipf and Stock, Eugene, Oregon, 1970, p. 66.
7. John Arquilla and David Ronfeldt, *Networks and Netwars: The Future of Terror, Crime and Militancy*, RAND, Santa Monica, California, 2001.
8. Audrey Kurth Cronin, "Cyber-Mobilization: The New Levee en Masse", *Parameters*, Summer 2006, p. 77.

# 8

# Recalibrating Law Enforcement to Keep Pace with New Technologies and Forms of Crime

*Madan M. Oberoi*

## Introduction

The future, moulded by advances in science and technology, looks very promising. The previously unimaginable advancements in the fields of communication technologies, computing, artificial intelligence, encryption technologies, virtual reality, robotics, genetics, neuroscience and biotechnology are radically transforming our world mostly for the better. These technological developments are shrinking the world and making it increasingly more open. This, for the most part has caused huge benefits for society, but at the same time there are some serious implications of these developments. In this constantly technologically upgrading world, everyone needs to keep up with technology in order to even conduct their daily functions effectively.

While looking at this "techno-utopian" vision of the future world, we have also to be conscious about the darker side of the emerging and new technologies. The organised criminal network is very adept as well as highly innovative in exploitation of these emerging technologies.

Law Enforcement Agencies are also trying, though in a very limited way and in some select jurisdictions only, to make use of these new technological innovations to prevent crime and improve the performance of the police. However, there is a need to systematically analyse how these innovations are being adopted by all the actors and to understand full implications, both desirable as well as undesirable, of these fast-paced technological developments.

A wide range of new technological innovations have very significant implications for law enforcement agencies in general, and particularly in the areas of crime prevention, crime control, criminal intelligence, crime detection, investigation, secure communications, forensics, anti-forensic applications by criminals, tracing the proceeds of crime, need for multi-jurisdictional and multi-stakeholder coordination, etc.

In order to understand these developments, an attempt has been made in following paragraphs to understand how various cutting-edge scientific and engineering transformations are redefining law enforcement environment. The analysis has been divided in five mega trends – Data Tsunami, Smart Everything, Digital Disruption, Cyber Insecurity, and Alternate Anonymous Cyber Economy. This model will help in understanding and distilling how revolutionary technology could spearhead major changes across different sectors such as transport, banking, communications, energy and defence, and their implications for law enforcement agencies, and also look at the kinds of new technologically advanced tools that could be developed in the near future.

## Data Tsunami

According to a report, "Around 40 per cent of the world population has an Internet connection today. In 1995, it was less than 1 per cent. The number of Internet users has increased tenfold from 1999 to 2013. The first billion was reached in 2005, the second billion in 2010, and the third billion in 2014."[1] Mobile growth is fuelling the high rate of growth of the Internet. Mobile phone connections are increasing more rapidly than purchases of any other consumer product. The Internet connections through mobile devices are following the growth curve of mobile devices.

According to Cisco's global Internet Protocol (IP) traffic forecast, "Annual global IP traffic will pass the zettabyte (1000 exabytes) threshold by the end of 2016, and will reach 2 zettabytes per year by 2019. By 2016, global IP traffic will reach 1.1 zettabytes per year, or 88.4 exabytes (nearly one billion gigabytes) per month, and by 2019, global IP traffic will reach 2.0 zettabytes per year, or 168 exabytes per month. Global IP traffic has increased fivefold over the past five years, and will increase threefold over the next five years."[2] According to this survey, "Content delivery networks (CDNs) will carry nearly two-thirds of Internet traffic by 2019. Two-thirds of all IP traffic will originate from non-PC devices by 2019." These predictions are in line with advent of technologies like Internet-of-Things (IoT).

According to an EMC study,

> The digital universe is growing 40 per cent a year into the next decade, expanding to include not only the increasing number of people and enterprises doing everything online, but also all the 'things' – smart devices – connected to the Internet, unleashing a new wave of opportunities for businesses and people around the world. The data coming from embedded systems (e.g. MP3 players, traffic lights, MRI scanners) has grown to a level where it's starting to challenge established practices in datacentres; the migration to digital entertainment – movies and TV – is almost complete; and metadata, once tightly coupled with the data it describes, has grown into a category in and of itself, the fastest-growing subcategory of the digital universe.[3]

According to Professor Patrick Wolfe, Executive Director of the University College of London's Big Data Institute, "The rate at which we're generating data is rapidly outpacing our ability to analyse it. The trick here is to turn these massive data streams from a liability into a strength. Only about 0.5 per cent of all data is currently analysed, and that percentage is shrinking as more data is collected. At the same time, big data has almost limitless potential."[4] With the concept of Smart Cities, collecting huge amount of data, not using big data analytics would no longer be an option.

The endless proliferation of technology users and their active engagement with the ever-ascending digital avenues unleash unstructured, consumer-generated content, or also referred to as Data Tsunamis. These rich sources of information stream in at astronomical velocity, volume, and variety. Steep-priced, big-data analytics are rigorously applied to study data tsunamis, though identifying and monitoring the sheer scale of material remain as one of the biggest challenges.

There is a limited range of data-mining tools that can keep up with the pace of the information explosion and are capable of producing useful leads. The monitoring of data from individuals and their continuously inflating network connections, over lengthy durations, as well as connecting the copious amounts of disparate pieces of details would the biggest challenges for law enforcement agencies.

## Smart Everything

A relatively new term, "Pervasive Computing", to an extent, captures the direction in which we are heading. The fast-paced developments and progress witnessed in technologies used in the fields of computing and communication are leading us towards an ecosystem of networks of networks. The resultant very high speed computing power is pushing us into a new era, where microprocessors will become so small and inexpensive that they can be embedded in almost everything used in our daily lives, including very ordinary things to digitise and capture almost everything. The concept of IoT in its full bloom looks at interweaving and connecting together all such devices by wireless networks. These networked devices together with powerful and cheap sensors and progress in processor and communication technology will lay the foundation for making everyday objects "smart", where these smart objects know where they are and they are able to adapt to the environment and they provide useful value-add services in addition to their original purpose.

The prospect of a world of things that virtually talk to each other, besides being fascinating, also raises concerns especially with regard to the political, legal, and social implications. Privacy is a prime concern. The full repercussions of extensive adoption of the new technological developments like IoT into our everyday lives are difficult to predict.

The rapidly evolving technology in the area of global digitalisation is leading to enhanced smart applications. The research has made much headway with IoT and in the field of robotics, with millions of industrial and service robots currently employed in different sectors of economy. Contrary to popular opinion, the intelligence of robots resides in their software which regulate their cognitive and affective abilities when interacting with humans or machines. Robotics, which used to be available exclusively for industrial use, such as for car manufacturing and precision engineering, today has found its way into consumer products in the form of vacuum cleaners and drones. Driverless cars have become a real possibility. Behind all of this technological innovation is the software development and big data in the cloud that's making the devices smart. The advent of intelligent technology has unlocked a Pandora's Box, where states, civilians, law enforcement agencies as well as felons vie to harness the collective intelligence of a cosmic web of interlinked devices processing and exchanging information in real time.

In this overall context it is of paramount importance that besides installing urban infrastructure to support smart appliances, law enforcement agencies should acquire an in-depth understanding of how to manage and leverage on such devices in order to perform their policing work effectively, as their stakeholders and criminals may exploit these automata to attain profits.

## Digital Disruption

The classic examples of 'Digital Disruption' are Uber and Airbnb, which have up-ended the taxi and hotel industries, respectively. However, the disruption caused by technology transformation is becoming far-reaching and dramatic. In the cargo transportation industry for example, where something as remotely related to 3D printing is now posing a serious threat to this traditional industry. With 3D printing, replacement parts and even entire components can be replicated on-site instead of buying and shipping from the original equipment manufacturer. Logistics demand will fall as a result. Similar disruption is being witnessed in the financial services sector. It is now possible to find more than 100 start-ups providing different services traditionally offered by a single bank. Banking services are fast becoming unbundled. The more progressive banks globally are learning to disrupt themselves proactively. They are experimenting with multi-modal operations and investing in Financial Technology (FinTech) start-ups in order to understand the emerging technological landscape and ride the wave of change. As new processes and business models emerge, jobs, skills and regulations must also change to suit the new way of doing things.

Faced with hundreds of competitors and external companies, forward-looking enterprises strive to avoid becoming a shell or shadow of themselves owing to disruptive innovations. A necessity to survive and a need to expand successfully are beginning to push these firms to invest in start-ups for their core businesses and engage in creative unbundling of services. Technological maturation and disruption to organisations could have a positive association, however the exploitation of these technologies by criminals and terrorists is worrisome. The Islamic State is a case in point, wherein the group has stymied businesses through its strong offline and online presence, and appeared to have revamped the traditional model of terrorism.

There could be value for law enforcement agencies to examine and develop business models that may serve them well. In identifying good models, rigid thinking patterns and cultural issues like ways of communication might be underlying challenges, which law enforcement agencies should overcome. The law enforcement agencies could capitalise on specialist units to look into the future impact of disruptive inventions. Such set-ups need to be self-driven where existing models can be questioned and disrupted to assess its drawbacks, and experts being given sufficient room and time to explore and stress-test different scenarios. It is then necessary to consciously bring the considerations of these evaluations into strategic foresight for the organisation.

## Cyber Insecurity

Global companies like J P Morgan Chase, Sony Pictures, Target, Anthem Healthcare and Fiat Chrysler have been hit by high-profile cyberattacks in the past few months. There is a strong perception of cyber insecurity amongst vast majority of cyber users. The perception may be over-hyped, but the threat is real, and few agencies/organisations/countries are in a state of readiness to fully combat this threat. This is not to imply that no efforts are being made in this direction, but there is a growing sense of vulnerability and a realisation that the efforts being made may not be adequate to meet the magnitude of the threat.

The weakest links perpetuating cyber threats might be the failures in human behaviour such as risky online disclosure, instant gratification, complacency and negligence, which incentivise criminals to keep to their illicit pursuits. Criminals who target financial and security institutions are capable of devoting months and years to build false relationships over social networks and masquerade as IT administrators before striking. Further complicating the surveillance of user-created data for law enforcement agencies, could be misinformation fed by criminals into public systems. Propagating fast within the IoT, the 50 billion interconnected devices projected to arrive by 2020, can potentially become points of vulnerability embedded in a massive attack surface for criminals to revel in.

There is a need for shared responsibility among consumers, manufacturers, and supervisory bodies, as legislative frameworks relevant to the accountability of ensuring full knowledge of the technological products are inadequate. A regulatory system is required to tackle issues on transparency when utilising the devices, as users may not be aware of the intricate effects of a software or a firmware update on the operating systems of their devices. To remain accustomed to gadgets that are complex in features and usage, law enforcement agencies may model an ecosystem, and work in synergy with external parties who may come up with multiple, fast and creative solutions. Law enforcement agencies will need to make their own well-informed choices on the risks and benefits new gadgets proffer, such as the type of information to transact when using unfamiliar WiFi and prioritising data visualisation as opposed to running indiscriminate analysis to generate interesting reports.

## Alternate Anonymous Cyber Economy

New products in alternate anonymous cyber payment systems are being witnessed every day. These products are leveraging the anonymity enabled through encryption technologies and models like blockchains for establishing parallel systems to central financial institutions. These systems are establishing trust models and criminal elements and others with malicious intent are using these alternate mechanisms for moving money around the world by circumventing anti-money laundering controls.

Decentralised cryptocurrency are the latest tools in this regard. These cryptocurrencies operate in an ecosystem, which is fundamentally different from the centralised banking and economic systems such as the Federal Reserve System and Reserve Bank, where a central authority controls the supply of money through various regulatory policies and by printing units of fiat money. These cryptocurrencies are available today in around 1,000 formats and are similar to or derived from bitcoin. These cryptocurrency systems work in an environment, where the safety, integrity and balance, etc. are maintained on open public ledgers (blockchains), which are maintained by an unrelated group of users referred to as miners. These miners are members of the general public using their computers to help validate and timestamp transactions adding them to the blockchain for a financial incentive. These unregulated anonymous payment systems deliver economic services without requiring users to verify who they are, thereby enabling anonymous transactions. Technology is enabling further layers of anonymity through use of tumbler services and darknet.

## Implications For Law Enforcement Agencies

These trends have huge implications on the working of law enforcement agencies. Most of these implications can be understood by asking questions like: Who is capable of enforcing laws/regulations? Who has the information to develop intelligence? Who has the expertise? How do we resolve issue of difference in framing regulations and enforcing them? What are the possible points where laws/ regulations can be enforced?

Before proceeding to examine the implications, it would be worthwhile to summarize major technology trends, which are Darknet; Cryptocurrencies; Blockchains; IoT; Embedded Devices; Encryption; Big Data Analytics; Social Media; Cloud Computing; Drones; Biometrics; 3D Printing; Bio-agents; Wearable Devices; Secure Communication Technologies; and Machine Enabled Decision-making.

The criminals are broadly exploiting these technologies for the purposes of secure and anonymous communication; attacking the computational resource for data; laundering of proceeds of crime; supply chain; leveraging technology for achieving a wider and bigger scale; hiding the footsteps – anti-forensics; and crowd funding for illegal activities.

The significant trends observed in cybercrime include Ransomware – India suffered more than 60,000 ransomware attacks last year; ATM Malware; Laundering of Proceeds of Crime through cryptocurrencies and other technology based methods; Bitcoin Thefts; Theft of Computational resources for bitcoin mining; Crime-as-a-service model, where service provided may not be illegal per-se; Privacy breaches; and Profiting by illegal access of legal information.

The major implications for law enforcement agencies are summarised as follows, under the broad headings of Intelligence Gathering, Investigation and Third-Party Policing.

## Intelligence Gathering

The main purpose of this law enforcement activity is to have access to varied sources of information and analysing them to produce actionable intelligence in order to

Figure 1. New Age Crime: Public-Private Collaboration Platform Framework

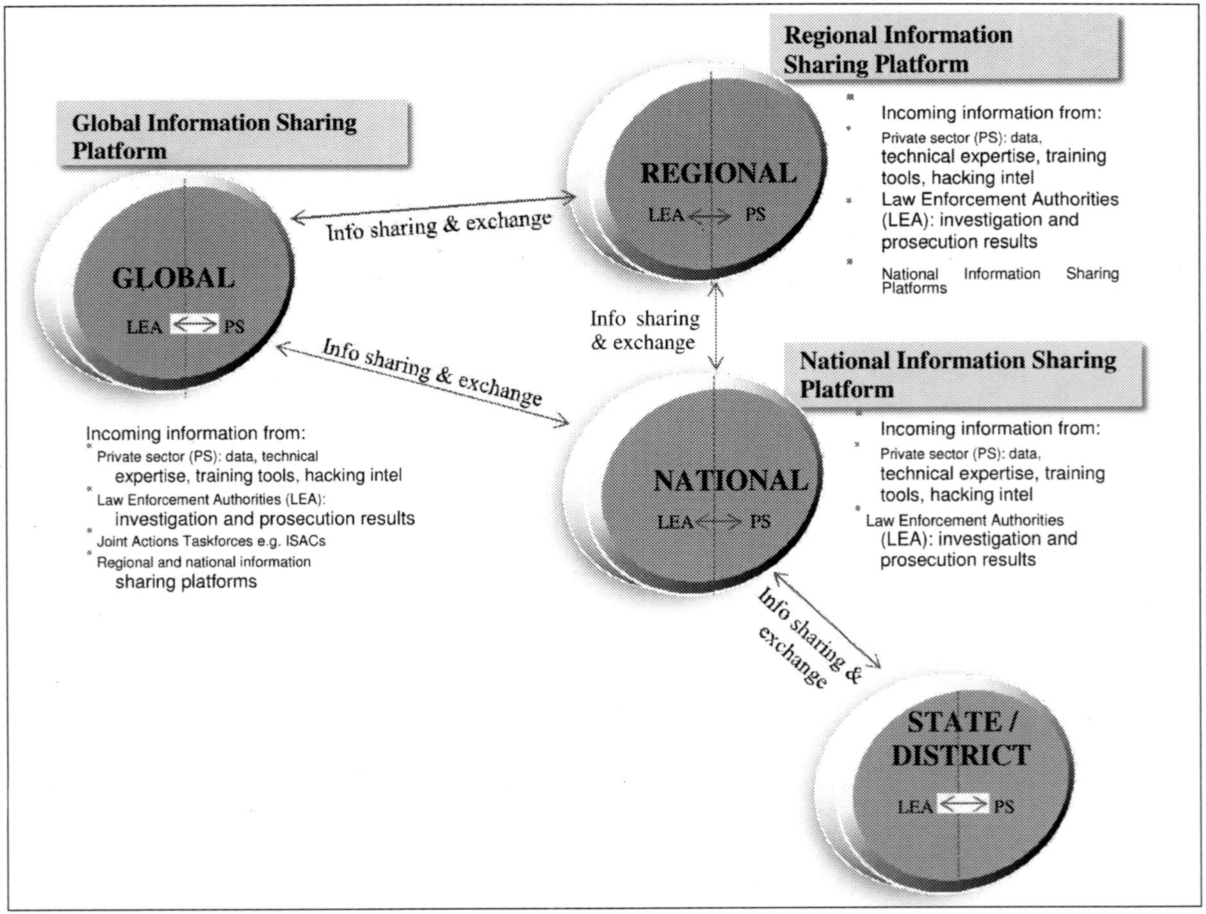

anticipate, prevent, detect and investigate crimes. The most important activity here is having access to all pieces of information, which in the cyberspace is the most challenging work. This is so because the new technology space has completely altered the information equations as most of the pieces of information lie outside the law enforcement domain with other stakeholders and also in other jurisdictions. Figure 1 describes a way forward in tackling this issue.

## Investigation

The major impediments to a successful investigation lie in:

- Multi-jurisdictional investigation information exchange.
- Obtaining investigation information from other stakeholders like ISPs.
- Legal and procedural harmonisation across jurisdictions.
- Technology-based challenges like encryption.
- Acquisition of evidence from newer kinds of space.
- Seizure/preservation and presentation of new kinds of digital exhibits.
- Forensic procedures for new kinds of digital exhibits.
- Forensic procedures for cryptocurrencies.
- Forensics on new platforms.
- Availability of adequate skills, infrastructure, and laws/procedures/rules.
- Internet governance policies especially those impacting attribution.
- Research support.

In light of these challenges we are seeing a trend where law enforcement agencies are relying more on a disruption-based strategy rather than a prosecution based strategy. Because of difficulties in accurate attribution, successful prosecution is becoming very challenging; as a result, law enforcement agencies are focussing more on disrupting criminal infrastructure. However, since the criminal actors are not targeted in this strategy, it only leads to criminals shifting the infrastructure in case of a successful disruption. This means they are back after a short time gap.

## Third-Party Policing

The Criminal Justice Systems in all jurisdictions is facing a new kind of challenge, where there is a huge gap in actual incidence of cybercrime and reported/registered cybercrime. The victims no more appear to be interested in reporting crime to law enforcement agencies. This is a very significant and dangerous trend. Let us take a concrete example, Ransomware cases in India. Symantec report in 2013 reported that 11 per cent of ransomware cases of the Asia-Pacific region took place in India. In the 2014 report, Symantec mentioned that there were more than 60,000 ransomware cases in India. However, if we look at the official crime statistics, there are negligible number of ransomware cases reported to law enforcement agencies.

A relevant additional trend is that many hardware repair shops in New Delhi are doing brisk business in re-formatting encrypted hard discs. There is a huge reluctance in reporting cybercrime to law enforcement agencies.

A careful analysis of the situation shows that this reluctance is not because of the often-cited reason of fear of reputation loss, the real reason is much more significant as the victims do not see any value in reporting cybercrime to police forces as they perceive that they would not be able to deliver anything worthwhile. This is also significant considering the victims are more comfortable going to private sector, including multi-national consultancy companies, who through the multi-jurisdictional footprint are apparently more effective in investigation and forensic services. This is leading to the trend of "Third Party Policing".

## Conclusion

The changing technologies are leaving a significant impact on the working of law enforcement agencies. The impact of various challenges is much more than normally perceived and can go to the extent of marginalisation of state actors in this domain, which used to be their exclusive area of operation. There is a need for a recalibration of law enforcement strategy and shift has to be towards a multi-stakeholder model involving private sector, academia, research bodies, inter-governmental bodies, civil society and law enforcement agencies.

## NOTES

1. Internet Live Stats, at http://www.internetlivestats.com/internet-users/.
2. "Cisco Visual Networking Index: Forecast and Methodology, 2014–2019", White Paper, May 26, 2015, at http://www.cisco.com/c/en/us/solutions/collateral/service-provider/ip-ngn-ip-next-generation-network/white_paper_c11-481360.html.
3. Vernon Turner, John F. Gantz, David Reinsel and Stephen Minton, "The Digital Universe of Opportunities: Rich Data and the Increasing Value of the Internet of Things", International Data Corporation (IDC), April 2014, at http://www.emc.com/leadership/digital-universe/2014iview/executive-summary.htm.
4. Lauren Browning, "The Mind-Blowing Growth and Power of Big Data", *Business Insider*, June 10, 2015, at http://www.businessinsider.sg/mind-blowing-growth-and-power-of-big-data-2015-6/?r=US&IR=T#.VpxJytIrLIU

## REFERENCES

Andrieu, Joe, "Beyond Data Ownership to Information Sharing", at http://blog.joeandrieu.com/2010/01/21/beyond-data-ownership-to-information-sharing/comment-page-1/.

Bruner, Casey, "Cyber Insecurity", The National Bureau of Asian Research, at http://www.nbr.org/research/activity.aspx?id=522.

Byrne, James and Gary Marx, "Technological Innovations in Crime Prevention and Policing: A Review of the Research on Implementation and Impact", *Cahiers Politiestudies*, 2011 (20), 2011, pp. 17-40, at https://www.ncjrs.gov/pdffiles1/nij/238011.pdf.

Dueweke, Scott, "Virtual Identity Fuels Anonymous Cyber Economy", Booz Allen Hamilton, at http://www.boozallen.com/insights/2012/01/virtual-identity-fuels-anonymous-cyber-economy.

*Firstpost,* "Bitcoin Extortion Group DD4BC Ramps up Attacks against Financial Sector", September 11, 2015, at http://www.firstpost.com/business/bitcoin-extortion-group-dd4bc-ramps-up-attacks-against-financial-sector-2430096.html

Goodman, Marc, "How Technology Makes Us Vulnerable", CNN, at http://edition.cnn.com/2012/07/29/opinion/goodman-ted-crime/.

Holmes, B, "DD4BC Expands Extortion Campaigns for Bitcoin", Brave New Coin, at http://bravenewcoin.com/news/dd4bc-expands-extortion-campaigns-for-bitcoin/.

IBNLive, "Why the Underworld Loves Bitcoin", March 16, 2014, at http://www.ibnlive.com/news/india/why-the-underworld-loves-bitcoin-674448.html.

Ilves, Luukas Kristjan, "Cyber-insecurity Is a Problem We Can Solve", *Diplomaatia,* 121, September 2013, at http://www.diplomaatia.ee/en/article/cyber-insecurity-is-a-problem-we-can-solve/.

Lipinski, Tomas A. and Johannes Britz, "Rethinking the Ownership of Information in the 21st Century: Ethical Implications", *Ethics and Information Technology,* 2 (1), March 2000, pp 49-71, at http://link.springer.com/article/10.1023%2FA%3A1010064313976.

Mattern, Friedemann, "The Age of Pervasive Computing – Everything Smart, Everything Connected?", Institute for Pervasive Computing, ETH Zurich, Switzerland, at http://www.vs.inf.ethz.ch/publ/papers/The-Age-of-Pervasive-Computing-2003.pdf.

Meng, Khoong Chan, "Five Megatrends Changing Our World in 2016", Institute of Systems Science, National University of Singapore, at https://www.iss.nus.edu.sg/AboutISS/NewsRoom/eNewsletters/FiveMegatrendsChangin gOurWorldin2016.aspx.

Mick2009, "Ransomlocked Files Using Built-In Windows Tools", Symantec, October 25, 2013, at http://www.symantec.com/connect/articles/recovering-ransomlocked-files-using-built-windows-tools.

Narang, Satnam, "Cryptolocker Alert: Millions in the UK Targeted in Mass Spam Campaign", Symantec, November 18, 2013, at http://www.symantec.com/connect/blogs/cryptolocker-alert-millions-uk-targeted-mass-spam-campaign.

National Crime Prevention Council, "Evolving with Technology", at http://www.ncpc.org/topics/fraud-and-identity-theft/evolving-with-technology.

Naveen, "Bit Coin Exaction DDOS Campaign by DD4BC Targeting Payment Industry", WebDefence, June 16, 2015, at https://webdefense.in/bit-coin-exaction-ddos-campaign-by-dd4bc-targeting-payment-industry/.

PTI, "Ransomware Attacks: India among 10 Worst Affected", *The Times of India,* August 7, 2015, at http://timesofindia.indiatimes.com/tech/tech-news/Ransomware-attacks-India-among-10-worst-affected/articleshow/48391766.cms.

Savage, Kevin, "Cryptolocker: a Thriving Menace", Symantec, October 22, 2013, at http://www.symantec.com/connect/blogs/ransomcrypt-thriving-menace. Recovering.

Savage, Kevin, Peter Coogan and Hon Lau, "The Evolution of Ransomware", Version 1.0, Symantec, August 6, 2015, at http://www.symantec.com/content/en/us/enterprise/media/security_response/whitepapers /the-evolution-of-ransomware.pdf.

Turner, Vernon, John F. Gantz, David Reinsel and Stephen Minton, "The Digital Universe of Opportunities: Rich Data and the Increasing Value of the Internet of Things", International Data Corporation (IDC), April 2014, at http://www.emc.com/leadership/digital-universe/2014iview/index.htm.

Wallace, Ian, "The Risks of Cyber Insecurity", *The Fletcher Forum of World Affairs,* August 17, 2014, at http://www.fletcherforum.org/2014/08/17/wallace/.

# 9

# Evolving Role of Government in Cybersecurity

*Kah-Kin Ho*

So far, the 21st century has been an era of wide-scale deregulation and privatisation, with much of the nation's critical infrastructure – in sectors such as energy, transport, finance and medicine – now in the hands of the private sector. Adversaries constantly target these critical infrastructure sectors, with security threats potentially causing both cascading and crippling effects regionally, nationally, and even internationally, as a result of the increased globalisation of society.

The difficulty in securing critical infrastructures is partly due to the divergence of interests between the private and public sectors. The private sector's primary focus is corporate efficiency: the main goal is profit making, so it does what it believes is merely "enough", implementing the minimum level of security, in order to achieve profit as quickly as possible. By contrast, the government is principally concerned with achieving social order, national security and economic prosperity for its population. A 2010 EuroStat survey reported that almost 70 per cent of European Union (EU) citizens believe it to be very important that governments ensure the safety of citizens against all threats. However, governments do not provide close supervision of, or operational control over, the critical infrastructures that are within the private sector. As a result, it has been argued that the role of government as a legitimate provider of security has diminished, and will continue to weaken moving forward.

Nevertheless, this matter is by no means straightforward, and governments are indeed grappling with the challenge of determining what their roles in cybersecurity could or should be, particularly with regard to the private sector. However, I would argue that the changing global landscape should not mean that the role of governments as legitimate providers of security be diminished, rather they should work to understand how the world has changed and is changing, and what their role(s) should be within this new environment of increasing interconnectedness. Furthermore, in order for governments to be successful in this new environment, their remit must transcend what their historical regulatory role has typically entailed. They must now tackle how they can best assist the private sector to invest in security (facilitation), and how public and private sectors can work together to improve the current state of security (collaboration). This is the Regulate, Facilitate, Collaborate (RFC) framework through which governments must strategise, and be ready to draw upon analogous lessons learnt from past strategies geared towards other areas of threat, such as pandemics and terrorism.

While it may be useful to have a framework made up of three components, the next, obvious question is: Under what circumstances and in what combinations can these three components be put to good use? This chapter is an attempt to identify the key factors that governments should consider when deciding on the best course of action for ensuring that the private sector increases their security investment and efforts.

## Private Sector Security Investment Decision

Risk management textbooks typically spell out an objective and easy-to-understand equation that describes risk. Put simply, security risk is a function of: 1) the likelihood of being targeted, 2) the vulnerability of the organisation, and 3) the impact of a security incident.

It is logical to assume that, given the high frequency and severity of cybersecurity incidents – which translates to high security risk – reported almost on a daily basis, companies, especially in the areas of critical infrastructure, are stepping up their cybersecurity investment accordingly. However, this is not true, and a straightforward Internet search shows that security vulnerabilities with critical infrastructure are on the rise. Why is it so? Why don't critical infrastructure companies invest more in security?

Certainly if we look at how companies are run, they face numerous competing risk factors, and cybersecurity risk is just one of the many they have to take into account in their risk mitigation plan. Additionally, some companies simply don't have an appropriate security culture, where everyone from the CEO to lowest ranking employees are involved in ensuring they all do their part to protect the company from security intrusion. Not having the right security culture creates a negative impact on the security budget, which affects the tools and security savvy individuals that need to be employed. A lack of security investment can sometimes be attributed to companies not having good examples to emulate, while there is also an argument that, because the adversaries are so skilful, increased security investment may not be of much use in the long term.

What governments can do is identify the different factors that influence how the private sector makes security investment decisions, and use a combination of tools within the RFC framework to prompt increased security investment. The rest of this chapter will outline these relevant factors and, where possible, provide lessons learnt.

## Market Forces

Market forces can cause severe repercussions for companies when they suffer from a security incident. Many examples, including Target, Home Depot, Sony, etc. have suffered from loss of market capitalisation, loss of revenue and profit, loss of brand reputation, and reduced market share. In reference to the risk equation described earlier, market forces affect the impact variable, and thereby increase the security risk considerably. Companies that believe they will be severely impacted by market forces tend to take security investment seriously. For example Cisco, realising the importance of security, not just to the company but also to customers, has made significant investments in key security initiatives, such as Secure Supply Chain, Secure Development Lifecycle and Trust programmes to harden equipment. Similar examples can be found with other major companies such as Microsoft, Apple and Google.

However, not all companies are affected by market forces, especially non-public listed companies or companies whose products are embedded in bigger systems usually sold by larger companies. Very often there is lack of incentive for these companies to invest in security, thus governments need further understanding on the factors holding them back and what could be done to ensure they step-up security investment.

## Tragedy of the Common

First brought to attention in 1833 by William Forster Lloyd to illustrate the deterioration that would occur in a pasture due to the innate traits of humans, the principles can be extended to cybersecurity. Companies, especially those with low profit margins, are extremely cost-conscious. Faced with stiff competition, if other firms are not making the same level of security investment as they are, they will be at a cost disadvantage, putting their survivability at risk. Furthermore, when one of their peers suffers from a security incident everyone is impacted by the consequences. For example: in the aftermath of a security incident, the government intervenes by imposing new regulations or security standards that companies must adhere to, thereby escalating the costs for companies who are already making security investment.

The main goal would be for everyone to be at an equal level of protection. One way of achieving this would be for governments to impose regulations on all companies, ensuring their compliance with a minimum standard of security. However, regulations should not be viewed as the cure-all solution, because they could be counterproductive when not implemented properly. A good example of this is the European Commission's proposed Network and Information Security (NIS) directive, as part of the EU's cybersecurity strategy, aimed at tackling network and information security incidents and risks across the EU. Key elements of the directive would require organisations to adopt risk management practices and report major security incidents on their core services. These organisations would include member states, key Internet enablers and critical infrastructure operators, totalling 42,000 organisations. On the surface, reporting major security incidents may seem a reasonable approach, however closer examination reveals potential issues: Could the mechanism scale to 42,000 organisations? Are all organisations on the highest end of the risk/threat spectrum and would the same security measures equally apply across all companies in different sectors? Given that the majority of these organisations are small and medium enterprises, would it impose unnecessary burden and cost? Another potential issue with reporting security incidents is it could hinder security professionals from actually preventing cyber attacks and mitigating their effects by distracting them with complying with the requirements and dealing with the reputation fall out of the security incident. A further complication is that certain sectors such as finance and banking are already required, by their individual governments, to comply with incident report obligations, so this would be yet another overlapping requirement that they must fulfil. Fortunately in late 2015 the European Parliament decided that only critical infrastructure falls within the scope of the directive, which allows the European Commission to provide proper oversight and enforcement, as the number of organisations affected by this directive has reduced significantly.

However, the question therefore still remains over what could be done to beef up security investment for the non-critical infrastructure players. The next section discusses and draws lessons learnt from a counterterrorism programme called Customs-Trade Partnership Against Terrorism (C-TPAT) – which was designed to "tip over the lead dog" – and examines whether these could be applied in cybersecurity.

## Tipping Over the Lead Dog

According to James F. Moore, in a *Harvard Business Review* article, "Predators and Prey: A New Ecology of Competition", innovative businesses don't evolve in a vacuum because they must attract resources of all sorts, drawing in capital, partners, suppliers, and customers to create cooperative networks. In his view, companies are part of a business ecosystem where they work cooperatively and competitively to support new products, satisfy customer needs and eventually incorporate the next round of innovation. Over time, they coevolve their capabilities and they tend to align themselves with the directions set by the ecosystem leader (also known as the lead dog).

Recognising the dynamics of a business ecosystem leader and members paved the way for the success of C-TPAT, which is a voluntary supply chain security programme led by US Customs and Border Protection, with a focus on improving the security of private companies' supply chains in regard to terrorism. In the early stages of this programme, the government worked closely with the so-called lead dogs of business ecosystems in four areas: site security, personnel, material movement and process control. After successfully "tipping over" these lead dogs to sign up to the C-TPAT programme, other members of the business ecosystem began to follow suit and they, under the direction of the lead dogs, started to comply with the security requirements. Today, C-TPAT membership boasts some 11,000+ companies. The main lesson to be learnt here is that governments don't always have the right answer as to what security measures need to be taken by a diverse set of companies in different sectors, but by working with different business ecosystem lead dogs – who have more know-how in their respective areas – to develop security requirements, the lead dogs can then "impose" these security requirements on members of their business ecosystem. As such, the government is indirectly able to help solve the tragedy of commons problem by ensuring all companies are at an equal level of protection through strong security partnership with the lead dogs.

## Security Interdependency within a Vertical

An important consideration for a government, when assessing how best to use the RFC framework to create a conducive environment for security investment, is whether there is heavy security interdependency within a particular industry vertical. Critical infrastructure, such as air traffic controls and electricity transmission companies, exhibit such characteristics, in that the impact of a disruption to one company will affect other companies. In a security interdependent world, risks faced by one company depend not only on its choices but also those of all others. Do companies then have adequate incentives to invest in security against a risk, the severity of which depends on the actions of others? The situation could go in one of two directions for a company: a decreasing trend in security investment if more and more companies are unprotected, or an increasing trend if more and more companies are protected. Fortunately, a government can play a critical role in preventing the former situation from deteriorating. To understand more about this, we will have a look at the lessons learnt from a European Commission directive for critical infrastructure protection.

In 2008, the European Commission issued a Directive called 2008/114/EC, for the identification and designation of European critical infrastructures and the assessment of the need to improve their protection. There are a certain number of critical infrastructures in the EU, the disruption or destruction of which would have significant cross-border impact. This may include trans-boundary cross-sector effects resulting from interdependencies between interconnected infra-structures. Member states forwarded suggestions as to what would enhance European prevention of and response to terrorist attacks involving critical infrastructures. In addition, member states were required to build and operationalise Operator Security Plans (OSPs), or equivalent measures, comprising the identification of important assets, a risk assessment and the identification, selection and prioritisation of counter measures and procedures.

At the outset, the directive was a success: member states worked cooperatively to fulfil the requirements. The lesson learnt from this is clear: a government can facilitate by creating a platform in order to bring together important stakeholders so they can collaboratively work out appropriate security measures. Stakeholders invest more when they can see that other stakeholders are committed to the process. Furthermore, they all understand the impact of security interdependency, and therefore that it is in everyone's interest to invest in security protection. One could argue this is an instance where the governing entity has achieved an appropriate balance between the regulatory, facilitative and collaborative elements in order to promote a positive outcome for everyone.

A further example of this is the Information Sharing and Analysis Centre (better known as ISAC). ISACs were created as a result of the US Presidential Decision Directive 63 (PDD-63) in 1998, which requested that the public and private sector create a partnership to share information about physical and cyber threats, vulnerabilities and events, in order to help protect the critical infrastructure of the United States. PDD-63 was updated in 2003, with the Homeland Security Presidential Directive (HSPD)-7, to reaffirm the partnership mission.

Today there are ISACs for 14 critical infrastructures, including Financial Services, Electric, Energy and Surface Transportation. Thus, it is not surprising that the ISACs are functioning very well, especially in sectors such as Financial

Services where there is heavy security interdependency between different firms. Membership of the FS-ISAC grew from 68 members in 2004 to 5,000+ members in 2014, so it is often touted as an example of a successful model for information sharing and collaboration.

## Exposure to Offensive Mindset

As previously discussed, the importance of having the proper security culture within an organisation is of paramount importance. Companies with a good security culture have a greater recognition of threats and invest in skilled security personnel in order to put security tools into effective use and protect the organisation. But what could government do to induce the right security culture within an organisation? One solution, which is both intriguing and possibly counterintuitive, is the training programme created by the Idaho National Laboratory in collaboration with Industrial Control Systems-Computer Emergency Response Team (ICS-CERT), under the purview of Department of Homeland Security. The programme trains the people who run ICS – typically used by critical infrastructure such as electrical power stations, oil refineries and water plants – to hack and attack their own systems. Towards the end of the training, participants are split into two groups: a large blue team and a smaller red team. The red team's task is to breach the ICS network and wreak havoc in the production system, while the blue team's job is to defend it. The exercise usually results in the red team gaining the upper hand and successfully creating a "disaster" (e.g. the spilling of toxic chemicals).

The training exercise serves a number of purposes. Firstly, to show critical infrastructure owners that cybersecurity incidents can have serious consequences, which may result in death and injury to humans and/or the destruction of property. Secondly, participants understand that it is not impossible to breach the ICS network when key vulnerabilities are discovered and exploited by adversaries. Hence they have to be more vigilant about, and better informed on vulnerability alerts and advice put out by ICS-CERT. Thirdly, by putting critical infrastructure operators into an offensive mindset, they are better able to anticipate the movement of adversaries and therefore improve and increase investment in their defence accordingly.

All three of these purposes contribute to building a proper security culture within the organisation. It is also worth noting that this strategy does not involve the regulatory element for the government; only the facilitative and collaborative elements are needed to bring about a positive change.

## Threat of Regulation

One powerful tool at the government's disposal is the threat of regulation.

Regulation often carries the notion of requiring private sector businesses to do certain things or face penalties. When faced with such prospects, companies tend to do "everything possible" to avoid this regulation. Thus it is not surprising that, when government agencies perform security assessments – which could be in the form of penetration testing – companies are open to scrutiny and as a result are responsive to these assessment findings, and employ proper counter measures to close security gaps. Such action provides a strong argument for the legitimacy and effectiveness of the private sector's self-regulation.

The threat of regulation without the need for a government to impose it could actually avoid potential challenges. In a rapidly evolving cybersecurity environment, regulation may improve cybersecurity by making companies address the threats of the previous generation but it is rarely equipped to address the constantly changing threats that will emerge from the current and future generations of technology. Regulation often creates a culture of compliance, and companies often seek the lowest-cost way of meeting these standards. Regardless of whether such actions would be beneficial for improving security, this mindset of compliance-over-security takes away precious private sector resources in order to focus on current and evolving cyberthreats.

## Trust

A key factor in private sector acceptance of regulation is the trust they place in the government. Trust provides the necessary lubrication to ease inherent frictions between the private sector and regulators. As trust is an expression of confidence between parties, a low-trust environment makes it much harder for the private sector to accept regulation. Hence, it is not uncommon that in a high-trust environment there is acceptance of regulation, even if parties do not entirely agree with a regulatory decision, because they perceive the regulators to be competent, fair and efficient. These three components are known as the three dimensions of trust, and we will discuss each one in turn.

By far, a regulator's competency is the most important component of trust. Numerous questions abound when regulators are evaluated based on competency: Do regulators have a proficient process? Do they possess the technical know-how? Do they have the practical experience to relate to challenges faced by the private sector? If the regulators are not seen as competent, additional expertise will be required, lest trust be compromised. On the subject of fairness, regulators will be evaluated on whether they take everyone's interests into account, not just those of powerful interest groups. Careful management of deliberation is usually required in order to demonstrate the impartiality of the regulators. The final component, efficiency, is often a sticky point with regulators, as it is notoriously perceived to promote bureaucratic red tape, thus crippling decision-making and causing the compliance procedure to be overly complicated and costly. A government should strive to demonstrate that they are able to process a regulatory decision and the eventual enforcement of the regulation as efficiently as possible.

Furthermore, trust is the foundation that underpins successful public-private partnerships; it takes a long time and a lot of effort to build, but it is easy and quick to destroy.

## Risk Transfer

Undoubtedly the consensus amongst security professionals is that it is both technically and economically impossible to design and protect critical infrastructure to withstand any and all disruptions, intrusions, and attacks, i.e., there is no such thing as perfect security. The logical consequence of this is that we have to manage residual risk. One way to do this is to enhance resilience, enabling the organisation to absorb the adverse impact of the security incident and re-establish itself quickly. Another way – albeit complementary – is for companies to utilise the insurance market for risk transfer. Unfortunately the cyber insurance market for critical infrastructure is underdeveloped; it is likely that insurers are either reluctant to provide coverage or they charge a high premium, making it too costly for anyone to sign up. Such concerns are not unreasonable when looked from the perspective of the insurers.

As addressed previously, cybersecurity incidents could result in catastrophic consequences, with death and injury to humans and/or the destruction of property. Often the impact is hard to quantify because there is a lack of historical data on which to base the actuarial analysis. Understandably, insurers are extremely concerned about black swan events – low frequency but very high impact – that would make them insolvent. Moreover, limited confidence in the accuracy of predictions on the likelihood of successful cyber attacks exacerbates the matter further.

Looking back at the events of 9/11 and the ensuing $ 40 billion in estimated insured loss, it is clear that the market for terrorism coverage, post 9/11, became dysfunctional. Unable to accurately model or price terrorism exposure, insurers and reinsurers withdrew from the market. This situation was a serious threat to industries where lenders and investors required terrorism protection for their investments. Subsequently, the US Congress stepped in and passed the Terrorism Risk Insurance Act (TRIA) at the end of 2002, which provided a federal backstop of up to $ 100 billion for private insurance claims related to terrorism. The government's role in inducing and sustaining a properly functioning insurance market is indispensable and such learning lessons could easily be applied to critical infrastructure protection against cyber attacks.

It is also important to note that a proper functioning insurance market provides strong incentives for companies to improve their cybersecurity posture. Insurers, seeking to avoid adverse selection – where they can't distinguish risky organisations – are going to offer insurance premiums and coverage limits that are proportional to the risk faced. Towards this end, insurers are already partnering with key security companies to better understand the risk exposure faced by different companies, and the mere process of applying for cyber insurance can help companies identify tools and best practices they may lack. Insurers may shun companies that have lacklustre security performance; hence companies with weak security posture will have to invest in security protection in order for them to get a proper cyber insurance coverage.

## Conclusion

The case for government to continue to be the legitimate security provider for the nation is ever more compelling, but governments have to strive to understand the changing world of increasing interconnectedness and interdependency in order to strike a proper balance on how best to use the RFC tool sets to affect a positive outcome for the private sector. It is inevitable that the security environment around us is complex and different organisations are more receptive to certain measures than others, hence we have to recognise that there is no one-size-fits-all solution. While governments can't control every aspect of cybersecurity, they can certainly shape the future of cybersecurity in part because there are important lessons to be learnt from other nations, other threats, etc., that could be drawn upon and applied to cyber threats in their respective nations. Cybersecurity is vital to the proper functioning and prosperity of economy and it is important for the citizens to realise that the economy has thrived not in spite of government, but in many ways because of government.

# 10

# Governance Challenges at the Intersection of Space and Cybersecurity

*Jana Robinson*

## Introduction

The disruption of capabilities that space assets provide would have immediate, far-reaching and devastating economic, political, and geostrategic consequences. Over the past two decades, space vulnerabilities have grown dramatically in a manner commensurate with terrestrial dependency on space-based capabilities and enablers. This is true for both the civilian and military activities. Purposeful interference with space systems could rather easily trigger a retaliatory spiral of actions that could compromise a safe and secure space operating environment. Accordingly, having available a range of measures to prevent or pre-empt an incident, or even full-up conflict, is of rapidly growing importance to an increasing number of countries.

The interruption of space services through a cyberattack could involve large, and possibly very complex, knock-on effects. As the space and cyberspace domains are linked operationally (i.e. space cannot exist without cyber and cyber, in some cases, without space) and they permeate all other warfighting domains (i.e. land, air and sea), cyber-related vulnerabilities of space assets are a major concern. Global effects would be virtually instantaneous.

Given these realities, space-dependent civilian governments are wise to be seeking new ways to engage in serious international discussions concerning how best to ensure responsible behaviour in these two connected domains. Meanwhile, space-dependent militaries are, to a lesser or greater degree, bracing themselves for the worst by the establishment of crisis management mechanisms to address fast-moving security threats emanating from cyber-related vulnerabilities embedded in space systems and operations. In some cases, this mechanism includes taking proper account of growing government dependency on commercial providers as key parts of both military and civilian missions.

This chapter examines various dimensions of space crisis management related to the vulnerabilities at the intersection of space and cyberspace. It first reviews how the space and cyberspace domains interact and later focuses on man-made (counterspace) threats to space assets and operations stemming from their dependency on cyberspace. It then assesses the present discourse concerning governance issues related to these domains. Finally, it offers several considerations for more effective crisis management preparedness and concludes that configuring transparency and confidence-building measures (TCBMs) in combination with the prospect of robust, and often asymmetric, responses to crises provide the proper ingredients to construct a common critical path for the management of militarily-sensitive space situations/incidents.

## Interaction of Space and Cyber Domains

Delineating the multifaceted interaction of the inherently global space and cyberspace domains can help detect space-related cyberspace vulnerabilities and configure the proper level of preparedness and responses should purposeful disruption of space operations occur. The International Telecommunication Union (ITU) describes cyberspace as "systems and services connected either directly to or indirectly to the internet, telecommunications and computer networks".[1] Cyberspace encompasses the hardware, software, data and information systems, as well as people and social interaction within the networks and the whole infrastructure.[2]

Space systems include not only the satellites themselves, but also the ground stations that operate and control them, and the links between them. Basic elements of satellites include the bus (a structural subsystem carrying other elements); a thermal regulation subsystem (cooling the active parts of the satellite, e.g. computer and receiver); a power source (e.g. arrays of solar cells); a computer control system (monitoring the state of the subsystems, controlling actions, and processing data); a communications system (the link between the satellite and its ground stations or other satellites, which includes a receiver, transmitter and radio antennae); an attitude control system (maintaining the correct direction of the satellite); a propulsion subsystem; and mission-specific equipment (e.g. radio receivers, transmitters, transponders, remote sensing systems and weapons system). Ground stations monitor and control satellites and links (e.g. uplinks, downlinks, crosslinks, telemetry, tracking and command (TT&C) link), as well as communicate with the satellite. TT&C is part of the uplink and downlink controlling a satellite's function and monitoring its health.[3]

As evident from above, space operations are entirely cyberspace dependent. In other words, space capabilities cannot be employed without cyberspace. Operators use specialised computers and computer programmes (themselves complex information systems) to transmit information to and from spacecraft over a computer network.[4] The US military doctrine regarding cyberspace, the *Joint Publication (JP) 3-12 (R)*, describes several layers of cyberspace relevant to space operations: the physical network layer (i.e. the information systems, the circuits, the ground equipment and space vehicles); the logical network layer (embedded in each piece of the physical layer, e.g. encryption or decryption of transmission, changing configurations and sending commands); and the cyber-persona layer (i.e. space operators who rely on the physical and logical network layers). It also highlights that a "critical portion of cyberspace can only be provided by space operations".[5]

As mentioned in the introduction, the interconnectedness of space and cyberspace is not only a military concern. A now-popular term "Internet of Things" (IoT) describes a concept of connecting any device that has an on and off switch to the Internet (and/or to each other), ranging from cell phones and washing machines to the drill of an oil rig. According to some estimates, there will be over 26 billion connected devices (and others say many more) by 2020.[6]

The choice to be part of this enormous network of connected devices has far-reaching implications concerning issues of vulnerability and the ability to mitigate them. Since space operators conduct cyber-dependent missions, they need to better understand not only their environment, but also the multiplicity of new threats that they must manage due to this coming IoT reality. The private sector is arguably better prepared operationally to face this new way of life as they are at the forefront of information technology development.[7] It is in the purview of governments, however, to ensure proper space governance. Hopefully, this establishes a logical basis for an intensified and expanded public-private sector partnership concerning the cyberthreats to space operations.

The dual-use nature of both cyberspace and space technologies complicates this calculus further. The ties between non-military and military cyberspace applications, as well as the use of commercial space assets for military operations blur the line between strictly civilian or military usage. A commercial software, for example, can be used to affect a country's critical infrastructure. Cyber-related vulnerabilities to space operations and services can, therefore, manifest themselves in often unexpected ways.

For example, an attack on location and timing information from a Global Navigation Satellite System (GNSS) may not be a result of jamming or spoofing of the system itself, but could be the result of the exploitation of the network-accessible systems. Accordingly, even if the GNSS receiver is working properly, the data can be false or compromised. Basically, any component of an integrated system can be manipulated, especially if it is connected to a network. Such modes of attack are attractive as they can be conducted from anywhere, do not require special hardware, and the perpetrators can more easily hide their identity.[8]

This reality makes it virtually impossible to employ traditional arms control approaches (including the verification aspects) for the governance of either domain, as it is how these technologies are used – rather than the technologies themselves (e.g. a computer and a GNSS system) – that needs to be addressed. Accordingly, in this uncertain, complex security environment, TCBMs stand out as an important policy tool in preserving global security, including in space.

## Cyberspace as a Counterspace Tool

An attack on a space asset through cyberspace has many advantages over a kinetic attack, not least of which is that it offers plausible deniability, in some cases; or can be masked as defensive even if conducted for offensive purposes. Implementing redundancy, back-ups, and design alternatives in constellations can, of course, help reduce vulnerability to a single component (not the whole system).[9]

Continuous innovation and transformation of information technology creates an unmatched set of challenges in configuring their adequate defences for both civilian and military space operators. Part of the problem is the difficulty associated with detecting and attributing an adversary action to its source. Even the US, which has arguably the most advanced thinking concerning cyberwarfare, (signing an information warfare directive in 1992, DoDD TS 3600.1, and the first doctrine on "information warfare" in early 1996[10]) struggles to configure proper safeguards.

The *JP 3-12(R)* referenced in the previous section, clarifies that cyberspace operations are not a subset of information operations, and perform three types of missions: offensive (projecting power by the application of force in and through cyberspace); defensive (defending US Department of Defense or other friendly cyberspace); and Department of Defense Information Network (DoDIN) operations. Offensive missions are authorised like all "operations in the physical domains, via an execute order". Defensive missions can be either passive or active (and can even create effects outside DoD networks that "rise to the level of use of force").[11] DoDIN operations are all actions that "create and preserve data availability, integrity, confidentiality, as well as user/entity authentication and non-repudiation".[12] The commander conducts these missions using four basic kinds of cyberspace actions: cyberspace defence,[13] cyberspace attack,[14] cyberspace Intelligence, Surveillance, and Reconnaissance (ISR)[15] and cyberspace operational preparation of the environment.[16,17]

As is evident from above, the US is acutely aware that its high dependence on space and cyberspace is exposing it to asymmetric risks of disruption. Back in 2001, the National Aeronautics and Space Administration (NASA) highlighted in its audit report that six computer servers associated with IT assets that control spacecraft and had critical data, contained vulnerabilities that could be exploited by a remote attacker.[18] If we look at China, for example, space capabilities enable the People's Liberation Army (PLA) to project military power to, and through, space. The PLA operates many of China's satellites and all terrestrial launch and support facilities. Civilian space applications are integrated into the country's more important military goals and strategies.[19] The most recent report by the Congressional US-China Economic and Security Review Commission concluded that China continues to develop a robust and comprehensive array of counterspace capabilities.[20]

Cyberattacks against satellite computer systems are of priority concern. The PLA understands, having observed the US (e.g. operations in the Balkans, Afghanistan and Iraq), that information-related technologies, including those space-based, are of unique importance to warfighting.[21] Cyber capabilities designed to achieve information dominance accomplished through counter-command, control, communications, computer, intelligence, surveillance, and reconnaissance (C4ISR) operations, are carefully examined.[22] Space-to-ground communications links and ground-based satellite control facilities also represent attractive targets for cyber exploitation.

Cyberattacks by China against the US systems have been periodically reported (with many remaining classified), including against the command and control system of Landsat-7 in 2007 and 2008, and the Terra Earth Observation System (EOS) in 2008.[23] The 2015 congressional report referenced above stated that hackers tied to China were likely behind a number of computer attacks against US space assets, including a September 2014 hack of a National Oceanographic and Atmospheric Administration (NOAA) satellite and weather service systems.[24]

It is evident that access to a satellite's controls could enable an attacker to damage or even destroy the satellite. The attacker could likewise deny, or degrade, or otherwise manipulate the satellite's transmission. High-level access could reveal the satellite's capabilities or information (e.g. imagery). Terrestrial or space-based networks can also be spied upon, or compromised, by a cyberattacker.

It is not just the major space powers that are vulnerable to cyberattacks. A

German ROSAT satellite was exposed to a cyberattack against the Goddard NASA Centre computer in September 1998, which caused it to orient itself toward the Sun eventually causing its shutdown. An alleged attack by Russia (never officially confirmed) happened against India's telecommunications satellite, INSAT-4B-S.[25] In short, any country, that relies heavily on satellites (including European countries, India and Japan), is exposed to these risks.

A stove-piped space systems-related defence would be ineffective, as a disruption could merely be a cyber intrusion in the information chain itself (i.e. data collection, processing and dissemination) without affecting satellites, but with severe consequences (e.g. unreliable imagery and wrong alarm).[26] The US Air Force, therefore, considered it proper to consolidate space and cyberspace domains under one command.[27]

## Governance Aspects

An overarching architecture for space and cyberspace governance has to cover both commercial and military activities and account for their global, strategic and dual-use nature, as well as their dependence on the electromagnetic spectrum and IT infrastructure.

Russia and China are at the forefront of promoting an arms control approach in the space domain. The notion they advance is that if countries do not engage in space arms control, the world will face unrestricted "weaponisation". This premise is difficult to accept, however, as space security in not a zero-sum game and targeted counterspace tool development and use, rather than an unrestricted one, to gain asymmetric advantages, has been underway for decades.

The Prevention of an Arms Race in Outer Space (PAROS) has been on the agenda of the Conference on Disarmament (CD) since 1985. In addition, Russia and China jointly presented a draft Treaty on the Prevention of the Placement of Weapons in Outer Space and of the Threat or Use of Force Against Outer Space Objects (PPWT) in February 2008, and a revised version in June 2014. The PPWT construct has been jointly sponsored and marketed by these two countries since 2002.

Interestingly, an integral, legally-binding verification regime for effectively monitoring compliance, has not been included. Moreover, Russia pledged at the United Nations General Assembly (UNGA) First Committee in October 2004, "not to be the first to place weapons of any kind in outer space". In 2014, Russia proposed such an approach at the UNGA First Committee in a Resolution on "No First Placement of Weapons in Outer Space".[28] This initiative is also aimed at supporting the draft PPWT.

The "no first placement", PAROS and PPWT initiatives are now being pursued proactively by Russia and China, while these same countries are stepping-up their offensive military space capabilities. This, however, does not prevent a number of UN countries to reflexively sign up to these proposed arms control schemes.

Russia and China took a different approach concerning cyberspace. They introduced a draft International Code of Conduct for Information Security (Cyber Code of Conduct) to the UN in 2011 (with an updated version introduced in January 2015), ostensibly to mitigate cyberspace conflict. The issue is, however, that the proposal discusses 'information weapon' which is another term for a restraint on free speech (a new version is talking about 'not using information and communication technologies to carry out hostile activities or acts of aggression'). The US and European countries are using the term 'cyberspace security'. Besides 'information security', the Code also discusses the principle of sovereignty in cyberspace which could both be interpreted as allowing for censorship and state control over the Internet.[29] Although seemingly nuances, they are of great importance, when countries with differing views on security concepts such as the US, Russia and China, are trying to find common ground. This reality argues for behavioural restrictions, rather than technology-related ones.

For space, the European Union (EU) proposed a different kind of Code of Conduct (the latest version of which is dated March 31, 2014), one that promotes behavioural norms to prevent irresponsible behaviour. The EU has taken the approach that without active space diplomacy, incidents and even conflicts involving the space domain are highly probable. In an environment of growing geopolitical tensions, however, the negotiations over the final version of the Code have slowed significantly – possibly even terminally.

Other initiatives related to TCBMs, including the United Nations Committee for the Peaceful Uses of Outer Space (UNCOPUOS) Scientific and Technical Subcommittee's Working Group on Long-Term Sustainability of Outer Space Activities, and implementation of a consensus report of July 2013 by the Group of Governmental Experts on TCBMs for Outer Space Activities, are, therefore, of great importance in helping advance space security in the long run.

TCBMs are also promoted in the context of cyberspace security, including in the reports of the Group of Governmental Experts on Developments in the Field of Information and Telecommunications of 2010, 2013 and 2015. The latest, consensus report (A/70/172) recommends behavioural rules, principles and confidence-building measures in cyberspace. The Group recommended that states cooperate to prevent harmful Information and Communications Technology (ICT) practices and should not knowingly allow their territory to be used for internationally wrongful acts using ICT, while respecting human rights, privacy and freedom of expression.[30]

Besides the UN, other organisations, including the Organisation for Economic Cooperation and Development (OECD), Organisation for Security and Cooperation in Europe (OSCE), North Atlantic Treaty Organisation (NATO) and Association of Southeast Asian Nations (ASEAN), actively seek to configure proper governance procedures for the cyber domain. The OSCE, for example, agreed in December 2013 on a set of TCBMs aimed at reducing the risks of conflict stemming from the use of information and communication technologies. The Council of Europe, with the participation of Canada, Japan, South Africa and the US, elaborated the 2004 Convention on Cybercrime (the so-called "Budapest Convention"), a multilateral treaty designed to address cybercrime matters and serve as a model for drafting national legislation.

International norm-building efforts in the areas of space and cyberspace reveal that the concerns of potential conflict in, and stemming from, these domains have already reached senior political levels. Granted, TCBMs for both domains are only politically, not legally, binding, and their success, therefore, is based on the premise of good will and voluntarism of states. Nevertheless, they represent a good foundation in trying to configure boundaries of what is permissible – a useful preventive tool and a barometer of political/diplomatic relations.

Real contingencies are also being discussed. When a cyberattack might be considered an armed attack by another state is, for example, discussed in the "Tallinn Manual on the International Law Applicable to Cyberwarfare" published in 2013. Issues that are difficult to address in these instruments are related to hostilities that are not ostensibly taking place under the direction of a state, namely the issue of attribution.[31]

Accordingly, situational awareness is required not only to operate effectively in space and cyberspace, but also to safeguard these domains and respond quickly to any contingencies. Yet, situational awareness in these domains is most challenging – in space because of the requirement to cover some 73 trillion cubic miles and in cyberspace because of the man-man nature of the domain itself and its less-well understood vulnerabilities.[32]

Accordingly, the establishment of strong communication links between relevant authorities, including hotlines, exchanges of information concerning policies and doctrines and regular dialogues among decision-makers remain the foundational TCBMs that states use to better understand each other and prevent misperceptions and miscalculations. At the same time, the ties between space and cyberspace require expanded contingency planning if the essential services offered by these domains are to be protected and preserved over the long haul.

## Crisis Management Considerations

The intersection of space and cyberspace is an integral component of broader security and foreign policy considerations in every space-dependent country. Crisis management related to this intersection must track the ever-changing nature of operational capabilities, and ensure the availability of effective organisational structures to facilitate sound processes for various contingencies. Should a serious incident occur, there would likely be little time for "dress rehearsals".

Internationally, it is important to not only establish norms of responsible behaviour, but also gain agreement on clear procedures to deal with escalatory spirals and other eventualities. While such a set of universal rules might prove elusive (as every situation will likely require a tailor-made solution), it remains a valuable exercise, particularly if substantial penalties are discussed for violators. For that to happen in an effective manner, however, the tool box must be defined and readily at hand.

Considerations when building the proper instruments include:

### *National Considerations*

- Establishment of a mechanism to acquire common 24/7 situational awareness.
- Education of space operators to understand cyberspace-related threats.
- Building collaborative arrangements between the space and cyberspace operators.
- Building a dossier of possible space vulnerabilities stemming from cyberspace, and possible impact (including potential for escalation).
- Understanding strategic-level implications of different contingencies.
- Configuration of smooth interaction among the relevant government authorities, and commercial and other actors to enable rapid reaction to unexpected events and shaping proper defences and damage control.
- Ensuring political-level preparedness through the establishment of a link between the operationally responsible entities/authorities and government authorities responsible for space security.
- Practising national tabletop exercises that involve government, commercial, and Non-Governmental Organisation (NGO) representatives to test how a comprehensive picture with possible political, economic, and social impacts, can be created in the event of "incidents".
- Understanding the benefits and challenges of establishing a separate "cyber command" within existing military and intelligence structures.

### *International Considerations*

- Discussion of possibilities of including a cyber operating picture into the current efforts to construct shared space situational awareness.
- Engaging in joint tabletop exercises with key space partners that address electromagnetic spectrum threats.
- Organising exchanges among and between government and commercial entities concerning various approaches to crisis management related to cyberthreats (including detection, classification and risk assessment) for space operators.
- Determining how to mutually reinforce efforts in various international organisations, including the UN, the OSCE, NATO and the OECD.

## Conclusion

As it would not be feasible to reduce the world's heavy – and increasing – dependence on space and cyberspace, the security of these domains stand equal with other national and international security considerations. The priority attention that safeguarding these domains has attracted internationally, creates a window of opportunity to formulate foundational governance concepts based on realistic, operational considerations and, at international level, a large (if incomplete) consensus.

A good start would be to find ways how to marry TCBMs with mature crisis management. The former can serve as a practical tool for bilateral, regional and global collaboration, while the latter acts as a necessary contingency should the identified rules of the road be violated. The implications of increasingly sophisticated counterspace systems in the hands of less-responsible actors are simply too far-reaching to ignore or minimise.

Acknowledging that actual capability will be developed by each state separately, the opportunities for political-level collaboration should be seized to develop a common critical path to manage militarily-sensitive space situations. Such collaboration is key as intentional acts against space assets, including those stemming from cyberspace, could jeopardise space stability systemically.

## NOTES

1. Frederick Wamala, *ITU National Cybersecurity Strategy Guide*, International Telecommunication Union (ITU), September 2011, at http://www.itu.int/ITU-D/cyb/cybersecurity/docs/ITUNationalCybersecurityStrategyGuide.pdf.
2. Martti Lehto, "Phenomena in the Cyber World", in Martti Lehto and Pekka Neittaanmaki (eds.), *Cyber Security: Analytics, Technology and Automation*, Springer International Publishing, Switzerland, 2015, p. 6.

3. David Wright, Laura Grego and Lisbeth Gronlund, *The Physics of Space Security: A Reference Manual*, American Acedemy of Arts and Sciences, 2005, pp. 109-115.
4. Chris Babcock, "Preparing for the Cyber Battleground of the Future", *Air and Space Power Journal*, November-December 2015, p. 62.
5. *Joint Publication (JP) 3-12 (R), Cyberspace Operations*, February 5, 2013, pp. v-vi and I-2.
6. Jacob Morgan, "A Simple Explanation of 'The Internet of Things'", *Forbes*, May 13, 2014, at http://www.forbes.com/sites/jacobmorgan/2014/05/13/simple-explanation-internet-things-that-anyone-can-understand/.
7. Ibid.
8. Logan Scott, "Spoofs, Proofs and Jamming", *Inside GNSS*, September/October 2013, at http://www.insidegnss.com/node/3183 (Accessed December 12, 2015).
9. Ibid.
10. Department of the Army, Department of the Navy and Department of the Airforce, *Joint Doctrine for Command and Control Warfare (C2W), Joint Publication 3-13.1*, February 7, 1996.
11. *Joint Publication (JP) 3-12 (R)*, No. 5, pp. II-2 and II-3.
12. Ibid., pp. vii.
13. Actions that protect, detect, characterise, counter, and mitigate DoD Information Network. See *Joint Publication (JP) 3-12 (R)*, No. 5, pp. II-2 and II-4.
14. Denying, degrading, disrupting, destroying, and manipulating actions. Ibid., p. II-5.
15. An intelligence action that includes ISR activities in cyberspace conducted to gather intelligence that may be required to support future operations, including offensive and defensive cyberspace operations. Ibid., p. II-5.
16. Non-intelligence enabling activities conducted to plan and prepare for potential follow-on military operations. Ibid.
17. Michael Warner, "Notes on Military Doctrine for Cyberspace Operations in the United States, 1992-2014", *The Cyber Defense Review*, August 27, 2015, at http://www.cyberdefensereview.org/2015/08/27/notes-on-military-doctrine-for-cyberspace/ (Accessed October 14, 2015),
18. Xavier Pasco, "Various Threats of Space Systems", in Kai-Uwe Schrogl et al. (eds.), *Handbook of Space Security*, Springer, 2015, p. 673.
19. *2008 Report to Congress of the US-China Economic and Security Review Commission*, US Government Printing Office, Washington, November 2008, p. 160.
20. Ibid., p. 284.
21. Dean Cheng, "Chinese Concepts of Space Security", in Kai-Uwe Schrogl et al. (eds.), *Handbook of Space Security*, Springer, 2015, pp. 431-448.
22. Cortez A. Cooper, "Chinese Perceptions of and Strategic Response to Threats in Cyberspace", in *China and Cybersecurity: Political, Economic and Strategic Dimensions*, University of California Institute on Global Conflict and Cooperation, April 2012, pp. 8-9 at https://www.usnwc.edu/Academics/Faculty/Derek-Reveron/Documents/China-and-Cybersecurity-Workshop-Report-final.aspx (Accessed December 12, 2015).
23. *2011 Report to Congress of the U.S.-China Economic and Security Review Commission*, U.S. Government Printing Office, Washington, 2011, pp. 215-217 at http://origin.www.uscc.gov/sites/default/files/annual_reports/annual_report_full_11.pdf (Accessed December 12, 2015).
24. *2015 Report to Congress of the US-China Economic and Security Review Commission*, US Government Printing Office, Washington, November 2015, p. 296.
25. Xavier Pasco, No. 18, pp. 673-674.
26. Ibid., p. 674.
27. John E. Hyten, "An Airman's Story", *The Air and Space Power Journal*, November-December 2015, p. 9.

28. Since 2005, some other countries have also formally accepted this policy, including, for example, Argentina, Armenia, Belarus, Brazil, Cuba, Indonesia, Kazakhstan, Kyrgystan, Sri Lanka, and Tajikistan.
29. "An Updated Draft of the Code of Conduct Distributed in the United Nations – What's New?", NATO Cooperative Cyber Defence Centre of Excellence, Tallinn, Estonia, February 10, 2015 at https://ccdcoe.org/updated-draft-code-conduct-distributed-united-nations-whats-new.html.
30. "UN Group of Governmental Experts: Developments in the Field of Information and Telecommunications in the Context of International Security", Council on Foreign Relations, July 22, 2015, at http://www.cfr.org/internet-policy/un-group-governmental-experts-developments-field-information-telecommunications-context-international-security/p36949.
31. Myriam Dunn Cavelty, "The Normalization of Cyber-International Relations", ETH Zurich, Switzerland, April 3, 2015, at http://www.isn.ethz.ch/Digital-Library/Articles/Detail/?id=189525.
32. John E. Hyten, No. 27, p. 8.

# 11

# Cybersecurity Threats to Critical Infrastructure: A Case Study of Nuclear Facilities

*Caroline Baylon*

## Introduction

Recent high-profile cyberattacks on nuclear facilities have raised new concerns about their cybersecurity vulnerabilities. This is of particular importance because of the potential – even if remote – for the release of ionising radiation as a result of a cyberattack. Given the sensitivities surrounding the nuclear industry, even a small-scale cybersecurity incident at a nuclear facility would be likely to have a disproportionate effect on public opinion and the future of the industry itself.

Growing recognition of the rapidly changing cybersecurity scenario led Chatham House to undertake an 18-month project exploring the potential impact on and implications for the civil nuclear sector. The project sought to assess the major cybersecurity challenges and risks posed to nuclear facilities and nuclear power plants in particular; identify international policy measures that could help to enhance cybersecurity at nuclear facilities; and increase knowledge and awareness among both industry practitioners and policy-makers of cybersecurity concerns in the nuclear sector. The research focused on the major cyberthreats to nuclear facilities, in particular on those that could affect industrial control systems, and suggests potential responses and solutions. It took a fourfold approach: a literature review; interviews with industry practitioners, policy-makers and academics; a series of expert roundtable workshops at Chatham House; and soliciting feedback from industry experts at international conferences.

## The Challenges

Meanwhile, cybercriminal activity is becoming ever easier to conduct, and more widespread: automatic cyberattack packages targeted at known and discovered vulnerabilities are now widely available, and search engines can readily identify nuclear facilities and other critical infrastructure that are connected to the Internet. As states and terrorist groups expand their online activities, the fear of a serious cyberattack is ever present as well.

At the same time, nuclear facilities are increasingly making use of digital systems, commercial off-the-shelf software and internet connectivity – all of which provide efficiency and cost-saving benefits but also make facilities more susceptible to cyber attacks. As these changes are currently under way, personnel at nuclear facilities may not realise the full extent of their cyber vulnerability. Some still cling to the myth that nuclear facilities are 'air gapped' – or completely isolated from the public internet – and that this protects them from cyberattack. Yet not only can air gaps be breached with nothing more than a flash drive but a number of nuclear facilities have Virtual Private Networks (VPN) or undocumented or forgotten connections, some installed by contractors.

The nuclear industry as a whole is currently struggling to adapt to these changes. Notwithstanding important recent steps taken by the International Atomic Energy Agency (IAEA), the industry's long-standing focus on safety and physical protection has meant that while these systems are now relatively robust, less attention has been paid to upgrading cybersecurity. In addition, its relatively late adoption of digital technologies means that it has less experience than other sectors in this area. As a result, exploiting weaknesses in digital technology may be a particularly attractive route for those seeking to attack nuclear facilities.

Other characteristics of the sector, such as the associated national security sensitivities, make disclosure of cybersecurity incidents that have occurred less likely, leading nuclear industry personnel to believe that cyberattacks are less of a threat than is actually the case. It also means that the sector's limited collaboration with others leaves it unable to learn from those with greater cybersecurity readiness. Furthermore, the rapid evolution of the threat means that regulatory standards are currently inadequate. As a result there is insufficient spending on cybersecurity, including a lack of funding for agencies poised to deal with the challenge.

All this suggests that the industry's threat assessment or risk calculation may be inappropriate, and that it is not investing as much as it should in cybersecurity.

Consequently the cost–security equation may be out of balance. Developing countries may be particularly at risk, because they have even fewer resources available.

There are also significant issues in the culture of the industry that contribute to the challenge. The different priorities and ways of thinking of nuclear plant personnel, who are Operational Technology (OT) engineers, and cybersecurity personnel, who are Information Technology (IT) engineers, frequently lead to misunderstandings. The problem is exacerbated by the fact that cybersecurity personnel are often located at a considerable distance from nuclear facilities and rarely visit the facility.

Furthermore, the level and quality of cybersecurity training at nuclear facilities are insufficient: in addition to a lack of cyber drills, nuclear personnel may have a poor understanding of key procedures, in part as a consequence of the cultural divide, since the training material is written by IT engineers. Thus, nuclear plants may lack preparedness for a large-scale cybersecurity emergency, particularly one that occurs after normal working hours.

There are numerous technical challenges too. Having been designed in the 1960s when the idea that a malicious actor would try to attack them was inconceivable, many industrial control systems lack basic security measures such as authentication and encryption, making them 'insecure by design'. Moreover, the flexibility of code means that any attacker who can get past network perimeter defences can make logic changes that are very difficult to spot. And standard cybersecurity solutions used in home or office IT environments, such as patching, are much more difficult to implement in nuclear facilities. Supply chain contamination is also a concern.

## The Solutions

### *Assessing the Risk – and Attracting Investment*

Given that many in the nuclear industry do not believe that cybersecurity poses a real risk to nuclear facilities, a first step is to raise awareness of the challenge. One way to do so would be through the development of guidelines on ways of measuring cybersecurity risks in the nuclear industry. Since at present there is no risk assessment methodology that would permit a nuclear facility to perform a combined safety risk and security risk assessment (only a safety risk assessment and a separate security assessment, which includes cybersecurity risk), such guidelines include the need for a combined risk assessment methodology for safety and security. Developing a methodology will require reflection within the industry, perhaps led by the IAEA's Interface Group, which was formed to address conflicting priorities between safety and security.

A greater understanding of the risk will also help to tackle the challenge of insufficient spending on cybersecurity in the industry. In addition to raising awareness of the need to invest in cybersecurity, it will make cyber security more commercially attractive and provide a clear economic rationale for CEOs and corporate boards to increase expenditure on it.

Since the insurance industry requires solid risk assessments, promoting the further development and adoption of cyber insurance in the nuclear sector might also be beneficial in helping develop these guidelines to measure cyber risk; cyber insurance may therefore be an important tool to enhance cybersecurity. The French Government has been conducting a major study on this question. An early conclusion is that to succeed (and to find the right level of underwriters' exposure when measured against the cybersecurity risk), a key need is the accurate calculation of that risk based on metrics agreed between insurers and the insured.

Insurance may also make cybersecurity more commercially attractive and drive the process of implementing appropriate measures, by providing the necessary financial incentives (in the form of lower premiums) to persuade owner-operators to invest in them.

### *Handling the 'human factor'*

Given that part of the challenge stems from the 'human factor' – such as engineers or contractors who set up rogue or unauthorised connections or those who plug their home laptops directly into nuclear facility networks – raising awareness among the personnel involved of the inherent dangers in doing so will be key.

There is also a need for nuclear facilities to establish rules where they are not in place already. For instance, in countries or facilities where personal devices are not already expressly forbidden within nuclear facilities, engineers should be required to hand in any personal devices such as laptops when they enter the facility; the devices should only be returned to the engineers when they depart.

There is also a need for rules requiring nuclear plant personnel to change the default passwords on equipment to secure passwords; this should apply to both existing equipment and to any new equipment installed.

In order to ensure that engineers actually follow such policies, enforcement is key. In particular, independent verification methods, in which multiple personnel check compliance with procedures, should be rigorously followed for cybersecurity issues. One interviewee suggested that if a device has been signed out, an assigned person should independently check that it is the correct device before it is hooked up to a nuclear plant; a person should also be assigned to run a virus scan on the device.

Technical means can also be used to help enforce compliance. For example, given that nuclear plant personnel may plug Universal Serial Bus (USB) devices into the nuclear facility computers even though this is not allowed, owner-operators may want to glue USB ports. Another option is to ensure that USB devices are checked for malware and cleaned before they are allowed into nuclear facilities.

### *Promoting Disclosure and Information-Sharing*

Since the industry's reluctance to share information about cyberattacks that have occurred stems partly from concern about potential damage to reputation, encouraging nuclear facilities to share threat information anonymously would promote greater disclosure. Anonymity could be achieved by asking facilities to share 'indicators of compromise', which are the traces left on a network or system indicating a malicious actor has been in the system. These might include phishing emails, the Internet Protocol (IP) addresses from which an attack was launched, or the malware code itself. In sharing indicators of compromise, nuclear facilities do not have to reveal their identity, nor what the impact of the attack has been.

Given that nuclear facilities tend to focus on reacting to attacks as they unfold, another benefit of sharing indicators of compromise is that it would encourage a proactive approach to prevent future attacks. In communicating valuable information about prevalent attacks – including the types of vulnerabilities exploited by hackers, attack pathways used to gain access, and systems targeted – sharing indicators of compromise would provide others with an early warning of such an attack. This would enable them to put defensive countermeasures in place, perhaps by increasing monitoring or by deciding to patch systems that are identified as particularly vulnerable. Anonymous sharing has been successful in other fields. Such mechanisms could be adapted in the nuclear industry.

Fostering personal contacts, which are central for the trust-building required for information-sharing, is also key for promoting the exchange of information at both national and international levels. People may not trust other companies – or governments, for that matter – but they do trust other individuals with whom they have developed strong personal relationships; they are therefore prepared to take the risk of sharing information with them. International conferences can be an important avenue for building these relationships, and more such initiatives in the nuclear industry (and critical infrastructure more broadly) should be encouraged.

Although governments are concerned that sharing threat information with other governments could jeopardise national security and thus are reluctant to collaborate at the international level, they recognise that at the national level such sharing is a key priority for defence. Governments can therefore play a key role in encouraging information-sharing within their own countries by leading the establishment of national Computer Emergency Response Teams (CERTs) specialised in industrial control systems.

The unique characteristics of industrial control systems mean that CERTs specifically dedicated to industrial control systems will be more effective. In fact, the United States has achieved success with its Industrial Control Systems Cyber Emergency Response Team (or ICS-CERT) established in 2009, which operates in addition to the United States Computer Emergency Readiness Team (referred to as US-CERT). Of course, for countries that are yet to establish national CERTs, doing so is a first priority and ICS can be handled as a division within these as a first step.

Some measure of government-backed international sharing can also take place between close allies. One avenue for this is the national CERTs; encouraging greater information-sharing between national CERTs could prove beneficial. At present, there is only limited information-sharing between CERTs on an informal, *ad hoc* basis. Even though some governments will take more information than they contribute, this will still strengthen cybersecurity. Many in the industry feel that any information-sharing, however limited, is still better than the current minimal situation.

Furthermore, given that owner-operators can be wary of disclosing cybersecurity breaches or incidents in case they are held liable, creating an environment where they feel they can speak candidly without fear of repercussions is key to increasing the level of reporting to ICS-CERTs (or CERTs more generally). The regulator should reassure owner-operators that they will not be penalised for any information they share – provided they show good faith – and that, if they disclose a cybersecurity problem or incident that arose because they violated the code, they will not be prosecuted.

Regulators thus need to understand that in order to foster a more proactive cybersecurity culture in the nuclear sector, they should be content to stay remote from some of the necessary dialogue between stakeholders; that their prime focus is on outcomes, rather than on the mechanics of delivering a minimum level of security. They also need to be aware of the difficulties of security in the electronic medium, and take a pragmatic approach to enforcement. Every system, whether it is air gapped, patched or otherwise protected, is liable to intrusion; as long as the root cause of a particular breach is not negligence or purposeful violation of rules, then regulators should only be concerned that the nuclear sector should learn from its experiences as the cybersecurity culture develops over time and corresponding capabilities are developed.

## *Developing International Policy Measures*

A number of policy measures would be beneficial as well. Given that only a small number of nations have implemented regulations regarding cybersecurity at nuclear facilities, the remaining countries should be encouraged to adopt regulatory standards. Since a large number of countries follow IAEA guidance, the agency's further development of its work on cybersecurity at nuclear facilities will prove beneficial. This can be encouraged by allocating more resources to the IAEA (and other agencies) to enable them to deal more effectively with cybersecurity threats.

Particular attention should be dedicated to helping developing countries improve their cybersecurity readiness in the nuclear sector, given their greater vulnerability. These countries are likely to require funding assistance as well to enable them to achieve this.

## *Bridging Communication Gaps*

In order to overcome the communication barriers between nuclear plant personnel (OT engineers) and cybersecurity personnel (IT engineers), fostering face-to-face communication between the two groups will be essential. For example, it is important that the cybersecurity personnel physically visit nuclear facilities on a regular basis. As cost-saving measures, they will be tempted to use methods of remote collaboration, but face-to-face contact is key to promoting mutual understanding between the two cultures.

In particular, encouraging nuclear plant personnel and cybersecurity personnel to work together on integrated projects would allow them to gain greater appreciation of each other's ways of thinking. This might involve working together on joint vulnerability analyses or risk assessments, for example. It would also help raise general awareness of cybersecurity risks among nuclear plant personnel.

It will be important to improve cybersecurity training at nuclear facilities. Given that one problem identified is that some of the training may be conducted by groups without sufficient qualifications, there may be a need for accreditation of training programmes. One source suggests that the IAEA would be the vehicle that could provide international accreditation.

In addition, training quality and frequency could be enhanced by holding integrated drills on a regular basis. This will also provide an additional avenue for communication between the two groups that will help reduce the cultural divide.

There is also an urgent need for more cross-disciplinary university and professional programmes. Interdisciplinary programmes on the cybersecurity of industrial control systems within the nuclear industry, which include both computer science and engineering disciplines, are now being established, and the creation of more such programmes should be promoted in order to help bridge the cultural gap and start to usher in cultural change within the industry.

Another initiative to improve communication, in view of the limited dialogue between cybersecurity companies and vendors, is to encourage more partnerships between cybersecurity specialists and vendors. Deeper knowledge of how vendors' propriety protocols work will enable cybersecurity companies to provide better security protection for these products. According to one interviewee, the cybersecurity company, McAfee has recently signed partnership contracts with vendors such as Alstom and Schneider for this purpose.

### *Enhancing Security*

Given that most industrial control systems were designed without considering cybersecurity requirements – and that, as noted above, it is difficult to 'add on' cybersecurity at a later date – it is essential that the designers of future generations of control systems take cybersecurity into account during the initial conception phase. For example, ICS should avoid the inclusion of non-essential digital features that could introduce cybersecurity weaknesses; otherwise, removing such features will require partial or complete redesign. In practical terms, this may mean that particularly important functions should not be digitised.

Additionally, given that the growing uptake of digital systems is leading to a reduction in redundancy, it is important for nuclear facilities to realise this and to ensure that sufficient redundancy is retained. This may involve, for example, making certain that there are manual backups for critical systems in the event of a failure.

Encouraging the greater adoption of authentication and encryption technologies in future generations of ICS will also be key, since their lack contributes to making Supervisory Control And Data Acquisition (SCADA) systems 'insecure by design'. Adding authentication when sending and receiving communications or commands means that the different parts of a SCADA system have to prove their identity to each other – and that the communication or command being transmitted is legitimate. It makes it harder to carry out cyberattacks that send unauthorised commands to a device which automatically accepts these commands, or falsifies communications (e.g., as happened with Stuxnet). And adding encryption to authentication would also make the contents of the communications or commands unintelligible to hackers, providing an even greater level of security.

Given that the unprecedented flexibility of the current generation of ICS also makes them 'insecure by design', it will be vital to restrict their malleability. While the specific nature of industrial environments means they face particular cybersecurity challenges that do not exist in everyday home or office IT environments – such as patching difficulties – these special characteristics also permit unique cybersecurity solutions that would not be possible in the latter. Promoting the adoption of 'whitelisting', for example, could therefore be an important way to bolster cybersecurity at nuclear facilities. As an information exchange protocol that only permits actions or traffic if they are on an authorised list known to be safe, whitelisting contrasts with traditional 'blacklisting' methods of cyber defence, a model under which all actions or traffic are permitted unless they are on a blocked list.

Whitelisting can be done both at the device level and at the network level. At a device level, the methodology involves authorising the device to carry out only a narrow set of actions that are necessary for its role. The computer would only be allowed to run certain types of pre-approved executable files, rather than, as now, any executable files on a USB key. This would reduce the risk of infection carried across an air gap by insertion of a USB device (which was the likely pathway used by Stuxnet).

Whitelisting at a network level involves only authorising traffic between specific points that are needed for its activities. For example, instead of allowing a computer to talk to all of the computers on the network, whitelisting would only allow it to talk to a small number of other previously identified computers with which it needs to communicate.

Industrial environments are particularly suited to whitelisting because they are predominantly static in functionality, making it possible to determine exactly what actions or traffic should be authorised. Most everyday home or office IT environments are in constant flux. At the device level, users regularly download new software on their computers – either new applications or software updates to existing applications. At the network level, computers are regularly added to or removed from parent networks. These computers also generate a lot of unfamiliar traffic, as they visit new websites and receive and send numerous emails to and from new people all over the world. This results in a high level of unpredictability.

By contrast, the industrial world is relatively fixed. At the device level, patching is rare (particularly in the nuclear environment), so device configurations change little. At the network level, industrial control systems primarily involve computers talking to computers; thus the communications and commands that different parts of such systems must exchange with other parts should follow relatively stable patterns. This predictability makes it possible to determine what actions and communications should be authorised in industrial environments.

Whitelisting can also provide a solution to the patching challenges experienced by nuclear facilities: by restricting the functionality of a device or network, it becomes less important to patch systems, and this in turn facilitates whitelisting.

In order to implement whitelisting, if the programmable logic controllers are modern, purchased within the last 10 years or so, and as long as they are digital, only a firmware upgrade to them or a new Ethernet card would be needed. The financial expense of an upgrade would be manageable. In fact, the largest share of the cost would be the additional testing and planning needed to make the upgrade safely.

If a system is older, perhaps 20–30 years old, then whitelisting may not be possible. In this case, other options that can add security include active management, the deployment of intrusion protection systems, and intrusion detection systems which monitor the electronic traffic within a nuclear facility for anomalous behaviour. Some of these are discussed further, as follows:

> Intrusion detection systems such as network monitoring, which involves examining the traffic within a nuclear facility for anomalous behaviour, would enable nuclear facilities to take a more proactive approach to cybersecurity. When the system detects unusual traffic that does not fit the established pattern, it alerts the owner-operator.
>
> For many facilities (nuclear and otherwise), the first step in network monitoring is to map the expected traffic between devices in order to establish a standard baseline. Many nuclear facilities are yet to do this, and others may not have undertaken the mapping at a sufficiently detailed level.
>
> The use of virtualisation – the creation of a virtual version of a device, operating system or network – may be a useful process in helping understand the data flows and serve as an effective way to map out those connections. By virtualising the entire network, it is possible to learn about the data flows without the degree of risk involved in actual experimentation.
>
> Furthermore, monitoring needs to be done on the entire industrial control system network, not just on the perimeter. Since personnel at nuclear facilities (and, in fact, critical infrastructure more generally) too often concentrate only on perimeter defence, allowing malware to operate undetected if it is able to get past the perimeter, they need to recognise that they must monitor all networks.
>
> In addition, encouraging the adoption of secure optical data diodes, where not already implemented, would significantly enhance cybersecurity. This is key given that there are some nuclear facilities that may have only a firewall to protect the industrial control system network.

With regard to supply chain challenges, the globalisation of manufacturing means that resolving vulnerability remains difficult. However, some countries are taking important steps towards the nationalisation of their supply chains (in the nuclear sector and beyond). Japan has had the greatest success here in enabling indigenous companies to build the entire product range for its nuclear power plants. Although microchips from foreign sources may be used, one interviewee states that Japanese power plants are "almost 100 per cent national; they make the products that they need".

The best option for countries that lack the required extensive national industry is to reduce their supply chain vulnerability to the maximum extent possible. Russia, for example, views the nationalisation of its supply chain as a priority, including in the nuclear sector. Given the difficulty of manufacturing all of its products domestically, in the short term Russia is seeking to reduce its dependency on components manufactured in countries that it considers 'less friendly'; instead, it is substituting them with components from China, which it considers a 'more friendly' country at present. Russia views this as an intermediate step while it continues to build up its own national industry. In the long term, it hopes to be able to replace the majority of components with Russian products.

Of course, for financial reasons it will be important for nuclear facilities to identify the most crucial parts of the plant from a cybersecurity perspective (notably, their critical cyber assets) in order to grant those the highest levels of protection. As one source states, "It needs to be a graded approach; we can't afford to do everything for every system." Prioritisation of the cyber risks is therefore key.

# 12

# CHALLENGES OF CYBERSECURITY: MALWARE AND AS-LEVEL STRUCTURE

*Ted G. Lewis*

## Challenges of Cybersecurity

Cybersecurity is a complex evolving challenge at both policy and technical levels. Table 1 summarises the policy challenges, and Table 2 summarises the technical challenges. Generally, policy challenges have to do with asymmetric exploits and society's inability to deal with them, quickly, and without trampling on civil liberties. Technical challenges, on the other hand, stem from the design of the Internet itself, and the self-organisation of the Autonomous System (AS)-level Internet. This survey presents the challenges in brief, and then provides a detailed examination of the last two technical challenges: the monoculture of the Internet, and the scale-free structure or "wiring diagram" of the AS-level network.

Anonymity protects perpetrators because it is difficult or often impossible to identify the source of an exploit. The dark network called TOR, for example, obfuscates source Internet Protocol (IP) addresses on purpose. Most exploits are asymmetric – they cost little to launch and may do damage out of proportion to the cost and effort of the attacker. Add to anonymity and asymmetry four more critical societal factors: inadequate IT operational skills; lack of cooperation across local, provincial, federal, and international levels of government; potential imbalances between privacy and security; and leaders unprepared to cope with new and rapidly changing technology – and the policy challenges mount. Large government and private sector exploits that have gained widespread attention, such as the Office of Personnel Management (OPM) and Target Store exploits, could have been easily avoided if the operators were equipped with better skills. Additionally, governments around the world are either ill-equipped to deal with necessary legislation to protect consumers due to civil liberty issues, or are ignorant of the technology. Without a basic understanding of computer and communications technology, it is unlikely that a balance between privacy and security will evolve from clueless politicians.

**Table 1: Policy Challenges to a Secure Cyberspace**

| | |
|---|---|
| Anonymity: | Hard to trace back to source of exploit. |
| Asymmetry: | Predator-prey relationship favours attackers. |
| Cooperation: | Local-Federal-International coordination does not exist. |
| Skills: | IT systems unable to compensate for unskilled system operators and consumers. |
| Civil liberties: | Where is the balance between privacy and security? |
| Clueless leaders: | Most policy-makers lack technical or analytic skills needed to understand the problem. |

The technical challenges are better defined and understood, but remain difficult. First and foremost, the Transmission Control Protocol/Internet Protocol (TCP/IP) was never designed to be secure. Rather, it was designed to be simple and efficient. In fact, the Internet is a simple and straightforward machine at all levels, which is where the problem begins. Almost anyone can hack the Internet and systems connected to it by simply reading the documentation online. The vulnerability of TCP/IP and other protocols such as the Secure Socket Layer/Transport Layer Security (SSL/TLS) and Border Gateway Protocol (BGP) would not be so critical if the Internet remained a backwater experiment for computer scientists. Instead, it has become wildly successful and with its success, entanglement with other systems continues to grow. Cyber systems link together water, power, telecommunications, transportation, medical, and industrial control systems of all sorts. This entanglement is a major source of vulnerability to the Internet and the systems it connects.

The technical challenges of cybersecurity are multiplied by volume, complexity and speed. Millions of exploits are launched every day, and the sophistication of exploits increases daily. Stuxnet was ten-fold more complex than the average exploit, and exploits continue to increase in complexity and power. At the same time, the number of consumers logging in to the Internet grows exponentially, adding more laptops, cell phones, and industrial Supervisory Control and Data Access (SCADA) devices to the global network. These devices are typically open and contain very weak edge protection. A cell phone, for example, may require a 4-6 digit Personal

Identification Number (PIN) to activate, and contain unencrypted passwords, contact lists, and bank account numbers. Edge devices are easy targets for hackers, especially when connected to the network by open Wi-Fi or Bluetooth.

Of particular interest in this chapter are the last two challenges listed in Table 2: the monoculture of the Internet, and its scale-free structure. Since the beginning of the global Internet, protocols and open source software has been widely reused as building blocks. Identical code and identical protocols make for a monoculture. As such, a flaw in one code or protocol is a flaw in them all. An exploit that works locally, also works, globally. Thus, malware spreads quickly and efficiently from server to server with little to stop it.

In addition to the Internet's weakness due to its monoculture foundations, the global Internet has self-organised into a scale-free network which accelerates the spread of contagious malware. A scale-free network is one in which the distribution of peering relations (connections or links) across nodes (Autonomous Systems) obeys a power law. That is, most nodes have modest numbers of connections while a handful of hubs have very large numbers of connections. Typically, large server hubs connect hundreds or even thousands of peers. For example, AS701 links together over 2,700 other AS-level nodes. Thus, large hubs serve as superspreaders of malware.

**Table 2: Technical Challenges to a Secure Cyberspace**

| | |
|---|---|
| Occam's razor: | TCP/IP and other protocols are simple and were never designed for secure operations. |
| Entanglement: | Internet convergence with all other sectors opens them to new vulnerabilities. |
| Volume: | Millions of exploits per day. |
| Complexity: | Recombinant DNA of malware evolves. |
| Speed: | Zero-day exploits can go undetected for years. |
| Weak edges: | Laptops, cell phones and SCADA systems have weak authentication/ encryption capability. |
| Technology monoculture: | A flaw in one system is a flaw in all systems. |
| Scale-free structure: | The Internet's "wiring" contains superspreaders, which accelerate the spread of exploits. |

## AS-level Structure

The scale-free structure of the AS-level Internet is apparent in Figure 1, which contains 13,579 AS-level nodes and 37,448 peering links. The layout of Figure 1 places peering nodes in a circle around highly connected nodes so the structure is made evident. As you can see, most peers belong to a hub, hence the high spectral radius. [Spectral radius, r is the largest eigenvalue of the connection matrix, and increases with both density of links and size of hubs. For large systems like the Internet, it is computed by the power method, which takes time proportional to $O(n^3)$].

Suppose we model the *vulnerability* of the monoculture nodes of AS13579 by assigning a probability *v* of being contaminated by a malicious code or worm. Then, the number of nodes contaminated by the spread of a malicious code is also probabilistic and defined by an exceedance probability, EP(q), where q is the power law exponent in $EP(q) = c^q$, and c is the number of contaminated nodes. Exponent, q is also called the *fractal dimension*, because power laws are very simple self-similar fractals. EP is called the *exceedance probability* because it is the probability that a randomly contaminated node will contaminate c or more AS-level nodes.

Malware spreads through network systems much like an animal virus spreads through a population of susceptible animals. However, in place of physical contact, network viruses travel through links to adjacent nodes. In turn, adjacent nodes spread the virus to their adjacent nodes and so forth. The rate of spreading is determined by vulnerability, v, which is the probability of contracting the virulent malware. This author developed the following (average) relationship between fractal dimension, vulnerability, and spectral radius for the spread of a contagion through any network:[1]

$\text{Log}(q) = b - kvr$; b, k are constants, and v, r are vulnerability and spectral radius, respectively.

This fundamental resilience relationship for cascading or spreading of malware through networks such as AS13579 defines a networked system's resilience in terms of fundamental factors v and r. Constants b and k depend on other network properties and are obtained by simulation of thousands of network epidemics.

The resilience of AS13579 was obtained by simulation, noting $r = 73.5$:

$$R = 1.1 - 0.5vr = 1.1 - 36.8v$$

A tipping point exists at $R = 0$, so that $v_0 = 3.0\%$. This is the point that separates a massive contamination from a minor one. In other words, malicious code spreads without limitation when the vulnerability of monoculture nodes exceeds $v_0 = 3.0\%$. The fact that tipping point vulnerability is very low indicates high susceptibility of the AS13579 network to malware.

Mitigation of malware spreading is achieved only by reducing v, r, or both. Vulnerability $v_0$ is already very low, so the alternative is to reduce $r = 73.5$. This can be done in several ways, described as follows.

**Figure 1: AS13579-level Internet and its 37,448 Peering Connections and Spectral Radius of 73.5; Mean Degree is 5.5 Links; AS701 is the Hub with 2,637 Links**

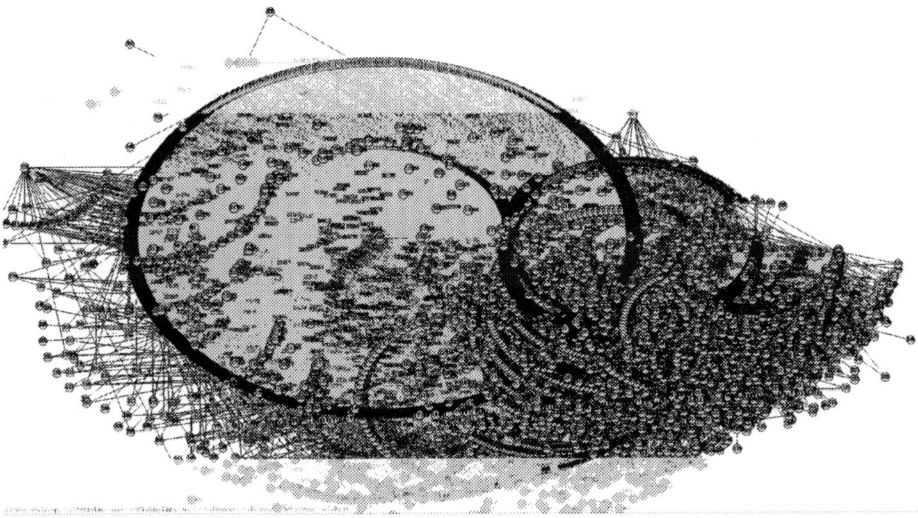

## Reducing Spectral Radius

Spectral radius is a function of number of links and size of hubs. Hub size is also known as node degree, which is defined as the number of links connecting the node to other nodes. Therefore, the obvious way to reduce r is to remove links or restructure the Internet to remove highly connected AS nodes. Both of these methods are technically and politically impractical.

Another approach is to find and protect the blocking nodes. A blocking node is a node that cannot be removed without segmenting the network into disjoint components. That is, a blocking node is an essential bridge between components of the network. As it turns out, AS13579 contains 944 (7 per cent) blocking nodes. Malware spreading drops to nearly zero when spreading is blocked by a blocking node. In effect, hardening of the 944 blocking nodes of AS13579 reduces its spectral radius from 73.5 to 25.6. In turn, the tipping point vulnerability increases to $v_0$ = 8.6%. In addition, the maximum number of contaminated nodes equals the largest component defined by blocking node removal, i.e., 2577 in Figure 1.

The blocking node algorithm can be applied again to identify more blocking nodes within components, so that the maximum number of AS-level nodes affected by malware can be reduced even further.

Blocking nodes can be identified by a brute-force algorithm, which examines each node one at a time to determine if its removal separates the network into islands. This algorithm is $O(n^2)$ in time complexity, requiring 185 million steps to find 944 blocking nodes in AS13579. For example, a second application of the blocking node algorithm finds 883 additional blocking nodes, with corresponding maximum component size of 263 nodes, and reduces spectral radius even further to 11.5. Malware spreading is virtually eliminated by protecting 13.5 per cent (944 + 883 = 1827) of all AS-level nodes.

## Open Shortest Path First (OSPF)

The Internet's OSPF algorithm maintains routing tables that foster heavy reliance on shortest-path routing. A natural question to ask is, "How many blocking nodes also lie on the most-likely routes?" Simulation of shortest-path routing of random messages originating and destined for randomly selected nodes shows that most blocking nodes are also OSPF nodes. Specifically, 93 per cent (819 of 944) of the blocking nodes identified in Figure 1 lie on shortest paths. Hence, the likelihood of a message passing through a blocking node is high, which implies that the likelihood of malware passing through a blocking node is also high. This suggests that a keen focus on blocking nodes will be rewarding in terms of malware resilience.

## A Note on Persistence

A persistent virus is one that recurs. That is, removal of an infected server without patching the vulnerability leaves an AS node susceptible to repeated contamination. Once again, spectral radius plays a role in determining persistence. That is, malware recurs under the conditions that AS-level servers are not patched quickly:

$$\text{Persistence: } v > \text{removal rate/spectral radius}$$

In the case of AS13579, $v > 0.5\%$ leads to persistence.

## Recommendations

Shrinking the spectral radius of the AS-level Internet increases malware blocking, and therefore resilience in the face of a malware threat. Given that it is unlikely that owners and operators of autonomous systems will voluntarily reduce their level of peering, the author prescribes the hardening or even *air gapping* of critical nodes called blocking nodes. By preventing further spreading through 7 per cent of the AS-level nodes, most malware spreading can be minimized. Blocking of 13.5 per cent of the AS-level nodes virtually eliminates spreading.

Given that blocking nodes correlate with OSPF routing, it is highly probable that a blocking node strategy will succeed in reducing contamination by malware. An alternative to blocking node hardening is to re-think and re-design the OSPF routing algorithm. Perhaps a randomized routing algorithm can be used. In either case, blocking nodes act as a kind of gateway between components of the Internet. Protecting these gateway nodes should be the highest priority.

## NOTES

1. Ted G. Lewis, *Critical Infrastructure Protection in Homeland Security: Defending a Networked Nation*, John Wiley & Sons, Hoboken, New Jersey, 2015, pp. 376.

## REFERENCES

Lewis, Ted G., *Critical Infrastructure Protection in Homeland Security: Defending a Networked Nation*, 2nd ed., John Wiley & Sons, Hoboken, New Jersey, 2015, pp. 376.

# 13

# Non-State Actors and Cyberspace: An Overview

*Sanjeev Relia*

On January 2, 2016, a heavily armed group of terrorists attacked the Pathankot Air Force Station. After over 48 hours of gun battle and seven casualties, (all security personnel) all six attackers were neutralised. As it appears from media reports, it was a group of fidayeen who infiltrated across the international border and executed the act. Mobile intercepts showed that they were reporting back and were controlled by their handlers based in Pakistan. We in India have been fortunate that the terrorists coming from across the borders are yet to adopt more devastating tactics and technologies that are being used in other theatres across the world.

Nevertheless, it is clear that the non-state actors operating in the virtual world today pose as dangerous a threat as non-state actors operating in the real world. For example, in the Pathankot case, while the gun battle was on at the Pathankot airbase, in the early hours of January 03, the Northern Indian electricity grid tripped, plunging the states of Punjab and Jammu and Kashmir (J&K) into darkness. Early morning on the same day, the Base Transceiver Stations (BTSs) in and around Pathankot started misrouting mobile calls, and communication via calls was mostly not possible. By seven in the morning, however, pictures started appearing on social media of burning aircraft and dead soldiers at the Pathankot airbase. The posts went viral in minutes. Investigations at a later date showed that all this was caused by cyberattacks launched from Pakistan. Obviously, the Pakistan Government denies having indulged in any such activity. Like the physical attacks, perhaps some non-state actors, on behalf of the government coordinated and implemented the cyberattacks, too.

In the *Arthashastra*, Chanakya has written that a state could be at risk from four types of threats, viz. internal, external, externally aided internal and internally aided external. While Chanakya wrote this in 350 BC, when no cyberspace existed, yet despite a passage of over 2,000 years, his words still hold good and apply to cyberspace, a phenomenon which is not even 35 years old.

What Chanakya referred to as externally aided internal and internally aided external threats are the threats originating from non-state actors. Though there is no universally accepted definition of who a non-state actor is, the Geneva Centre for the Democratic Control of Armed Forces, in its article, entitled, "Armed Non-State Actors: Current Trends & Future Challenges", defines non-state actors as any organised group with a basic structure of command operating outside state control that uses force to achieve its political or allegedly political objectives.

While the above definition applies particularly to non-state actors in the real world, it is important to ask, Who could be the non-state actors in the virtual world and how? Would they be different from the ones who operate in the real world? Cyberspace is an arena where non-state actors of various kinds thrive. Once again, there are no universal classifications of such players. I have identified two distinct categories of groups that could be termed as non-state actors operating in the cyberspace.

The first type of non-state actors is organisations that have created cyberspace. Critical infrastructure of most nations is created and largely owned by the government of the country. Cyber infrastructure on the contrary has been created and is owned mostly by private business houses and organisations. Thus, they belong to the first category. They also are amongst the most targeted elements of cyberspace. But they are not our focus of attention today.

The second category is the bad guys who pose a threat in cyberspace. I further categorise this variety of non-state actors into two distinct subcategories:

(a) The first subcategory is of those non-state actors who form part of radical organisations such as the Islamic State of Iraq and Syria (ISIS) and Al-Qaeda. They are also referred to as cyberterrorists. These are the people who indulge in the use of the Internet and networks against critical national infrastructures in order to hit or intimidate a society and its peoples causing casualties or injuries for ideological, political or religious reasons. Though not many examples of major cyberattacks by such non-state actors can be cited, yet that does not imply that these groups are not capable of undertaking such acts. These organisations are exploiting cyberspace in other manners too. Recruitment, generation of funds, projection of power and propaganda using the social media are some of the activities being undertaken by cyberterrorists. For example, consider the manner in which ISIS is targeting the US Government websites. As per some news reports that appeared in the US media in December 2015, ISIS is looking to expand its online capabilities in order to carry out cyberattacks against US targets, including the federal government and airplanes.

(b) The second important subcategory of non-state actors in cyberspace is Cyber Militia. As per my analysis, this is the category of non-state actors we in India presently need to worry about and guard against.

## What is Cyber Militia?

A cyber militia can be defined as a group of volunteers who are willing and able to use cyberattacks or other forms of disruptive cyber actions on behalf of a nation state in order to achieve a political goal. In the words of Chanakya, they belong to the group of externally aided cyber threat. Cyber Militia are men, not in uniform but motivated enough to be employed in covert government-orchestrated campaigns with the purpose to further the strategic political or military objective of the instigating state. It is said that China has established the People's Liberation Army (PLA) Unit 61398 staffed by thousands of computer professionals as "Cyber Troops" acting on direct orders of the PLA. Unit 61398 is supposed to be responsible for all major cyberattacks and cases of cyberespionage against the US and other countries including India. China on the other hand completely denies even the very existence of any such unit leave alone its involvement or connection with any other state machinery. If a country like China can rely on non-state actors, it is only a matter of time before even Pakistan creates such capabilities and uses them against India in line with its policy of state-sponsored terrorism, if it is not already doing so. Pakistan specifically is influenced by two of its policies: first, disruption of opposing conventional operations while maintaining credible nuclear threat, and second, peacetime use of disruptive provocation in the form of terrorism. Disruptive provocation using the cyberspace will be an essential ingredient of the overall plan. Non-state actors will play a key role in the implementation of such a policy.

When a nation state covertly adapts the policy of use of non-state actors to sponsor terror though cyberspace, it indeed would lead to serious security concerns for our government. For this reason alone, the threat of cyber militia is higher than the threat of cyberterrorists.

But then the question arises, why use non-state actors? Why not just employ the government machinery for such acts? Employing cyber militia in place of regular cyber units/organisations has tremendous advantages. Some of these are as follows:

(a) **Freedom to Attack from Anywhere**. Cyberspace knows no boundaries, and hence non-state actors become non-territorial players. That is, they need not be based in the same country which is sponsoring them. The attack can be carried out with the same precision and impact with the attacker based in a third country, or perhaps from Indian soil in our case. This makes the task of the attacking nations easier as attribution becomes even more difficult in such cases.

(b) **Cost Factor**. To raise a well-organised cyber wing as part of the government or defence forces would cost a lot of money as such a force will have to be funded and manned by uniformed personal. By recruiting suitably motivated and technically competent non-state actors, the same task can be achieved at little or no cost.

(c) **Counterstrike**. In cyberspace, lack of any hard evidence which can establish attribution protects the attacker against any political ramifications. Thus, the threat of a counterstrike is negligible. In 2007, while all evidence showed that the distributed denial-of-service (DDoS) attacks on Estonia originated from Russia, Estonia or the North Atlantic Treaty Organisation (NATO) could not retaliate due to lack of attribution. While Russia completely denied any involvement, the execution may have been carried out by patriotic cyber militia on behest of the Russian Government. In India, we still have not been able to work out a clear cut policy for a counterstrike on state-sponsored terrorism.

(d) **Laws of War Do not Apply**. Even if an indisputable link is established between a non-state proxy and a nation-state, no laws of war apply to these cyber militias.

Having established that a non-state actor group in the cyberspace today is a potent force to reckon with, let us see how they could threaten the security fabric of India.

## What Threat Does India Face from Cyber Non-state Actors

The Stuxnet attack and the crisis it created for the Iranian nuclear programme in 2010 has been considerably analysed by experts all over. But many would not be aware that Stuxnet was detected in Indian hardware too. Stuxnet was designed to attack systems using Windows Operating System, Siemens PCS 7, WinCC and STEP7 industrial software applications that run on Windows and one or more Siemens S7 Programmable Logic Controls. In other words, Stuxnet was designed to target computers specifically associated with the SCADA systems, as these software are found in SCADA/industrial control systems. Based on a study of the spread of Stuxnet conducted by Symantec, the most affected countries in the early days of the infection were Iran, Indonesia and India. As per its report of September 2010, 8.31 per cent computers in India were found infected with Stuxnet.

So, how did the worm manage to reach well-protected hardware in Iran and India and what damage was caused by it in India? Obviously no nation state was directly involved in perpetrating the Stuxnet attack. The sophistication with which the worm's code was written and the lethality with which it carried out its task indicates that it was not a handiwork of some novice hacker. As no money or information was stolen by the exploits of the worm, it is unlikely that some motivated cybercriminals created and planted it. The precision with which Stuxnet attacked the SCADA systems indicate that it took a lot of planning and effort in the implementation of the attack. Such a task could have been done either by a nation state or non-state actors acting on behalf of some state. The same people who perpetrated the worm attack in Iran, also perhaps infected the Indian systems too. While the damage caused by Stuxnet in Iran is well documented, unfortunately no survey is available in the public domain which could establish the nature of damage that may have been caused in India by it. Though some reports in the media indicate that INSAT-4B, a communication satellite launched by India in 2007 and which effectively went dead in 2010 due to failure of its transponders affecting 70 per cent of DTH services in India, was a handiwork of Stuxnet, the same was denied by ISRO, and not confirmed by Siemens too, whose software the satellite was using. Whether the satellite went dud because of Stuxnet or not, the mere fact that such a deadly computer worm was able to penetrate unnoticed into control systems of our satellite network (if the Forbes report is to be believed), is an indication of the penetration capabilities of offensive cyber tools available today with rouge elements.

**Sabotaging the Critical Info-infrastructure:** Sabotage is an inherent part of Cyberwarfare. Malicious software and cyberattacks are ideal instruments of sabotage. This is especially applicable for sectors which provide direct services to consumers such as telecom, banking and power. As systems become more complex, the knowledge required to attack them also becomes more complex. Unless the attacker is backed up with full financial and knowledge support, sabotaging industrial control system will be a difficult task. While a nation state may undertake sabotage through the cyber domain in war as part of its strategic offensive cyber policy, in peacetime, non-state actors are the most suited cyber adversaries who can achieve such a task with ease as they have all the necessary backing of the state.

India is in the process of moving from being a "partially networked state" towards being a networked state. The digital India programme ridding on the massive telecom network of the country is making inroads into the remotest corner of the country. We are in the process of creating 100 smart cities. A large number of other such projects are either already in the process of being implemented or are on the anvil. With the state of automation and networking improving, India's critical information infrastructure will become vulnerable to cyberattacks. Cyber non-state actors therefore would pose a constant threat to our critical information infrastructure during both peacetime and wartime.

**Subversion:** Another activity which non-state actors can undertake effectively through the cyberspace against India is subversion – subversion of human resource as well as system subversion. Let us first deal with human subversion. Many a times, an adversary can achieve much more by internal subversion than going to war. As per Thomas Rid, famous British scholar and writer, information technology has enabled proliferation of subversive causes and ideas. Because of cyberspace, subversion has become more cause driven, it is seeing higher levels of membership mobility and is now characterised by lower levels of organisational controls. One common tool of all subversion activity is media, be it print, visual or online media. The exponential rise and infinite reach of social media today has made it a perfect tool for subversive activities. The kind of influence social media has on the society has got our government thinking about the impact it can have on the internal security of the country. Today politicians, senior government officials and scholars can often be heard voicing their concern about the negative and subversive impact of social media. Whether it is the Pak-based agents' honeytrap involving defence personnel or the recent arrest of ISIS operators in India, subversion seems to be high on the list of activities of our neighbouring nation.

System subversion is the covert and methodical undermining of internal and external controls over a system lifetime to allow unauthorised or undetected access to system resources and information. Network or system subversion, unlike a cyberattack is a very specialised operation. Subversion requires the subverter to have sufficient access to the system at one or more points during its life cycle to exert influence on its design, implementation, distribution, installation, and/or production in a way that can later be used to bypass the protection mechanisms. Both software and hardware can become victims of system subversion. Almost 100 per cent of hardware used in our critical infrastructure and public networks is being produced in China. With virtually no test facilities available in the country, supply chain vulnerability of our hardware is extremely high. In such circumstances, system subversion is a very likely disastrous event waiting to happen. This is also where the first category of non-state actors – the good variety of non-state actors – comes into picture. It is unlikely that the country producing the equipment would indulge in a policy of introducing supply chain vulnerabilities. It would be non-state actors backed by the nation and in connivance with the industry who could perhaps indulge in something like that.

**Espionage:** From nation states to novice hackers, everyone seems to be stealing information using cyberspace. While espionage was always a part of any military campaign and study, private lives of ordinary citizens were generally not much affected by such activities. Today cyber surveillance technologies range from malware which infects a target computer to record every keystroke, to systems for tapping undersea fibre-optic cables in order to monitor the communications of the entire population. Most of the world is inadequately prepared for defending against these new types of surveillance and espionage techniques. Nation states using their state machinery do indulge in all kinds of cyber espionage activities. Non-state actors can assist the government in undertaking this task more effectively.

The above are just some of the ways a nation can employ non-state actors in cyberspace. While sabotage, subversion and espionage would be the main motives behind employing cyber militia, there could be many other ways to use them in spreading terror in India using cyberspace. Our armed forces and other governmental organisations have mastered the ways to counter state-sponsored terrorism in J&K and the North East; we will now have to learn innovative methods for fighting actions perpetuated through the cyber domain.

## What Could We Do to Counter the Threat of Non-State Actors in Cyberspace?

To be completely secure, a nation needs economic security, energy security, environmental security and cybersecurity, besides securing the borders. But can the government achieve all this on its own? Perhaps no. Security today has become a multi-stakeholder function. Cybersecurity cannot be ensured just by the government of the nation. While the government can lay down policies with a broad framework of establishing a secure cyber ecosystem and provide a legal framework for the implementation of such policies, it is the responsibility of every organisation which has a stake in CII to follow these guidelines to make their infrastructure safe and secure against cyberattacks. All stakeholders therefore will need to come together to respond to the threats and challenges posed by non-state actors in cyberspace.

The ways and means of countering threats from non-state actors in cyberspace are not very different from the way threats from others sources have to be handled. Whether the threat to the critical infrastructure is from nation states, cyberterrorists or cyber militia, the ways to counter the threat would by and large remain the same. To handle this threat, the best method would be perhaps to create an environment where cyberattacks cannot take place on national assets. In other words, creating a cyber ecosystem in the country where each device can work with immunity. The focus of strengthening the cyber ecosystem is towards driving fundamental change in the way people and devices work together to secure cyberspace. This evolutionary change in the computing environment will be achieved by empowering individuals and organisations to operate securely, making and using more trustworthy cyber protocols, products, services, configurations, and architectures, building collaborative communities and establishing transparent processes. All this is not so easy to achieve and considerable time, money and effort will have to be spent to establish such an ecosystem in the country. Along with this, strengthening of the regulatory framework of the country to ensure that legal cover is available to this framework will also be required. The IT Act amended in 2008 in its present form is not adequately armed to meet the challenges of the cyberspace.

## Conclusion

Being a high-tech country in a volatile environment, India faces a different reality than most of the developed world. A large part of our nation is coping with problems of terrorism, left-wing extremism or communal violence. If these radical elements take the cyber route to spread terror, they could cause devastating damage to the country impacting our national security. It is time that we in India took a serious view of this and addressed the issue of cyber conflict with non-state adversaries, in order to establish a secure and resilient cyberspace in the country.

# 14

# NON-STATE ACTORS AND CYBERSPACE: A NORTH AFRICAN PERSPECTIVE

*Gillane Allam*

With the advent of the 21st century, the concept of it being the "Asian Century" was in full swing. In theory, the concept of an "Asian Century" – the world's political and economic centre of gravity is shifting to Asia in the 21st century – is correct. Indeed, one in every three people on Earth is Chinese or Indian. More importantly, Northeast Asia – particularly China, Japan, South Korea and Taiwan – has enjoyed astonishing levels of economic growth since the latter part of the 20th century with booming export sectors and emerging wide middle-class consumer bases. The global production and supply networks that have taken shape in Asia produce much of what the world consumes. The Northeast Asian countries also became major consumers of energy imported from the Middle East countries (West Asia), thus making them geopolitically important power centres.

Simultaneously, the blooming 21st century was arduously turning onto digitalisation and the use of cyberspace. Expansion of digital empowerment was sought to become an integral part of almost every aspect of institution management, including critical infrastructure and national security establishments. It is estimated that 1,400 million in Asia are cyber users. Presenting both opportunities and vulnerabilities, digitalisation entails a compulsory need for securing cyberspace from malicious uses and intents.

Another variable affecting the world geo-political scene at the time and taking significant root in Asia was the end of the Cold War. It was against the formal withdrawal of the former Soviet Union troops from Afghanistan in February 1988 and the widespread uprisings in the former republics of the then Soviet Union, that Francis Fukuyama, a US political scientist, published in 1989 his famous essay, "The End of History", relishing the triumph of the Western style liberal democracy combined with capitalism over communism and socialism. Fukuyama claimed that the advent of that system of governance may signal the end of humanity's socio-cultural evolution and thus set the final form of human government. As such, Fukuyama totally discarded any impact of culture and socio-economic realities in different societies in favour of democracy coupled with capitalism.

So the question arose: 'Is this the end of War?' As a response, in 1989, a US military thinker, William S. Lind co-authored, with a few high-ranking US officers, another famous essay, "The Changing Face of War: Into the Fourth Generation".

Lind indicated that his essay was a possible response to an often-asked question in the US military: 'How will the next war look like?' The next war, the Fourth Generation War (4GW) as Lind put it, is characterised by the loss of the state of its monopoly over war. All over the world, state militaries find themselves fighting non-state opponents such as Al Qaeda, Hamas, Hezbollah, etc. Almost everywhere, the state is losing. Lind acknowledged later that copies of his essay were found by US troops in the caves of Tora Bora, the hideout of Osama bin Laden in Afghanistan.

In a subsequent article in 1992 commenting on the US involvement in the war in Iraq, Lind indicated:

> "4GW is obviously marked by a return to a world of cultures, not merely states in conflict. And at the end of a war like the one in Iraq, it becomes inevitable that the local state attacked vanishes leaving behind a stateless region (Somalia) or a façade of a state (Afghanistan) within which more non-state elements rise and fight."

In brief, the rise of 4GW relates to a number of realities on the ground:

- The loss of the nation-state monopoly on war, thus forcing all open conflicts into the 4GW mould.
- The rise of cultural, ethnic, and religious conflicts.
- Globalisation leading to open societies and economies, and modern and new technologies, thus allowing an enhanced role of global networking including the role of media.

Along a similar notion, also in 1992, another US political scientist, Samuel P. Huntington, published in *Foreign Affairs* his famous essay entitled, "The Clash of Civilizations: The Next Pattern of Conflict". Huntington states,

> "World politics is entering a new phase, and intellectuals have not hesitated to proliferate visions of what it will be; the end of history, the return of traditional rivalries between nation states, the decline of the nation state from the conflicting pulls of tribalism, globalism, among others. Each of these visions catches aspects of the emerging reality. Yet they all miss a crucial, indeed a central aspect of what global politics is likely to be in the coming years ...
>
> It is my hypothesis that the fundamental source of conflict in this new world will NOT be primarily ideological or primarily economic. The great divisions among humankind and the dominating source of conflict will be cultural. Nation states will remain the most powerful actors in world affairs, but the principal conflicts of global politics will occur between nations and small groups of different civilizations. The clash of civilizations will dominate global politics. The fault lines between civilizations will be the battle lines of the future. Conflict between civilizations will be the latest phase in the evolution of conflict in the modern world."

The question arises: Have the 9/11 attacks on the US been envisaged and executed by Al Qaeda from that ideological vantage point? The stark reply is no. They were an act of defiance and vengeance.

The ensuing 2003 US invasion of Iraq under the false pretence of its acquisition of Weapons of Mass Destruction (WMDs), nuclear and chemical, brought about dramatic changes not only for Iraq but for the Arab world as a whole and the world at large. The New Middle East project found its moment at last.

Actually, the term "New Middle East" was introduced to the world in June 2006 concurrently with the Constructive Chaos Theory turned Concept during a visit by then US Secretary of State, Condoleezza Rice, to Tel Aviv.

The "Constructive Chaos" theory is based here on the conception that conditions of violence and warfare throughout the New Middle East could result in partition of its countries from the southern and eastern Mediterranean shores, from Lebanon and Syria to Anatolia (Asia Minor), up to Saudi Arabia, the Persian Gulf, and the Iranian Plateau up to Afghanistan in a structure that would respond to broad economic, geo-strategic and military objectives of the US and its allies and partners.

So where does all this history lead in relation to the rising of Non-State Actors (NSAs) or rather Armed NSAs (ANSAs) in the Middle East, hence in Asia? It is now obvious that ANSAs are not a novel phenomenon in Asia as Lind mentioned in his essays. But we shall address DA'ESH the Arabic acronym for the Islamic State in Iraq and EL Sham; also known as the Islamic State of Iraq and the Levant (ISIL); Islamic State of Iraq and Syria (ISIS); or simply the Islamic State (IS) as the most recent, cyber novel and notorious example in this respect. DA'ESH is also of particular relevance to an understanding of the ongoing process of political fracturing of Iraq, Syria, Yemen and Libya, attempts at destabilization of oil producing Gulf countries' and Iran's deep involvement in those and other conflicts in the Middle East.

The Centre for Preventive Action of the US Council on Foreign Relations in its Preventive Priorities Survey 2015 refers to 26 conflicts in the world of various levels of impact on the US interests, among them 16 in Asia, with 10 in the Middle East. The war against DA'ESH was classified among those as having a "critical impact on the US interests".

But how did DA'ESH come into existence? The Coalition Provisional Authority (CPA) Head, Paul Bremer's fateful May 23, 2003, Order Number 2 (Order Number 1 was the dis-establishment of the Al Baath Ruling Party in Iraq) to dissolve the Iraqi army robbed Baghdad's post-invasion military of some of its best commanders and troops. Instead of giving Iraq a fresh start with a new re-organised army on a secular non-discriminating basis, it helped create a vacuum that DA'ESH filled. Many of the Sunnis in the military who were chased out ended up on the other side. Combined with sectarian strains that still persist more than a decade later, it also drove many of the suddenly out-of-work Sunni warriors into alliances with a Sunni insurgency that would eventually mutate into DA'ESH. Many former Iraqi military officers and troops, trained under Saddam, have spent the last decade in Anbar Province battling both US troops and Baghdad's Shi'ite-dominated security forces. It is estimated that more than 25 of DA'ESH's top 40 leaders once served in the Iraqi military. Abu Bakr Al Baghdadi announced the establishment of DA'ESH on June 28, 2014. (Ezzat El Douri, Saddam's Deputy, indicated in an interview with the Egyptian Daily, *Al Watan*, February 2, 2016 that "Al Qa'eda and DA'ESH have liberated the Sunnis from Iran's tyranny. Indeed, the US has delivered Iraq to the hands of Iran".)

A shocking report by the European Union (EU) released after DA'ESH self-attributed the Paris terrorist attacks of November 2015 confirmed the DA'ESH control of large swathes of territory, approximately the size of Belgium (around 30,000 km$^2$), stretching in Syria from near the Turkish border to close to the Lebanese border, and to the east in Iraq close to the capital Baghdad. This territorial expansion gave it a significant strategic advantage, and a haven where it could function freely.

According to the same EU report, DA'ESH has recruited and continues to recruit experts with degrees in chemistry, physics and computer science indicating the possibility of waging a war involving WMDs against the West. The report further claims that DA'ESH has already smuggled into Europe chemical, biological, radiological, or nuclear (CBRN) material accessed in Iraq and Syria, and most recently in Libya.

The report also indicated that through activities centred in the parts of Syria and Iraq that are under its control, DA'ESH has access to 'extraordinary' levels of funding. A Reuters study published in October 2014 estimated that DA'ESH possessed assets of more than US$ 2 trillion, with an annual income amounting to US$ 2.9 billion. It earned about US$ 40 million a month from illicit oil sales. In addition, it reportedly taxes minorities, farmers and lorry drivers; confiscates property and livestock; sells foreign fighters passports; kidnaps civilians for ransom payments; and loots and sells antiquities. To that list the United Nations Secretary General in a recent United Nations Security Council statement added humans and arms trafficking, cash couriers, illicit payments, etc.

Furthermore, DA'ESH's adoption and effective implementation of an untraditional professional Cyberspace strategy allowed it a significant edge in its war assets. A recently published research by RAND concludes that "the threat ISIS poses today is graver than ever for two reasons: its war chest and its ability to attract foreign recruits are both at their peak. Redoubling international efforts to cut off ISIS from these two pillars of its war machine is necessary to sap the group's strength". Yet, cyberspace in the hands of the "Cyber Caliphate" has been used to achieve much more far reaching results in what could be fairly qualified as Asymmetric Warfare. For example:

- Ensuring sustainable financial capabilities, thus increasing its governance abilities and long-arm stretch over the extended territories and affiliate groups under its control.
- Disseminating its ideology based on its own politically motivated and ideologically biased and distorted views and concepts at no real substantive cost.
- Increasing the span and pace of radicalisation of targeted marginalised groups in Muslim and other countries around the world, thus luring the youth aspiring to personal glory and esteem to join their ranks.
- Recruiting foreign fighters through offerings of material and ideological enticements, leading to estimates of 30,000 to 50,000 recruits with different qualifications from over 100 countries as per a statement of the present Iraqi Minister of Defence.
- Communicating, exchanging and forwarding information among its elements in different areas and territories under its control as well as with affiliate organisations and cells in different countries.
- Providing guidelines for its jihadist supporters using cyber manuals.
- Gathering Intelligence and information.
- Establishing numerous footholds in different parts of the world through declarations of allegiance of insurgent groups and organisations to the Caliphate – in the Sinai Peninsula, the Philippines and Indonesia in Asia and Libya, Nigeria and Somalia in Africa.

- Globalising the organisation by enhancing synchronisation and spreading coordinated terror attacks within targeted countries and around the globe: Affiliate groups, returnee fighters and Lone Wolves from the Sinai to Jakarta, Istanbul to Grozny, Benghazi to Nairobi and Paris to San Bernardino USA, undertook attacks and suicide bombings that have kept the world on edge.
- Glorifying the organisation through news and images of its terrorist activities, thus consolidating its position empowerment and influence.
- Engaging the international media in following news about its acts, assessment of its power, tactics, effect on neighbouring and attacked countries, etc., turning it into an icon of terror (glorification of terror).
- Expanding the DA'ESH cyber territory by hacking into different domains of cyberspace. For example, in 2015, the Cyber Caliphate took control of 54,000 Twitter accounts, thus creating fear, chaos and confusion in anticipation of a potential digital meltdown. Twitter ended up deactivating those accounts.

Naturally, the question arises: How can the 4GW against DA'ESH be won? The answer simply is, by using, national, regional and international extended and concerted multilayered hybrid approach and strategy.

In this context, the opening statement of the Founder and Executive Chairman of the World Economic Forum, Klaus Schwab, in Davos January 2016 rings a continuing danger bell:

> "We stand on the brink of a technological revolution that will fundamentally alter the way we live, work, and relate to one another ... The possibilities of billions of people connected by mobile devices, with unprecedented processing power, storage capacity, and access to knowledge, are unlimited. And these possibilities will be multiplied by emerging technology breakthroughs in fields such as artificial intelligence, robotics, the Internet of Things ...
>
> More than 30 percent of the global population now uses social media platforms to connect, learn, and share information ... In an ideal world, these interactions would provide an opportunity for cross-cultural understanding and cohesion. However, they can also create and propagate unrealistic expectations as to what constitutes success for an individual or a group, as well as offer opportunities for extreme ideas and ideologies to spread ...
>
> The Fourth Industrial Revolution will also profoundly impact the nature of national and international security, affecting both the probability and the nature of conflict. The history of warfare and international security is the history of technological innovation, and today is no exception. Modern conflicts involving states are increasingly 'hybrid' in nature, combining traditional battlefield techniques with elements previously associated with non-state actors. The distinction between war and peace, combatant and non-combatant, and even violence and nonviolence is becoming uncomfortably blurry.

> As this process takes place and new technologies such as autonomous or biological weapons become easier to use, individuals and small groups will increasingly join states in being capable of causing mass harm. This new vulnerability will lead to new fears."

Further, in an attempt at presenting the US vision of the threats facing the world and the way to their mitigation, President Obama indicated in the State of the Union Address on January 13, 2016:

> "In today's world, we're threatened less by evil empires and more by failing states. The Middle East is going through a transformation that will play out for a generation, rooted in conflicts that date back millennia. And the international system we built after World War II is now struggling to keep pace with new realities. It's up to us, the United States of America, to help remake that system. And to do that well, it means that we've got to set priorities.
>
> Priority number one is protecting the American people and going after terrorist networks. Both al Qaeda and now ISIL pose a direct threat to our people, because in today's world, even a handful of terrorists who place no value on human life, including their own, can do a lot of damage. They use the Internet to poison the minds of individuals inside our country. Their actions undermine and destabilize our allies. We have to take them out.
>
> For more than a year, America has led a coalition of more than 60 countries to cut off ISIL's financing, disrupt their plots, stop the flow of terrorist fighters, and stamp out their vicious ideology. With nearly 10,000 air strikes, we're taking out their leadership, their oil, their training camps, their weapons. We're training, arming, and supporting forces who are steadily reclaiming territory in Iraq and Syria.
>
> Our foreign policy has to be focused on the threat from ISIL and al Qaeda, but it can't stop there. For even without ISIL, even without al Qaeda, instability will continue for decades in many parts of the world – in the Middle East, in Afghanistan, parts of Pakistan, in parts of Central America, in Africa, and Asia."

On the multilateral level and upon the joint initiative of the US and Russia after the November 2015 Paris attacks, the UNSC under Chapter VII of the UN Charter unanimously adopted the Resolution 2253 (2015). It is a 28-page resolution with more than 100 operative paragraphs. For the first time in the history of the UNSC, the Ministers of Finance of the P5 and 10 of their counterparts in the countries non-permanent members adopted a resolution that aims basically to suppress the financing of terrorism, through expanding and strengthening its already approved Al Qaeda sanctions framework to include a focus on DA'ESH . Yet, it must be noted that unlike the Al Qaeda, DA'ESH does not depend on external finances. It relies basically on its own generated sources of financing. The UNSCR 2253 reiterated support for special ways and means of cutting off even these sources of financing in a world where money moves around in seconds as indicated by the UK Chancellor of the Exchequer.

Egypt is engaged in wars with DA'ESH affiliates on two fronts, in its Asian territory in the Sinai and on its western borders with Libya where DA'ESH is establishing strong bases in the Eastern Libyan oil fields. Egypt is also faced with a strong hostile presence of radical elements acting on behalf of the Muslim Brotherhood (MB). Egypt is adopting a multi-track, multilayered strategy to face these threats:

- The country is party to the International Coalition against DA'ESH and the Arab Coalition fighting the insurgency in Yemen.
- The Egyptian Army has been engaged for the last two years in a real war of attrition scenario in the Sinai Peninsula.
- The country boasts of developmental efforts targeting the less privileged and the more marginalised sectors of the population.
- The Egyptian government has also launched several initiatives to improve the quality of university and vocational education and training of youth to help boost their career and employment opportunities.
- The Egyptian President on the New Year's Day in 2015 made an extraordinary speech to the Al-Azhar University and Awqaf (Muslim Endowments) Ministry calling for what he indicated as a long overdue virtual ecclesiastical revolution in the modernisation of the Islamic preaching and discourse.
- The Al Azhar University, which is the highest ecclesiastical authority on religious affairs for Muslims, has over the years undertaken many initiatives to help the youth, through its instructive programmes and educational courses; a number of its programmes are also shown on state media. Its role is not only limited to Egypt, but also extends to other countries. Many Al Azhar scholars annually visit countries around the world in order to explain the correct religious provisions to combat ideological extremism on ground and spread moderate thought.
- The Fatwa Authority (Dar Al Iftaa) established a multi-lingual blog on different cyber domains to reply and redress distorted or extreme religious assumptions propagated by DA'ESH.
- "Dar Al-Iftaa" also established an observatory for "Takfiry" fatwas issued by DA'AESH and other terrorist groups. The observatory detects all extremist cyber fatwas and seeks to pinpoint their intellectual and religious flaws. It is also working on a multi-lingual project to translate thousands of fatwas. The EU adopted a decision to consider the Fatwa Authority of Egypt as the sole source of Islamic rulings according to the Islamic Sharia.

- Egypt, a non-permanent member of the UNSC for the period 2016-2017, is now heading the Security Council Counter-Terrorism Committee (CTC). The CTC has been established in accordance with UNSCR 1373(2001) unanimously adopted in the aftermath of the 9/11 attacks. It works to bolster the ability of United Nations Member States to prevent terrorist acts both within their borders and across regions.
- Egypt has also won the membership of the Peace and Security Committee of the African Union (AU) for the next three years.

## Conclusion

Addressing the future of economic prosperity and security of Asia compels its countries to harness their collective political will to face up to the security challenges they are encountering. Violent, extremist and armed non-state actors must be totally contained and possibly eliminated. Winning the war against distorted ideas and extremism is of the essence. Use of Cyberspace should be monitored and positively directed through internationally agreed upon instruments. Asia is home to over 4.3 billion people in the world, that is, 60 per cent of the world population. It has the largest Muslim population in the world (over 1.3 billion); the largest oil reserves in the world; and the majority of users of cyberspace (over 1.4 billion). It is also the hub of the world's largest producers of digital and cyber technology (China, India, Japan, South Korea). Ernest & Young Megatrends 2015 report confirms that "A New knowledge World Order" is emerging with Asia as its hub.

In addition to the national and international military campaigns against terrorism, there is a pressing need at the international level to wage a committed digital counter-insurgency campaign as well. It should chip away the financial capabilities, cripple the governance abilities and strike with knowledge at the core moral grounds that ANSAs build their strength upon so that the realities on the ground prevent validation of their narratives.

Fair and sustained development, social justice, promoting greater tolerance within and between nations, deepening understanding between religions and gaining moral territory on the digital frontlines might be the most assured ways of destruction of such actors, their ideas and ill ambitions. Ensuring continuous and sustainable economic prosperity, stability and security for Asia should be the highest motivation to engage all available assets in these endeavours.

# 15

# REGIONALISING CYBERSECURITY GOVERNANCE IN AFRICA: AN ASSESSMENT OF RESPONSES

*Uchenna Jerome Orji*

## Introduction

The 20th century witnessed the end of colonialism in most states of the African continent. Following this development, the independent states began to embrace regional economic and political integration as a core component of their foreign policy objectives and development strategies.[1] The pursuit of regional integration was underscored by the need to facilitate wide regional cooperation as well as economic objectives such as free trade, and the development of common markets. Consequently, the quest for regional integration has led to the establishment of several regional intergovernmental organisations across the five geographical sub-regions that constitute the African continent (Southern Africa, Central Africa, East Africa, North Africa and West Africa). Africa currently comprises of 55 sovereign states and it is classified as the world's second largest and second most populous continent after Asia, with a terrestrial mass of 30,2044,049 million square kilometres and a human population of over one billion.[2] The African Union (AU) is the most prominent regional intergovernmental organisation that unities the African states, and it comprises 54 sovereign states, with Morocco being the only sovereign state that is not a member of the union.[3] Some of the notable intergovernmental organisations that operate within Africa's sub-regions include: the Common Market for Eastern and Southern Africa (COMESA)[4] which comprises 19 Member States, the Economic Community of West African States (ECOWAS)[5] which comprises 15 Member States, and the Southern African Development Community (SADC)[6] which comprises 15 Member States.

Africa's Internet user population has continued to grow in phenomenal proportions since the beginning of the 21st century. Statistical data indicates that the population of Internet users in Africa grew from 4,514,400 million in 2000 to 330, 965,359 million in November 2015, representing about 28.6 per cent of the Africa's entire population estimate.[7] This phenomenal growth which still continues into the foreseeable future[8] is linked to factors such as the liberalisation of telecommunications markets in African states, the widespread availability of mobile and wireless Internet technologies, and the increasing availability of broadband capacity.[9] However, the spread of Internet penetration in the African states has also raised serious concerns regarding the need to strengthen cybersecurity and also prevent Africa from becoming a 'safe harbour' for cybercrime.[10] More importantly, there are serious concerns about negative impact of cybercrime on African economies. For example, Nigeria which has the largest Internet user population in Africa is estimated to annually lose over US$ 13 billon due to cybercrime.[11] South Africa is also reported to annually lose over R 5.7 billion due to cybercrime,[12] while Norton also reports that 70 per cent of South Africans have fallen victim to cybercrime compared with a global average of 50 per cent.[13] It is also foreseeable that the impact of cybercrime on African economies will continue to increase with increasing availability and dependence on ICTs and the availability of broadband capacity.

In order to generally address cybersecurity and cybercrime concerns across Africa, several African intergovernmental organisations have developed binding and non-binding legal frameworks to promote the regionalisation of cybersecurity governance and also facilitate the harmonisation of cybersecurity laws in Member States. At the sub-regional level, the ECOWAS adopted a Directive on Fighting Cybercrime in August 2011, while the COMESA adopted a Model Cybercrime Law in October 2011. In March 2012, the SADC also adopted a Model Law on Computer Crime and Cybercrime. At the regional level, the AU adopted the Convention on Cyber Security and Personal Data Protection in June 2014.

## Cybersecurity and Cybercrime

Cybersecurity is an information age terminology that was derived by merging the prefix – 'cyber' with the concept of 'security'. The term is defined as "the collection of tools, policies, guidelines, risk management approaches, actions, training, best practices, assurances and technologies that can be used to protect the cyber-environment and organisation, as well as users' assets".[14] Cybersecurity governance measures include technical, organisational, policy and legal aspects.[15] The technical aspects of cybersecurity governance deal with the development and implementation of technical protection measures for computer systems and network infrastructure, while the organisational aspects deal with the development of institutional capacities to promote cybersecurity such as the establishment of law enforcement organisations as well as the development of institutional capacities such as the establishment of Computer Emergency Response Teams (CERTs) to provide critical services including prevention, early warning, detection and management of cybersecurity incidents.

On the other hand, the legal aspects of cybersecurity governance deal with legal measures that aim to promote cybersecurity. Legal measures are usually considered as probably the most relevant aspect of cybercrime control.[16] Such measures include the establishment of laws prohibiting acts that violate the security or integrity or availability of computer data and systems or networks and attacks against critical information infrastructure. They also include measures to facilitate cross-border cooperation on cybersecurity as well as measures for the prevention, investigation and prosecution of prohibited acts. The scope of cybersecurity laws may also extend to the criminalisation of acts that do not affect the security of computers or data or networked information infrastructures such as online child pornography or online xenophobism.[17] Malicious acts that are prohibited by cybersecurity laws are commonly referred to as 'cybercrime' or 'computer crime'. These terms are often used interchangeably to refer to instances where computer technologies are the target of a malicious or unlawful activity or the instrument for facilitating a crime or malicious activity. However, there is no universally accepted legal definition of cybercrime or computer crime[18] and cybersecurity laws generally tend to avoid such explicit definitions.[19]

## *Regional Legal Responses to Cybersecurity in Africa*

### *The ECOWAS Directive on Fighting Cybercrime*

The ECOWAS was founded by the Treaty of Lagos on May 28, 1975. Its aims include promoting regional cooperation and integration that will lead to the establishment of an economic union in West Africa and also fostering economic stability and relations amongst Member States.[20] The ECOWAS Treaty requires Member States to ensure "the harmonisation and co-ordination of national policies and the promotion of integration programmes" in areas including communications, technology and legal matters.[21] On the strength of the above mandate, the ECOWAS Council of Ministers adopted the Directive C/DIR.1/08/11 on Fighting Cybercrime at its Sixty Sixth Ordinary session at Abuja in August 2011.[22] The adoption of the Directive was underscored by the need to curb cybercrime within the ECOWAS region as some Member States were gaining global notoriety as major sources of email scams commonly referred to as the *West African Letter Scam*.[23] Accordingly, the Directive requires Member States to criminalise cybercrime[24] including: unauthorised access to a computer system;[25] unauthorised interference with the operation of a computer system;[26] unauthorised modification of computer data;[27] unauthorised interception of computer data;[28] computer fraud; [29] unauthorised manipulation of personal data;[30] and online child pornography.[31] The Directive also establishes a framework to facilitate international cybersecurity cooperation[32] and such cooperation is required to be carried out in line with relevant ECOWAS international instruments and mechanisms on international cooperation in criminal matters"[33] such as the ECOWAS Convention on Mutual Assistance in Criminal Matters[34] and the ECOWAS Convention on Extradition.[35]

In order to facilitate the development and harmonisation of national cybersecurity laws in Member States, the Directive establishes binding positive obligations on Members to implement its provisions. Accordingly, Article 35 of the Directive declares that "Member States shall adopt the necessary legislative, regulatory and administrative measures in order to comply with this Directive not later than 1st January 2014".[36] However, some Member States have not complied with the obligations under the Directive. As of January 2016, seven ECOWAS Members including Cape Verde, Gabon, Guinea Bissau, Liberia, Sierra Leone, Mali and Togo were yet to establish national cybersecurity laws, although there were ongoing initiatives to develop laws in some of those states.

### *The COMESA Model Cybercrime Bill*

The COMESA is a free trade union that was formed in December 1994. It aims to achieve regional integration by reducing barriers to cross border trade amongst Member States.[37] In line with its objectives, the COMESA developed a Model Cybercrime Bill in October 2011,[38] with a view to providing a uniform framework that would serve as a guide for the development of cybersecurity laws in Member States. Thus, the Model Cybercrime Bill provides a guide for the criminalisation of offences against computer systems and data such as unauthorised access, data interference; data interception; misuse of digital devices; digital forgery; digital fraud and cyber extortion.[39] However, the Bill does not establish any binding obligations on Member States to criminalise cybercrimes. The Bill also establishes an elaborate guide for the development of national frameworks to facilitate international cooperation,[40] as well as extradition[41] and mutual assistance.[42] However, despite its provisions, the Bill only serves as a mere guide or model for development of national cybersecurity laws in Member States. The Bill does not establish any international cooperation obligations on Member States and neither can it be used as a legal instrument for cooperation amongst Member States. As of

January 2016, eight COMESA Member States including Djibouti, Eritrea, Libya, Comoros, Malawi, Swaziland, Democratic Republic of Congo and South Sudan did not have cybercrime laws.

### *The SADC Model Law on Computer Crime and Cybercrime*

The SADC was founded in 1980 with the objectives of achieving economic union and cooperation among Member States.[43] In March 2012, the SADC adopted the Model Law on Computer Crime and Cybercrime[44] to serve as a guide for the development of cybersecurity laws in SADC Member States. However, the model law does not impose any binding obligations on Members to establish cybersecurity laws. It does not establish any provisions to guide the development of international cooperation regimes in Member States and neither does it establish any international cooperation obligations on Member States. However, Member States that have used the Model Law as framework for developing their cybersecurity laws may rely on the SADC Protocol on Mutual Legal Assistance in Criminal Matters[45] and the Protocol on Extradition[46] to obtain international cybersecurity cooperation from other Members. As of January 2016, six SADC Members including Angola, Democratic Republic of Congo, Lesotho, Malawi, Mozambique and Swaziland did not have cybercrime laws.

### *The AU Convention on Cyber Security*

The AU was originally founded as the Organisation of African Unity on May 25, 1963, and later assumed its current name and structure in 2002.[47] The aims of the AU include to "accelerate the political and socio-economic integration" of the African continent[48] and also to coordinate and harmonize the policies between the existing and future Regional Economic Communities.[49] In line with its mandate the AU developed a Cybersecurity Convention which was adopted by AU Heads of State and Government during the 23rd Ordinary Session of the AU Assembly in Malabo on June 27, 2014. The Convention which is known as the AU Convention on Cyber Security and Personal Data Protection[50] aims to harmonise the laws of African States on electronic commerce, data protection, cybersecurity promotion and cybercrime control. The Convention will to enter into force after it has been ratified by 15 AU Member States.[51]

The Convention recognises that cybercrime constitutes "a real threat to the security of computer networks and the development of the Information Society in Africa".[52] Under the Convention Member States have obligations to establish national legal, policy and institutional governance mechanisms for cybersecurity. This includes the establishment of a National Cybersecurity Framework that comprises of a National Cybersecurity Policy, a National Cybersecurity Strategy[53] and National Cybersecurity Governance Structures.[54] In addition, the Convention requires Member States to establish laws to criminalise offences such as attacks against computer systems[55] and data[56] and online child pornography,[57] and also establish procedural measures for the control of cybercrime.[58] The Convention also establishes some provisions to facilitate international cooperation on cybersecurity.[59] In particular, Member States are required to "encourage the establishment of institutions that exchange information on cyberthreats and vulnerability assessment such as CERTS or Computer Security Incident Response Teams (CSIRTS)",[60] and also make use of existing channels of international cooperation (including intergovernmental or regional, or private and public partnerships arrangements) for the purpose of promoting cybersecurity and tackling cyberthreats.[61] Thus, to a large extent, the Convention adopts a holistic cybersecurity governance approach that apparently goes beyond that of the Council of Europe Convention on Cybercrime which focuses on only the criminalisation of cybercrime and the establishment of procedural mechanisms for law enforcement and international cooperation.[62]

However, despite its holistic approach to cybersecurity governance, the AU Cyber Security Convention does not appear to provide an elaborate framework for international cooperation amongst Member States when compared to the Council of Europe Convention on Cybercrime.[63] For example, the AU Convention emphasises the need for States to adopt the principle of double criminality (dual criminality)[64] when rendering cross-border assistance on cybersecurity issues[65,66] without creating any mechanisms for Member States to fulfil extradition and mutual assistance requests in the absence of an extradition treaty or mutual assistance arrangement on the basis of dual criminality. Also the AU Convention does not enshrine the *aut dedere aut judicare* (extradite or prosecute) doctrine, as such it does not establish any obligations on Member States to extradite or prosecute cybercriminals where a request for extradition is refused. This state of affairs is further compounded by the absence of an AU legal instrument for rendering extradition or mutual assistance requests between Member States. The apparent problem here is that an AU Member State that may have adopted and ratified the Convention into its national laws may not have an extradition or mutual assistance treaty with another AU state that is also a party to the Convention. As such, a request for extradition or mutual assistance may not be successful between two Member States to the Convention even where the requirements of the double criminality principle have been fulfilled. This implies that the states after establishing 'uniform' national laws that would guarantee the application of the double criminality principle would then have to individually establish mutual legal assistance treaties amongst themselves. As such, each Member State of the AU will have to establish mutual assistance treaties with the other 53 sovereign states of the AU. This will require each state to engage in tedious and expensive negotiation processes which may not always be successful. For example, under the Convention a small AU state such as Cape Verde may only be able to obtain a regional wide guarantee for mutual assistance and extradition where it has entered into extradition or mutual legal assistance arrangements with all the 53 other sovereign states within the AU. The above state of affairs could technically create an enabling environment for forum shopping by cybercriminals within Africa because an AU Member State may not likely honour extradition or mutual assistance requests from other Member States with which it has no extradition or mutual assistance treaty. This would further be compounded where such State does not have capacity to investigate or prosecute cybercrime or where it is reluctant to prosecute. Thus, the absence of a broad AU framework to facilitate mutual assistance and international cooperation would limit the effectiveness of the Convention by making it difficult to hold a cybercriminal accountable for offences in a particular AU state even where such criminal is still located within the borders of another AU state.

While it is agreed that cyberthreats that affect African states may also emanate from outside the continent, which also underscores the need for wide international cooperation amongst all states, however the development of a framework for such global cooperation is beyond the AU and also beyond the scope of this chapter. This notwithstanding, AU Member States should at least be able to obtain international cooperation amongst themselves to the widest possible extent. More, importantly, given that the AU Cyber Security Convention is meant to serve as a treaty for the promotion of cybersecurity within Africa, the ideals of African unity and cooperation which inspired the founding of the AU[67] would not have been fulfilled if there is no explicit and elaborate AU framework to facilitate international cooperation and mutual assistance amongst Member States.

To address the above state of affairs, it may be necessary for the AU to establish an additional Protocol that would create provisions enabling all Member States to the AU Cybersecurity Convention to adopt the protocol as a legal basis for rendering international cooperation such as extradition requests or mutual assistance in accordance with the principle of dual criminality where there is an absence of applicable treaties between Member States. The proposed Protocol should also establish obligations on Member States to extradite or prosecute cyber criminals where a request for extradition is refused. Such measures exist under the Council of Europe Convention on Cybercrime.[68,69] The AU may also consider the establishment of explicit extradition and mutual assistance instruments to facilitate the rendering of extradition and mutual assistance requests within the African region and also extend the application of such instruments to cybercrime offences established under the AU Convention. This type of mechanism already exists in the ECOWAS region in form of the ECOWAS Convention on Mutual Assistance in Criminal Matters[70] and the ECOWAS Convention on Extradition[71] which are also applicable to cybercrime offences. It also exists in Europe in form of the European Convention on Extradition[72] and the European Convention on Mutual Assistance in Criminal Matters,[73] which are also applicable under the Council of Europe Convention on Cybercrime.[74]

Unlike the ECOWAS Directive on Cybercrime, the AU Convention does not establish any timeline by when Member States are required to implement its provisions in their national laws. As such, Member States are entitled to implement its provisions at their individual pace. However, notwithstanding, the fact that no AU Member State has ratified the Convention,[75] some Member States have already established national cybersecurity governance frameworks. As of January 2016, out of the 55 States of the African continent, 30 states had established cybersecurity laws, while 14 states had established national cybersecurity policies; on the other hand, 16 States had established CERT frameworks (see Table 1).

**Table 1: A Summary of National Regulatory Responses to Cybersecurity in Africa**

| | *Country* | *Cybersecurity Legislation* | *National Cybersecurity Policy* | *Computer Emergency Response Teams (CERTS)* |
|---|---|---|---|---|
| 1. | Algeria | ✓ | None | None |
| 2. | Angola | ✓ | None | None |
| 3. | Benin | ✓ | None | None |
| 4. | Botswana | ✓ | ✓ | ✓ |
| 5. | Burkina Faso | ✓ | In progress | ✓ |
| 6. | Burundi | ✓ | None | In progress |
| 7. | Cameroon | ✓ | None | ✓ |
| 8. | Cape Verde | In progress | None | None |
| 9. | Central African Republic | None | None | None |
| 10. | Chad | None | None | None |
| 11. | Comoros | None | None | None |
| 12 | Côte d'Ivoire | ✓ | None | ✓ |
| 13. | DemocraticRepublic of the Congo | None | None | None |
| 14. | Djibouti | None | None | None |
| 15. | Egypt | ✓ | ✓ | ✓ |
| 16. | Equatorial Guinea | None | None | None |
| 17. | Eritrea | None | None | None |
| 18. | Ethiopia | In progress | None | None |
| 19. | Gabon | In progress | ✓ | In progress |
| 20. | Gambia | ✓ | None | None |
| 21. | Ghana | ✓ | ✓ | ✓ |
| 22. | Guinea | ✓ | None | None |
| 23. | Guinea-Bissau | None | None | None |
| 24. | Kenya | ✓ | ✓ | ✓ |
| 25. | Lesotho | None | None | None |
| 26. | Liberia | None | None | None |

| | *Country* | *Cybersecurity Legislation* | *National Cybersecurity Policy* | *Computer Emergency Response Teams (CERTS)* |
|---|---|---|---|---|
| 27. | Libya | None | None | None |
| 28. | Madagascar | ✓ | None | None |
| 29. | Malawi | None | None | None |
| 30. | Mali | None | None | None |
| 31. | Mauritania | None | ✓ | None |
| 32. | Mauritius | ✓ | ✓ | ✓ |
| 33. | Morocco | ✓ | ✓ | ✓ |
| 34. | Mozambique | None | None | None |
| 35. | Namibia | ✓ | None | None |
| 36. | Niger | ✓ | None | None |
| 37. | Nigeria | ✓ | ✓ | ✓ |
| 38. | Republic Arab Saharawi Democratic | No information | No information | No information |
| 39. | Republic of the Congo | None | None | None |
| 40. | Rwanda | ✓ | ✓ | ✓ |
| 41. | São Tomé and Príncipe | None | None | None |
| 42. | Senegal | ✓ | None | None |
| 43. | Seychelles | ✓ | None | None |
| 44. | Sierra Leone | None | None | None |
| 45. | Somalia | None | None | None |
| 46. | South Africa | ✓ | ✓ | ✓ |
| 47. | South Sudan | South Sudan | ✓ | None |
| 48. | Sudan | ✓ | ✓ | ✓ |
| 49. | Swaziland | None | None | None |
| 50. | Tanzania | ✓ | None | ✓ |
| 51. | Togo | None | None | None |
| 52. | Tunisia | ✓ | ✓ | ✓ |
| 53. | Uganda | ✓ | ✓ | ✓ |
| 54. | Zambia | ✓ | In progress | None |
| 55. | Zimbabwe | ✓ | In progress | None |

## Prospects of Regionalising Cybersecurity Governance in Africa

One of the advantages of establishing either non-binding regional model frameworks or binding regional legal instruments on cybersecurity is that they enhance regional awareness on cybersecurity and provide a model framework of minimum standards that will guide Member States in the development of national cybersecurity regimes. This contributes to a large extent in facilitating the development and harmonisation of national cybersecurity regimes and also helps in promoting regional cybersecurity cooperation. More importantly, binding regional cybersecurity instruments such as the ECOWAS Cybercrime Directive and the AU Cyber Security Convention carry more significant legal implications for their Member State because they impose positive legal obligations on their Member States to establish national cybersecurity frameworks that enshrine the minimum standards under those instruments. Thus, under international law, the general principle of *Pacta Sunt Servanda* which is expressed in Article 26 of the Vienna Convention on the Law of Treaties declares that "every treaty in force is binding upon the parties to it and must be performed by them in good faith."[76] The Vienna Convention further declares that "a party may not invoke the provisions of its internal law as justification for its failure to perform a treaty".[77] Apparently, the existence of positive obligations under the ECOWAS Cybercrime Directive and the AU Cyber Security Convention may provide a basis for holding a Member State accountable where its failure to fulfil such obligations by establishing national cybersecurity frameworks has encouraged the perpetration of cybercrime that caused a violation of human rights such as the rights guaranteed under the African Charter on Human and Peoples' Rights.[78] The possibility of holding a Member State accountable for a failure to fulfil its positive obligations under a Treaty has been illustrated in some decisions of the African Commission on Human and Peoples' Rights. For example, in *Social and Economic Rights Action Center for Economic and Social Rights v. Nigeria*,[79] the Commission found the Nigerian Government liable for not fulfilling its positive obligations under the African Charter on Human and Peoples' Rights as a result of its failure to take measures to prevent environmental pollution and promote sustainable development of natural resources in *Ogoni land*.[80] The possibility of holding a state accountable for the non-fulfilment of its Treaty obligations has also been illustrated outside Africa by the decisions of the European Court of Human Rights in the cases of *K.U. v. Finland*[81] and *I. v. Finland*.[82] In both cases, the Court found the State of Finland liable for not taking adequate measures to fulfil the positive obligations that are attached to the right to a private life under Article 8 of the European Convention for the Protection of Human Rights and Fundamental Freedoms (1950) due to the State's failure to establish adequate cybercrime and data protection frameworks.

Another significant implication that arises from the adoption of binding regional cybersecurity instruments such as the ECOWAS Cybercrime Directive and the AU Cyber Security Convention is that they also create a legal basis for imposing intergovernmental sanctions on Member States that fail to fulfil their Treaty obligations. For example, all AU Member States are generally bound to comply with the 'decisions and policies' of the AU including those made by the AU Executive Council and the AU Assembly of Heads of States and Government. In this respect, Article 23.2 of the Constitutive Act of the AU provides that "... any Member State that fails to comply with the decisions and policies of the Union may be subjected to other sanctions, such as the denial of transport and communications links with other Member States and other measures of a political and economic nature to be determined by the Assembly."[83] Article 77 of the ECOWAS Treaty also provides for the imposition of sanctions on Member States that fail to fulfil their obligations to the ECOWAS Community. However, in practice, the sanction mechanism appears not to have been invoked by the AU or ECOWAS against any Member State that failed to implement its positive obligations under a Treaty.[84]

## Challenges of Regionalising Cybersecurity Governance in Africa

There are several peculiar challenges that have hindered the effective regionalisation of cybersecurity governance measures in Africa. A major challenge is the slow pace that characterises the development of legal and policy frameworks on cybersecurity in many African states. This slow pace may be traced to factors including lack of awareness amongst policy makers and legislators, lack of capacity in terms of expert personnel to drive the development of national cybersecurity frameworks, poor national Internet and ecommerce penetration and lack of funding. In particular, most African States have not dedicated adequate financial resources to promoting cybersecurity initiatives. To some extent, the poor government funding of cybersecurity initiatives has been caused by the fact that cybersecurity is not really considered as a national security priority in many African states. This is also not unconnected with the fact many African states face physical national security challenges such as terrorism which is usually considered a more pervasive cybercrime and other cybersecurity challenges.

Another major challenge is the absence of dedicated regional institutional governance mechanisms to monitor and facilitate the development of national cybersecurity frameworks resulting in a poor regional coordination and harmonisation of cybersecurity frameworks. The absence of such governance mechanism also appears to have limited prospects for regional cybersecurity cooperation and the dissemination of best practices. There is also the challenge of the non-application of sanctions against Member States that failed to fulfil their obligations under binding regional cybersecurity instruments such as the ECOWAS Cybercrime Directive, thus reducing the need for the timely implementation of the Directive by Member States. The large size of the African continent with its 55 sovereign states and their diverse legal traditions and how they receive and implement international Treaties is also a challenge to effective national harmonisation of regional cybersecurity measures.

Addressing the above state of affairs would require the implementation of measures including:

(a) Promoting cybersecurity as a core national security priority.
(b) Funding cybersecurity capacity building initiatives to enhance the

development of skilled personnel for law enforcement and also promoting awareness amongst policymakers and legislators.

(c) Improving funding for national cybersecurity initiatives including the operation of National CERTs/CSIRTS and law enforcement institutions.

(d) Promoting a national culture of cybersecurity awareness amongst the citizenry.

(e) Establishing dedicated regional cybersecurity institutional governance such as an African Network Security Agency within the AU framework to monitor and facilitate the development of national cybersecurity frameworks and also assist Member States in the development of cybersecurity frameworks. The establishment of such regional Network Security Agency would also serve as a platform for disseminating best practices and enhancing cybersecurity cooperation in terms of information sharing and the regional coordination of cybersecurity incidents.

(f) Exploring the option of sanctions against Member States that fail to fulfil their obligations under regional cybersecurity instruments.

## Conclusion

The absence of a widely accepted global approach to cybersecurity governance has given rise to the proliferation of bilateral and regional governance arrangements. However, given the global nature of the Internet, bilateral and regional cybersecurity governance arrangements would never be able to replace a widely accepted global cybersecurity governance arrangement.[85] Nevertheless, despite their jurisdictional limitations, bilateral and regional arrangements may provide platforms for building global consensus on cybersecurity governance. In particular, regional governance arrangements can provide a basis for facilitating legal harmonisation and promoting cooperation to the widest possible extent among Member States. However, achieving these objectives is dependent on how the regional governance arrangement is structured in the first place and the ability of Member States to timely undertake the domestic implementation of the binding obligations that arise from such arrangement, as well as the ability of the relevant regional body to coordinate and monitor its implementation by Member States. These are some lessons that other developing regions that are contemplating the establishment of binding regional cybersecurity governance arrangements may draw from the African context.

In concluding this chapter, it can be said that the proliferation of non-binding model laws and binding regional cybersecurity laws in Africa indicates the continent's awareness of cybersecurity concerns and also signals its interest in promoting the development of a secure global information society. However, achieving the desired outcomes will require sustained efforts towards addressing the issues that have been highlighted in this chapter.

## NOTES

1. Trudi Hartzenberg, "Regional Integration in Africa", Working Paper ERSD – 2011-14, World Trade Organisation – Economic Research and Statistics Division, Geneva, Switzerland, October 2011, p. 2; also see: Malebakeng Forere, "Is Discussion of the 'United States of Africa' Premature: Analysis of ECOWAS and SADC Integration Efforts", *Journal of African Law* 56 (1), 2012, p. 33.
2. Matt Rosenberg, "Continents Ranked by Area and Population", at http://geography.about.com/od/lists/a/large.continent.htm (Accessed January 28, 2016).
3. http://www.au.int/en/member_states/country profiles.
4. http://www.comesa.int/.
5. http://www.ecowas.int/.
6. http://www.sadc.int/.
7. Miniwatts Marketing Group, "Internet Usage and Population Statistics for Africa", November 30, 2015, at http://www.internetworldstats.com/stats1.htm (Accessed January 28, 2016).
8. ITU Telecommunication Development Bureau, *The World in 2014 – ICT Facts And Figures,* at http://www.itu.int/en/ITU-D/Statistics/Documents/ICTFactsFigures2014-e.pdf (Accessed January 28, 2016).
9. GSMA, *The Mobile Economy Report 2013*, A.T. Kearney, London, United Kingdom, 2013, p. 16.
10. Loucif Kharouni, "Africa: A New Safe Harbour for Cyber Criminals?", Trend Micro Research Paper, Trend Micro Inc., USA, 2013, pp. 1-26.
11. Gbenga Sesan et al., *Economic Cost of Cybercrime in Nigeria*, Paradigm Initiative, Nigeria, 2013, p. 11, at https://pinigeria.org/download/download/cybercost.pdf (Accessed January 28, 2016).
12. Trust Mastile, "South Africa Loses R.5.7 Billion Annually to Cybercrime", CNBC Africa, February 12, 2015, at http://www.cnbcafrica.com/news/special-report/2014/06/10/safrica-loses-r57-billion-annually-to-cybercrime (Accessed January 28, 2016) .
13. Tom Jackson, "Can Africa Fight Cybercrime and Preserve Human Rights?", BBC News, April 10, 2015, at http://www.bbc.com/news/business-32079748 (Accessed January 28, 2016).
14. High-Level Experts Group (HLEG), *ITU Global Cyber-Security Agenda (GCA) High-Level Experts Group (HLEG) Global Strategic Report,* ITU, Geneva, 2008, p. 27; also see: Uchenna Jerome Orji, *Cybersecurity Law and Regulation*, Wolf Legal Publishers, the Netherlands, 2012, pp. 10-16.
15. Ibid. (Uchenna Jerome Orji), pp. 17-42.
16. Gercke Marco, *Understanding Cybercrime: A Guide for Developing Countries*, ITU, Geneva, 2009, p. 84.
17. However, some countries regard the criminalisation of the online dissemination of xenophobic materials as an impediment to free speech. See Kristin Archick, "Cybercrime: The Council of Europe Convention", *CRS Report for Congress*, September 28, 2006, p. 3.
18. Uchenna Jerome Orji, No. 14, pp. 17-19.
19. See, for example: The African Union Convention on Cyber Security and Data Protection, Malabo, 2014 and the Council of Europe, Convention on Cybercrime, 41 I.L.M. 282, Budapest, November 23, 2001.
20. Article 3, Treaty of ECOWAS, May 28, 1975, 14 ILM 1200, revised July 24, 1993, 35 ILM 660, 1996.
21. Articles 3(2)(a), 33(2) and 57(1), Treaty of ECOWAS.
22. ECOWAS Directive C/DIR.1/08/11 on Fighting Cybercrime, adopted at the Sixty Sixth Ordinary Session of the ECOWAS Council of Ministers at Abuja, Nigeria, August 2011.

23. Uchenna Jerome Orji, "Curbing Advance Fee Fraud in Nigeria: An Analysis of the Regulatory Framework and Contemporary Challenges", *International Company and Commercial Law Review*, 12, November 2011, pp. 408-421; Andrews Atta-Asamoah, "Understanding the West African Cybercrime Process", *African Security Review*, 18 (4), pp. 106-114.
24. ECOWAS Directive on Cybercrime, Article 2.
25. Ibid., Article 4.
26. Ibid., Article 6.
27. Ibid., Articles 7 and 9.
28. Ibid., Article 8.
29. Ibid., Articles 10 and 11.
30. Ibid., Article 12.
31. Ibid., Article 16.
32. Ibid., Article 33(1).
33. Ibid., Article 33(2).
34. ECOWAS Convention on Mutual Assistance in Criminal Matters (A/P1/7/92), Dakar, Senegal, July 29, 1992.
35. ECOWAS Convention on Extradition (A/P1/94), Abuja, Nigeria, August 6, 1994.
36. ECOWAS Directive on Cybercrime, Article 35 (1).
37. Treaty Establishing the Common Market for Eastern and Southern Africa, 1994, Articles 3 and 6.
38. *Official Gazette of the COMESA*, 16 (2), October 15, 2011.
39. COMESA Model Cybercrime Bill, Part VI.
40. Ibid., Section 41.
41. Ibid., Section 42.
42. Ibid., Sections 43 and 52.
43. http://www.sadc.int/.
44. SADC Model Law on Computer Crime and Cybercrime, Version 2.0, adopted on March 2, 2012.
45. SADC Protocol on Mutual Legal Assistance in Criminal Matters, Luanda, October 3, 2002.
46. SADC Protocol on Extradition, Luanda, October 3, 2002.
47. African Union, "African Union in a Nutshell", at http://www.au.int/en/abut/nutshell.
48. Constitutive Act of the African Union, adopted the Thirty-Sixth Ordinary Session of the Assembly of Heads of State and Government, Article 3 (c), Lome, Togo, July 11, 2000.
49. Ibid., Article 3(i).
50. African Union (AU) Convention on Cyber Security and Personal Data Protection [Hereafter, AU Convention on Cyber Security], EX.CL/846(XXV) adopted at the 23rd Ordinary Session of the Assembly of the African Union, Malabo, June 27, 2014.
51. Ibid., Article 36.
52. Ibid., Preamble.
53. Ibid., Article 24.
54. Ibid., Article 25.
55. Ibid., Article 29:1.
56. Ibid., Article 29:2.
57. Ibid., Article 29:3(1).
58. Ibid., Articles 29:3(4), 31:3(a).
59. Ibid., Article 28.
60. Ibid., Article 28:3.
61. Ibid., Article 28: 4.

62. Uchenna Jerome Orji, "Examining Missing Cybersecurity Governance Mechanisms in the African Union Convention on Cybersecurity and Personal Data Protection", *Computer Law Review International*, 5, October, 2014, pp. 131-132.
63. Uchenna Jerome Orji, "Multilateral Legal Responses to Cybersecurity in Africa: Any Hope for Effective International Cooperation?", in M. Maybaum et al. (eds.), *Architectures in Cyberspace - 7th International Conference on Cyber Conflict*, NATO CCD COE, Tallinn, Estonia, pp. 110-112.
64. "Double criminality" or "dual criminality" exists where a conduct in issue have been criminalised in the laws of both the State requesting for assistance or extradition and the State from whom such assistance or extradition is requested. Under this principle, an extradition request can only be granted in accordance with an extradition treaty between two countries where both countries have criminalised the criminal conduct for which an extradition request is sought and the crimes are punishable by one year imprisonment or more. See HLEG, No. 14, pp. 14 and 56; *The Black's Law Dictionary*, 8th Edition, West Group, 2004, p. 537.
65. See AU Convention on Cyber Security, Article 28: 1, which provides that: "State parties shall ensure that the legislative measures and/or regulations adopted to fight against cyber-crime will strengthen the possibility of regional harmonization of these measures and *respect the principle of double criminal liability.*" Article 28: 2 of the AU Convention on Cyber Security, which provides that: "State parties that do not have agreements on mutual assistance in cyber-crime *shall undertake to encourage the signing of agreements on mutual legal assistance in conformity with the principle of double criminal liability*, while promoting the exchange of information as well as the efficient sharing of data between the organizations of State Parties on a bilateral and multilateral basis."
66. Constitutive Act of the AU, July, 2000, Article 3.
67. See Council of Europe Convention on Cybercrime, Article 24 (3), which provides that: "If a Party that makes extradition conditional on the existence of a treaty receives a request for extradition from another Party with which it does not have an extradition treaty, it may consider this Convention as the legal basis for extradition with respect to any criminal offence referred to in paragraph 1 of this article." Also, Council of Europe Convention on Cybercrime, Article 24 (6).
68. ECOWAS Convention on Mutual Assistance in Criminal Matters, No. 34.
69. ECOWAS Convention on Extradition, No. 35.
70. The European Convention on Extradition, Paris, December 13, 1957 [ETS No. 24].
71. The European Convention on Mutual Assistance in Criminal Matters, Strasbourg, April 20, 1959 [ETS No. 30]; Additional Protocol to the European Convention on Mutual Assistance in Criminal Matters, Strasbourg, March 17, 1978 [ETS No. 99].
72. Council of Europe Convention on Cybercrime, Article 39.
73. Tom Jackson, "Can Africa Fight Cybercrime and Preserve Human Rights?", BBC News, April 10, 2015, at http://www.bbc.com/news/business-32079748; Daniel Finnan, "Lack of Laws Governing Cybercrime Making Africa a Safe Haven for Cyber Criminals (Interview)", Radio France Internationale, February 16, 2015, at http://www.english.rfi.fr/africa/20150215-lack-laws-governing-cybercrime-making-africa-safe-haven-cybercriminals-interview.
74. Vienna Convention on the Law of Treaties, concluded at Vienna on May 23, 1969 and entered into force on January 27, 1980, Article 26.
75. Ibid., Article 27.
76. OAU DOC. CAB/ LEGAL/67/3 rev. 5, 2ILM – 68 (1982).
77. Communication No. 155/96, African Commission on Human and Peoples Rights.

78. Fons Coomans, "The Ogoni Case before the African Commission on Human and Peoples' Rights", *International and Comparative Law Quarterly*, 52, July 2003, pp. 749-760.
79. [2008] ECHR No. 2872/02, Judgment of December 2, 2008.
80. [2008] ECHR No. 20511/03, Judgment of July 17, 2008.
81. Constitutive Act of the AU, July 2000, Article 23.2.
82. Chidebe M. Nwankwo (Jr), *Legitimation of the Economic Community of West African States (ECOWAS): A Normative and Institutional Inquiry*, a Ph.D. Thesis submitted to the Brunel University of London, College of Business, Arts and Social Sciences, June 2014, p. 191.
83. Solange Ghernaouti-Hélie, "Need for a United Nations Cyberspace Treaty", WISIS Forum 2010-High-Level Debate on Cybersecurity and Cyberspace, ITU, Geneva, May 10-14, 2010, p. 2.

# SECTION II

# Asian Perspectives on Cybersecurity

# 16

# Challenging Opportunities for the Asia-Pacific's Digital Economy

*Liam Nevill*

## Introduction

The Asia-Pacific is currently powering growth in the global economy. The region is expected to host two thirds of the global middle class by 2030, including 1 billion people in China and India alone.[1] However, this growth cannot be sustained without engaging new and more productive economic growth opportunities. Continued economic growth is vital to reduce the number of people in the region living below the extreme poverty line, currently numbering 2 billion. The digital economy offers significant potential to unlock this growth, and enable more efficient economic and business practices.

The digital economy is rapidly becoming a larger percentage of national and regional Gross Domestic Product (GDP) in the Asia-Pacific. As digital infrastructures expand their reach, particularly access to internet connected mobile devices, new services are being made accessible to more people. The region encompasses a spectrum of economies, ranging from some of the most advanced to the least developed, and the most connected to the least connected. Digital commerce can help overcome some of this divide, lowering traditional barriers to development and providing access to health and education resources in a manner not previously possible. The diversity of markets and levels of development mean that there is no 'one size fits all' solution to the challenges that face further digital economic growth, and each country must assess their own particular challenges, and seeking assistance where necessary to overcome them.

Continued growth of the region's digital economy is underpinned by cybersecurity. Without confidence in the security of personal data and financial information, consumers will hesitate to engage with digital commerce. Cybercrime is a global business, but greater regional cooperation will assist in overcoming well organised criminal groups who take advantage of their victim's digital naivety, or bad luck. The rapid growth of connectivity in some countries comes with additional security vulnerabilities caused by the use of counterfeit hardware and software.

The other barrier to greater digital economic growth is the legal and regulatory frameworks that have failed to keep pace with new types of business services and models. Innovative and even disruptive practices often clash with regulatory frameworks, disincentivising start-ups and their investors. This can cause these businesses to fail due to unfavourable regulatory and taxation frameworks, or see them move offshore to seek more favourable business environments. Some countries are now working to create an environment that encourages the growth of digital start-ups, and others now need to catch up to realise the full potential of the digital economy.

## Growth Opportunities for the Digital Economy

Far more than just a "marketplace of the Internet", the digital economy describes "the global network of economic and social activities that are enabled by digital technology, such as the internet, mobile and sensor networks".[2] The Asia-Pacific is the engine room of global economic growth, accounting for one third of global growth during a period of economic uncertainty, and increasing pressure to boost productivity. McKinsey estimates that the region's digital economy will be worth between US$ 220-625 billion by 2030, which is 4-12 per cent of the region's total projected GDP.[3] In 2014 the World Economic Forum forecasted that major technology trends could create between US$ 9.6-21.6 trillion in value for the global economy by 2020.[4]

Cyberspace is seen as a key enabler in unlocking growth and productivity. Beyond providing access to markets outside the local or national economy, the Internet and associated technology such as data analytics and mobile Internet can increase the productivity of businesses, open up new business opportunities or models, and lead to better jobs. For economies seeking to transition from industrial to service based economy, the digital economy compresses the time frame needed to create the conditions necessary for such a transition. For countries such as Brunei and Australia that are critically reliant on resource exports for economic growth, the digital economy offers an opportunity to diversify their economic base and build greater resilience to boom and bust price cycles.

Asia-Pacific states are looking to cyberspace to build well-educated and healthy populations, and digital business models to generate jobs to accommodate those entering the workforce.[5] The Asia-Pacific faces a variety of demographic pressures. Regionally the working age population has peaked and aged societies will soon proliferate, however this is not uniform at the national level.[6] China and Vietnam are enjoying a window of opportunity with large working populations, but this window may shortly close due to a slowing birth rate. Other countries are challenged by a combination of growing working age population and weak economies.

Cyberspace and the digital economy has the potential to enhance the economic potential of the least developed countries in the region. The Internet, when aligned with favourable conditions such as linguistic similarities has seen the growth of new industries, for instance, the call centre industry in India and the Philippines. In the Philippines, the call centre industry generated an estimated US$ 8.4 billion in revenue and employed about 500,000 people in 2012. Seeking lower cost skilled labour offshore shifts demand for labour, and provides better paid jobs to some.[7] Vietnam is seeing rapid growth in its digital start-up sector and an e-commerce market that was estimated to be worth US$ 4 billion in 2015.[8] Regionally, mobile network providers and related industries generated an estimated US$ 395 billion, constituting 1.6 per cent of the GDP.[9]

Continued growth of the digital economy will not come without issues for policymakers and others such as central banks with access to economic and fiscal policy levers. Technological change and automation will inevitably cause some jobs to cease to exist, or undermine the competitiveness of old regulated monopolies such as the taxi and hotel industries. The structure of the labour market will change as the digital economy assumes a greater proportion of overall output, and jobs that can be automated will decline in number while other jobs will transform, requiring a new type of worker.[10] The movement of jobs offshore challenges policy makers to identify the right balance between economic efficiency, job creation, and what capability and information should be protected for national security or personal privacy reasons.

Cybersecurity skills and services themselves provide a significant opportunity to develop new local and export industries. Globally, the cybersecurity market was worth US$ 75 billion in 2015, and is estimated to reach US$ 170 billion by 2020.[11] Extra-regional examples such as Israel and Britain, where cybersecurity has become a £ 2 billion a year export industry, provide an indication of the opportunity for development: from prioritising the development of cybersecurity skills, funding cybersecurity research, and cybersecurity-focussed trade promotion.[12] For example, China has significantly invested in the development of a national Information and Communications Technology (ICT) industry, driven by economic and national security concerns.[13] This is likely to continue in the region, building a competitive market for regionally developed digital services and products.

Overall the contribution of the digital economy to the Asia-Pacific economy is a net benefit. Continued growth of the digital economy will enhance the region's role as the powerhouse of the global economy, and possibly assist in overcoming the deep divisions between the most developed and least developed economies in the region.

## Enabling Digital Start-ups, Innovation and Disruption to Build the Digital Economy

Overcoming challenges to the growth of the digital economy must occur at the sub-national, national and regional level. Multilateral agreements, such as the Trans Pacific Partnership (TPP), and national efforts to enable digital start-ups to get off the ground are good starting points on this path. Efforts to liberalise regional trade and support economic growth and diversification will build resilience to boom and bust cycles linked to agriculture, tourism, or resources. South Korea and Singapore are regional leaders in creating the conditions for strong start up 'ecosystems' to emerge, and other countries such as Australia and India are now following suit.

The TPP has been heralded as "the most ambitious trade policy ever designed for the Internet and electronic commerce".[14] The TPP's overarching intent to liberalise regional trade is implemented in the digital space by requiring non-discriminatory treatment, a prohibition on duties for electronic transactions, prohibiting local computing use requirements and better enabling cross-border information transfers.[15] The TPP also requires states to implement legislation and policy that supports confidence in the digital economy including electronic transaction laws, protections for online consumers and the protection of personal information. There are concerns about how these requirements balance privacy and security with liberalisation of trade, and implementing the agreement will not be straightforward. However, the potential of the TPP to catalyse further growth of the region's digital economy is significant.

The South Korean Government has committed substantial funds to develop its digital economy over many years. Realising that there was a cost advantage to creating a skilled workforce for domestic research and development, and creating partnerships for the research and manufacture of high technology goods, successive governments have adopted strategies to harness digital business opportunities. The creation of the Information and Communication University in 1998 and the promotion of IT-related research and development, sponsored research projects and other scholarships have supported South Korea to become one of the most connected societies in the world and a leader in the development of new digital goods.[16] The South Korean Government has also played a critical role in supporting venture capital for digital business. In 2013, the Ministry of Science, ICT and Future Planning was established to implement President Park's vision of a 'creative economy'. In 2014, the Ministry committed US$ 2 billion to support the start-up sector, and eliminated restrictions on the venture capital industry.[17]

Singapore is consistently rated as one of best places to do business globally, and this remains the case for the digital economy. Strong support from the government has created a pro-innovation environment, supported by a world class education system, particularly in mathematics and science, which the World Economic Forum considers the best in the world.[18] Singapore's location makes it ideally suited as a regional hub for communications, an advantage it has embraced through the development of excellent digital infrastructure using public-private partnerships. This high quality infrastructure and its attractive environment for digital start-ups, private equity and venture capital means Singapore is a global leader in the digital economy.[19]

In Australia, the digital economy accounts for 5 per cent of the GDP, making it a bigger contributor to the overall economy than both agriculture and the retail industry. This is in large part due to the surge in mobile phone markets and the take-up of cloud services.[20] Prime Minister Malcolm Turnbull made his private fortune in the early days of the Internet, and upon assuming the prime ministership he quickly announced more than 20 policy changes to support the emergence of new digital start-ups. These policies, by encouraging a more productive, innovative and disruptive digital commerce sector in Australia seek to overcome Australia's reliance on resource exports for economic growth, diversifying the economy and building a stronger basis for future growth.

Included in this package of policy changes are tax incentives for early investors including tax offsets and exemptions; increases in the maximum size limits of early stage venture capital funds; reduction of the default bankruptcy period from three to one year; protections for directors if they seek restructuring assistance; allowing intangible assets such as patents to depreciate; and creating a digital market place where start-ups can bid for government tenders. The government will also invest in the creation of five 'landing pads' for Australian start-ups seeking global expansion, the first two in Tel Aviv and Silicon Valley. The government has committed AU$ 22 million for international research collaborations, and will provide new visas for financially backed international entrepreneurs and pathways to permanent residency for science, technology, engineering and mathematics (STEM) and ICT graduates. Further investment in women in STEM (AU$ 13 million) and primary and high school digital literacy (AU$ 51 million) are intended to grow the pool of potential employees and entrepreneurs in the digital economy.

The Indian Prime Minister, Narendra Modi, has also recently announced a package of incentives for India's start-up sector. Modi's 'Startup India Action Plan' includes tax exemptions on income, capital gains, and investment above fair market value; reduced regulatory requirements such as self-certification with labour and environment laws; a mobile app to enable the establishment of new businesses; fast tracked business closures; relaxed public procurement standards; a new start-up India hub; patent reforms; a variety of funds to support digital enterprises; new incubators and research hubs; and an innovation programme for school students.

Conversely, countries such as Bangladesh and Pakistan offer an example of how burdensome taxation regimes can inhibit digital growth. Deloitte notes that in Bangladesh, 18 per cent of the cost of owning and using a mobile device is due to taxation. In Pakistan, taxes represent nearly 30 per cent of the cost of owning a mobile device, including taxes on SIM card sales, provincial and federal excise duties on services including SMS and data, and an additional withholding tax.[21] Taxation regimes such as this deter lower income mobile users who could benefit most from digital technology and access to basic information that it enables such as health, agricultural and financial services that have previously been out of reach.

Embracing risk, and accommodating failure through more flexible regulatory and tax frameworks is an approach that is increasingly being applied by countries in the Asia-Pacific to fully benefit from the digital economy. The diversity of approaches to removing impediments to digital businesses and the capital they require will determine to some extent how successful these countries are in developing their digital sectors.

## Challenges for the Growth of the Digital Economy

Several challenges remain for the sustained growth of the digital economy in the Asia-Pacific. A lack of connectivity in some countries; consumer concerns about cybersecurity and cybercrime that reduce confidence in digital commerce; and regulatory and policy frameworks that obstruct new businesses and discourage investment, need to be addressed to unlock greater digital growth.

The limited ability of many states to invest in adequate infrastructure to harness the potential of the digital economy remains an impediment to growth. In the Asia-Pacific, about 8 per cent of the population has access to fixed broadband services, and these are often financially out of reach to lower income populations.[22] However, there may be benefits in bypassing the development of fixed infrastructure. Mobile phones have provided online access to a new generation in the region, and growth has been considerable. In 2014, 89 in every 100 inhabitants of the region had mobile Internet access, a rise of 387 per cent compared to 2005 when only 23 of every 100 had mobile Internet access. It may be that the absence of legacy infrastructure better enables them to adopt disruptive business models based on mobile technology in a manner that more developed economies cannot.[23]

Cybersecurity, and the confidence on effective measures to protect business and consumers from cyberthreats, underpins the development of the digital economy. Widespread cybercrime continues to undermine confidence in the adoption of digital commerce. Without a reliable and safe cyber environment, business will hesitate to invest in new business markets and models. The World Economic Forum estimates that if national and multilateral cybersecurity efforts are not effectively implemented, and cybercriminals retain their advantage, up to US$ 1.02 trillion in the value of the global digital economy would not be achieved. This would increase to US$ 3.2 trillion if the growth of digitisation is slowed by a lack of confidence in cybersecurity.[24]

As the number of people with access to the Internet grows every year in the region, first-time users who do not have an adequate understanding of basic cybersecurity measures and practices provide easy prey for cybercriminals. In particular the use of pirated or unlicensed software, for example, 84 per cent of all software in Indonesia and 81 per cent in Vietnam, creates a critical vulnerability. Cybercrime, and the difficulty of prosecting offenders across borders without a harmonised legal structure and other law-enforcement capacity constraints, such as digital forensics skills, are shared challenges in the region.[25] Cooperation on cybersecurity efforts between countries, and through multilateral organisations such as Interpol's Cyber Fusion Centre or the Asia Pacific Computer Emergency Response Team (APCERT) are critical to making a more reliable and secure cyberspace in the region. Cooperation between national private, public and academic sectors is critical to addressing cybersecurity vulnerabilities and challenges in the Asia-Pacific.

Further challenges remain in the ability of emerging economies to develop suitably skilled people for high tech industry. Literacy is essential for understanding user interfaces for digital devices, and countries with very low levels of literacy such as Papua New Guinea are slow to embrace digital technologies. Further, where there is little available content in local languages, there is less incentive for users to access and utilise digital devices for communication or commerce. Without this base, digital literacy is more difficult to generate in these less developed countries. The challenges of education are not constricted to less-developed countries, with technical skills emerging as a critical skills gap in most countries. Scholarships and training centres are emerging throughout the region, to boost domestic skills and increasingly to export cybersecurity skills and capability to the region.

## Addressing Challenges to the Growth of the Digital Economy

Achieving the full potential of the digital economy requires strategies to support the growth of the digital economy and overcome barriers to its development. Nationally and internationally, there are several challenges to that could be overcome with concerted effort and cooperation by regional countries.

Addressing cybercrime and cybersecurity vulnerabilities is key to increasing confidence in the digital economy. Initiatives such as Interpol's Cyber Fusion Centre in Singapore that aid information sharing between national cybersecurity agencies and organisations, as well as private sector and academia are key to overcoming cyberthreats. Enforcement of cybercrime across the region should be a priority supported by efforts to develop the capacity of less-developed countries police and law enforcement agencies to detect and prosecute cybercriminals wherever they may be. The Association of Southeast Asian Nations (ASEAN)'s work on Transnational Crime and the ASEAN Regional Forum both offer opportunities to pursue greater cooperation in addressing cybercrime threats.

Increased affordability of access to cyberspace is critical to the growth of the digital economy. The emergence of more low-cost handset manufacturers will make mobile devices more accessible, but taxation on mobile services usage will discourage lower income new users. Programmes to enhance digital literacy and provide access to local content should be supported by developed countries to boost the region's overall digital economy and cybersecurity. There is growing support for apps and content in local languages that is also necessary to grow access to cyberspace and the digital economy. Programmes such as Google's Indian Language Internet Alliance, which is working to develop news content in Indian languages to attract more users, and Indian device manufacturer Micromax that has developed 10,000 apps in Indian languages in 2015, should be supported and supplemented with similar programmes in other countries.[26] Large multinational technology firms, such as YouTube and Mozilla, are also working to increase the accessibility of digital platforms in languages other than English.

Each country should review their policy and regulatory and taxation frameworks as they begin to understand the potential that can be realised if new and disruptive business models are embraced by a more flexible approach to digital economic activity. The competitive advantage that countries can gain by attracting overseas talent through an accommodating business and regulatory environment should drive further innovation and change in how governments approach these reforms. For the region's less developed countries however regulatory and taxation reform without the development of a skilled workforce will be ineffective in stimulating digital start-ups at a large scale. Harmonisation of international regulatory frameworks that encourage cross border trade allows regional countries and consumers to benefit from international economies of scale. Multilateral organisations such as the ASEAN could potentially coordinate this important task, but effort at the national level to ignore protectionist impulses and engage in multilateral efforts is required to underpin this.

Achieving the full potential of the digital economy will further enable the Asia-Pacific to overcome economic and demographic challenges, and continue to be an engine room of global growth. Cooperation between public and private sectors at the national level and in multilateral organisations will be key to addressing these challenges to growth towards a mutually beneficial end.

## Conclusion

The Asia-Pacific has enormous potential to reap the benefits of digitally enabled business. Unevenness in development and capacity to embrace digital business will challenge some countries in implementing appropriate responses to the policy, economic and infrastructure challenges that are posed by the digital economy. However, cooperation between states and within multilateral institutions to develop harmonised frameworks for digital trade and increase confidence in the security of cyberspace will support sustained growth across the region. The capacity of the digital economy to diversify economies and bring prosperity to less-developed countries will become increasingly necessary as demographic changes make increased productivity vital in the Asia-Pacific while the region powers global economic growth in the 21st century.

### NOTES

1. Ernst & Young, *Hitting the Sweet Spot: The Growth of the Middle Class in Emerging Markets*, April 25, 2013, at http://www.ey.com/GL/en/Newsroom/News-releases/News_By-2030-two-thirds-of-global-middle-class-will-be-in-Asia-Pacific.
2. Rowena Barrett, "Digital Economy: Our Perspective", Working Paper no. 1, PWC Chair in Digital Economy, Brisbane, April 23, 2013, at http://www.chairdigitaleconomy.com.au/wpcontent/uploads/2015/04/RPT_DimensionsDigitalEconomy_WP01_201504231.pdf.
3. Jonathan Woetzel, Oliver Tonby, Fraser Thompson, Penny Burtt, and Gillian Lee, *Three Paths to Sustained Economic Growth in Southeast Asia*, McKinsey Global Institute, November 2014, at http://www.mckinsey.com/insights/energy_resources_materials/three_paths_to_sustained_economic_growth_in_southeast_asia.
4. Benat Bilbao-Osorio, Soumitra Dutta, and Bruno Lanvin (eds.), *The Global Information Technology Report 2014*, World Economic Forum, Geneva, 2014, at http://www.weforum.org/reports/global-information-technology-report-2014.
5. United Nations Economic and Social Commission for Asia and the Pacific, *Population Trends in Asia and the Pacific*, November 2013, at http://www.unescap.org/sites/default/files/SPPS-Factsheet-Population-Trends-v3.pdf.
6. Asian Development Bank, *Key Indicators for Asia and the Pacific*, Philippines, 2014, at http://www.adb.org/publications/key-indicators-asia-and-pacific-2014.

7. Organisation for Economic Cooperation and Development (OECD), "Skills and Jobs in the Internet Economy", OECD Digital Economy Papers, No. 242, OECD Publishing, at http://dx.doi.org/10.1787/5jxvbrjm9bns-en.
8. Vietnam's E-commerce and Information Technology Agency, "E-Commerce - Vietnam: a snapshot (2013)", July 28, 2014, at http://www.vecita.gov.vn/Home.
9. GSM Association (GSMA), *The Mobile Economy: Asia Pacific 2015,* at https://gsmaintelligence.com/research/?file=fba9efc032061d5066b0eda769ad277f&download.
10. OECD, No. 7.
11. Steve Morgan, "Cybersecurity Market Reaches $75 Billion In 2015 ; Expected To Reach $170 Billion By 2020", *Forbes*, December 20, 2015, at http://www.forbes.com/sites/stevemorgan/2015/12/20/cybersecurity%E2%80%8B-%E2%80%8Bmarket-reaches-75-billion-in-2015%E2%80%8B%E2%80%8B-%E2%80%8Bexpected-to-reach-170-billion-by-2020/#2715e4857a0b4f46d21a2191.
12. Richard Gluyas, "CBA Calls for strong cybersecurity Strategy", *The Australian*, January 19, 2015, at http://www.theaustralian.com.au/business/financial-services/cba-calls-for-strong-cybersecurity-strategy/news-story/ba59b4a40876cab074791e6effeb2ffa.
13. Tobias Feakin, "Cyber Capacity-building through the Lens of Techno Nationalism", *The Strategist*, August 17, 2015, at http://www.aspistrategist.org.au/cyber-capacity-building-through-the-lens-of-techno-nationalism-2/.
14. David Fidler, "The Top Five Cyber Policy Developments of 2015: The Trans-Pacific Partnership", *Net Politics*, December 24, 2015, at http://blogs.cfr.org/cyber/2015/12/24/the-top-five-cyber-policy-developments-of-2015-the-trans-pacific-partnership/.
15. David Fidler, "The TPP's Electronic Commerce Chapter: Strategic, Political, and Legal Implications", *Net Politics*, November 9, 2015, at http://blogs.cfr.org/cyber/2015/11/09/the-tpps-electronic-commerce-chapter-strategic-political-and-legal-implications/.
16. Economic Commission for Latina America and the Caribbean, *Korea's Informatization Promotion Strategies,* www.cepal.org/noticias/noticias/3/12743/kijoolee2.pdf.
17. Alan McGlade, "Why South Korea will be the Next Global Hub for Tech Startups", *Forbes*, February 6, 2014, at http://www.forbes.com/sites/alanmcglade/2014/02/06/why-south-korea-will-be-the-next-global-hub-for-tech-startups/.
18. Benat Bilbao-Osorio et al. (eds.), No. 4.
19. Bhaskar Chakravorti, Christopher Tunnard, and Ravi Shankar Chaturvedi, "Where the Digital Economy Is Moving the Fastest", *Harvard Business Review*, February 19, 2015, at https://hbr.org/2015/02/where-the-digital-economy-is-moving-the-fastest.
20. Tobias Feakin, Jessica Woodall and Liam Nevill, *Cyber Maturity in the Asia-Pacific Region 2015*, Australian Strategic Policy Institute, Canberra, 2015, at https://www.aspi.org.au/publications/cyber-maturity-in-the-asia-pacific-region-2015.
21. GSMA, No. 9.
22. Ibid.
23. Tobias Feakin et al., No. 20.
24. Benat Bilbao-Osorio et al. (eds.), No. 4.
25. Tobias Feakin et al., No. 20.
26. GSMA, No. 9.

# 17

# Economic Dimensions of National Cybersecurity Strategies in the Asia-Pacific Region: At the Nexus of National Security, Innovation Capability and Commercial Interests

*Candice Tran Dai*

## Introduction

Many Asian governments have integrated the information and communication technologies (ICT) into their socio-economic development strategy in a way that the rapid emergence of information societies in Asia involves an increased corollary dependence upon the ICT in terms of economic prosperity. In this regard, several countries in the region belong to the leading global exporters of ICT goods and services and, for a number of them, the contribution of ICT and digital technology to Gross Domestic Product (GDP) is relatively important.[1] Moreover, Asia-Pacific is the largest Business-to-Consumer (B2C) e-commerce region in the world, it accounts for nearly 55 per cent of global mobile phone ownership, and for around 40 per cent of global mobile data traffic.[2]

Hence, from a regional perspective, the expansion of digital economies and the development of information societies result in stronger vulnerability and greater exposure to cybercrimes and cyberthreats, both internally and externally. One thing is clear; Asian countries have become a favourite target of cyberattacks worldwide. The prevalence of cybercrime in the region reflects without doubt the economic vitality of the Asia-Pacific region, which has become very attractive for potential gains from cybercriminal activities. However, the boundaries between what constitutes an act of crime, economic or strategic espionage are still difficult to determine.

Cybersecurity has undoubtedly become a priority component in the formulation of national cyber-strategy roadmaps of Asia-Pacific countries. At a national level, Asia-Pacific countries have heterogeneous means and capabilities for prevention and fight against cybercrimes and cyberthreats, and they have to deal with challenges of different nature with regards to cybersecurity. As a matter of fact, each country in the region seeks to overcome its cybersecurity shortcomings, according to its objectives and ambitions, financial and human resources, and according to its needs and priorities in this domain. The result is a wide disparity in approaching the question of cybersecurity and in the formulation or reformulation of policies and/or *ad hoc* national strategies.

At the global level, cybersecurity policymaking has been evolving in recent years from a technology-focused issue towards a more holistic issue, encompassing economic, social, educational, legal, technical, diplomatic, and military, intelligence aspects. Two major dimensions tend to be more systematically integrated into recent cybersecurity strategy roadmaps: national sovereignty and economic policy, both aspects being often tied up into the dual concept of national independence and indigenous innovation. The Asia-Pacific region is no exception to this growing trend.

The objective of this chapter is first to introduce some of the economic dimensions of cybersecurity as a growing visible component of several national cyber strategies in the Asia-Pacific region. Moreover, the chapter shall question the main incentives behind the quest for indigenous cybersecurity capability and technology and show how it is intertwined with a broader quest for innovation capability and a pragmatic quest for commercial objectives. Finally, as governments' responses to cybersecurity challenges have been mostly designed along national lines and tend to involve a growing national sovereignty concern, the international dimension of cybersecurity from global trade and cyber supply chain perspectives must be taken into consideration.

## The Economic Dimension of Cybersecurity as a Growing Visible Component of National Cyber Strategies

Considered from an economic and business perspective, cybersecurity is becoming globally an ever-growing market, and an ever-competitive market, including in the Asia-Pacific region.[3] As such, it may be noted that several Asian countries tend to fully consider this economic dimension of cybersecurity, whether as a policy objective, rather reflecting an ambition, or as an integral part of their national cybersecurity strategy, hence reflecting an assertive and proactive strategy in this domain, and even as an industrial policy, reflecting international competitiveness objectives. Cybersecurity as an engine of economic policy can be implemented in different ways, as exemplified by the various strategies several Asian States have worked out, thus revealing a diverse view of the role of cybersecurity in the economy and more widely for the country. Without seeking to be exhaustive, we may draw the line between some distinctive types of strategies and highlight some significant initiatives.

India outlined a National Cybersecurity Policy in 2013,[4] which lays the stress on developing indigenous security technologies not only to protect national critical infrastructures but also to enable economic development. Section H of the document, "Promotion of Research & Development in Cyber Security" reads as follows: "To encourage Research & Development to produce cost-effective, tailor-made, indigenous security solutions meeting a wider range of cyber security challenges and target for export markets.... To facilitate transition, diffusion and commercialisation of the outputs of Research & Development into commercial products and services for use in public and private sectors." India takes a pragmatic approach, consistently encompassing the protection of economic interests. Another policy initiative appears to be quite compelling. On June 13, 2012, the Department of Telecommunications issued the National Telecommunications Policy which clearly promotes domestic telecommunications equipment development and which shows that government policy is to prefer domestically manufactured electronic products in procurement, due to security considerations.[5] Further to this policy position, we may note that India's Defence Research and Development Organisation (DRDO) is currently developing India's own Operating System (OS) with the aim of reducing the country's dependence on foreign imported OSes based on Windows and Linux, in a bid to better secure India's networks. Of particular interest, India's strategy is aiming at spurring knock-on technological effects and providing opportunities for innovation by the domestic industry.

Like India, China belongs to the countries where the ICT has contributed to their positioning in the global economy, and acted as a catalyst for their economy. The imperative of securing cyberspace is much more crucial for the countries, which have emphasised the role of the IT sector in their economy. China fits perfectly into this dynamic but it pushes the logic even further. "Information security"[6] has always been considered an issue of national security for the Chinese Central Government, especially within the context of a country that has expressly been encouraging the development and promotion of the IT sector combined with a strict control and censorship system on network contents and which vows to become a cyber power. China's specific positioning and objective in the cyber domain has had – and will continue to have – tangible impact on its cybersecurity policy making. Chinese Central Government's first formal initiatives regarding cybersecurity date back at least to what is commonly referred as "Document 27", i.e. "Opinions for Strengthening Information Security Assurance Work",[7] and which already stated the idea of "building an indigenous national assurance system, under firm domestic control". Since then, and as reinforced by the most recent policy pertaining to cybersecurity,[8] China has been consistently focusing on the promotion and development of the domestic information security industry coupled with an ever expanding discourse on indigenous innovation and more stringent government procurement laws. The ultimate goal is to nurture a domestic cybersecurity industry capable of significantly reducing the country's dependency on foreign IT hardware and software but also possibly to gain a competitive edge.

Japan and South Korea undoubtedly belong as well to leading players in the global IT economy and have, as such, designed tailored cybersecurity policies, specifically taking into account business parameters. In the case of Japan, it should be first noted that the country has been working out cybersecurity measures since the early 2000s and that cybersecurity policy making has been constantly evolving over the years and growing in strategic and operational sophistication. After rolling out a first National Strategy on Information Security in 2006, a second one in 2009 and an Information Security Strategy for Protecting the Nation in 2010, together with several packages of information security measures notably for critical infrastructures, Japan came up with a comprehensive Cybersecurity Strategy in June 2013. It has been revamped into a new Cybersecurity Strategy as approved by the Japanese Cabinet in September 2015.[9] This new document gives a prominent place to the economic dimensions of cybersecurity as illustrated in Section 5.1.3, "Improvement of Cybersecurity Business Environment". It is clearly stated that Japan aims at "promoting cybersecurity-related businesses", "developing fair business environment" and "improving environment for Japanese enterprises' global operations". Of particular interest is the fact that Japan is highlighting the role of the Internet of Things (IoT) industry and the cybersecurity industry as economic growth engines. The wording is unequivocal as the aim is "for Japan's IoT Industry and other ICTs-based digital businesses to become internationally competitive and subsequently become the engines of the national economy and for Japan to build capacities for the self-reliant assurance of cybersecurity [...]". Additionally, it is worth noticing that Japan clearly considers that cybersecurity measures shall not hinder innovation as a driver of economic growth.

As far as South Korea is concerned, the country has set strong ambitions not only regarding the development of the domestic information security industry but also regarding the export of domestically-developed cybersecurity products and services. South Korea has a very business-oriented view of cybersecurity as an economic driver both at home and abroad, hence demonstrating willingness to strengthen international competitiveness. The year 2015 has marked an increased drive towards this aim, though this dynamic had already been initiated for some time. For instance, in May 2015, the Ministry of Science, ICT and Future Planning (MSIP) set up the Global Cyber Security Partnership Council which gathers about 50 members ranging from domestic information security firms, IT service providers, to telecoms operators, and relevant authorities. The purpose of the Council is without ambiguity to support the overseas growth of the Korean cybersecurity sector. The MSIP expects Korean cybersecurity exports to increase threefold to US$ 4.1 billion by 2019.[10] On the domestic side, the Korea Internet and Security Agency expects the size of the Korean information security industry to double by 2017.[11] In February 2016, the MSIP announced an ambitious plan to sell cybersecurity technology worth US$ 3.7 billion to countries in the Middle East, Southeast Asia, Africa as well as South and Central America. South Korea's ambitious stance towards cybersecurity, considered as an engine for economic policy, has to be put in perspective with regards to the very specific status ICT has always had for the country.

## The Quest for Indigenous Cybersecurity Capability and Technology: Incentives and Challenges

A distinctive feature shared by several national cybersecurity strategies in the Asia-Pacific region takes the form of what we would call indigenisation. This promotion of indigenous cybersecurity competence is not disassociated from economic and business considerations. Altogether, the pursuit and development of indigenous competence is seen as an enabler for the joint benefit of innovation headway, national security, and economic interests. This seemingly genuine quest for indigenous cybersecurity capability and technology is not unchallenging and is often not deprived of political concerns, as there is a risk that it may move closer to a protectionist position in the name of national security.

The political will of several countries in the Asia-Pacific region to promote indigenous cybersecurity capability needs first to be assessed within the broader spectrum of the promotion and development of indigenous technologies in the overall field of ICT. As in other parts of the World, domestic innovation in the field of ICT and the digital economy has become a leitmotiv in the region. Some countries, like Singapore or South Korea, have already triggered for some time a strategic shift towards this direction. China has meanwhile developed a comprehensive politico-strategic discourse on the subject. The ultimate goal would be to get freed from an increased dependency on foreign IT technologies. For instance, the idea of being able to evade the obligation of paying substantial fees for the use of foreign technology licences, particularly for key technologies related to the development of the digital economy, is appealing for a number of countries in the region. Moreover, nurturing a domestic innovation ecosystem has clearly become a strong incentive for a new economic drive within the context of a sluggish global economy.

Within this framework, cybersecurity, strictly considered from the angle of a segment of the IT sector, is thus subject to specific R&D programmes in several countries. For instance, Singapore has established a National Cybersecurity R&D (NCR) programme in 2013,[12] which aims at "[developing] R&D expertise and capabilities in cybersecurity for Singapore" and which is fully integrated into Singapore's National Cyber Security Masterplan 2018 (NCSM2018).[13] Malaysia has, for its part, designed a National Cybersecurity Policy encompassing "Eight Policy Thrusts", among which Thrust 5, "Research & Development towards Self Reliance", aims "to Nurture the Growth of Local Cyber Security Industry" and "Promote the development and commercialisation of intellectual properties, technologies and innovations through focused research and development".[14] Cybersecurity has actually developed into an industry which, as it is the case for other industries, needs products R&D and commercialisation and which is equally subject to national and international competition.

The quest of several Asian countries for indigenous cybersecurity capability and technology is also to be assessed within the context of the predominance of foreign and especially US cybersecurity technologies. One can as such wonder about the impact of the Snowden revelations in this area. Everything suggests that one of the inevitable consequences is nothing else but the strengthening of the following conviction: the necessary development of indigenous capacities in the field of information technology, and all the more in the area of information security technology. The general idea would be that, compared to foreign imports, and beyond the question of ICT licence fees, indigenous products would be safer, or at least more controllable in terms of security hence more reliable in terms of national security. As a matter of fact, the hazard of foreign-installed backdoors on foreign IT products remains a key concern, especially as cyberespionage has gained momentum in recent years.

Indigenous cybersecurity technology shall nevertheless not be considered as an end in itself. At first hand, it appears to provide safeguarding, but it cannot be considered as the ultimate solution. Even for a domestically created product, potential risks do exist such as unintended design flaws and programming flaws. Besides, a product may potentially still be tainted during the various and numerous phases of its existence and running. Indigenous innovation policies shall thus still be accompanied with overall risk management policies. As far as indigenous OSes are concerned, India and China being notably both engaged in the development of their own domestically-made OSes, the relatively high cost and investment for such projects has to be put in balance with realistic and achievable results with regards to effective contribution to overall cybersecurity enhancement. Furthermore, when dealing with the specific issue of OSes, one should bear in mind that software applications have to be taken into account, especially as many Asian countries suffer from a relatively important prevalence of pirated software. According to the latest report by the Business Software Alliance (BSA), in 2013, Indonesia reached an unlicensed PC software prevalence rate of 84 per cent which was 81 per cent for Vietnam, 74 per cent for China, and respectively 71 per cent and 60 per cent for Thailand and India.[15] There is no doubt that unlicensed software is definitely not the sole and unique reason for malware encounter, however, there is a link between the two, and as such pirated software tends to contribute to cyber insecurity.[16] What is more, one daunting issue regarding the journey to indigenous cybersecurity capability and technology has to do basically with the truly effective capability of the countries promoting indigenisation to achieve tangible results in this domain. Far beyond the requirement of sufficient investment capacity, a dedicated workforce is critically needed as well as adequate R&D&I frameworks. Moreover, one may consider that time is not a negligible factor, especially in the context of a rapidly evolving and complexifying threat landscape. Today's cybersecurity solutions may well not be adapted to tomorrow's threats and soon become obsolete. Cybersecurity strategies need to remain flexible and agile and cannot rely solely on a tech-based approach.

From a wider point of view, we have to consider that one of the inevitable corollaries of the institutionalisation of domestically-made technologies preference is potential market access restrictions for foreign vendors. Regulatory translation of the preference for domestic suppliers may not only impose strict market access conditions for foreign companies but may also potentially lead to the emergence of multiple markets with various technical, standard and certification requirements as well as interoperability issues. As a matter of fact, national cybersecurity policies may have an impact on the global cyber supply chain and cannot be entirely decorrelated from international trade considerations. This aspect is particularly obvious when keeping in mind that most IT products are composite, globally sourced and part of a global cyber supply chain.

## The International Dimension of Cybersecurity from Global Trade and Cyber Supply Chain Perspectives

The insertion of economic dimensions into national cybersecurity strategies cannot evade the truly global nature of the cyber supply chain and the economic interdependency among countries in an ever liberalising global economy.

From a global political economy perspective, cybersecurity has evolved from an almost invisible and understated issue towards a more visible and challenging concern. What is called the digital revolution is fundamentally based on global networks, cross-border flows of data, and network infrastructures and contents platforms largely owned by global companies. Within this overarching framework, coupled with the concomitant global growth of economic liberalisation and technological innovation, the cybersecurity imperative has been mostly designed along national lines, and is hence subject to tangible difficulties in creating shared cybersecurity norms at the international level.

As a matter of fact, cybersecurity issues are now clearly taken into account in trade and investment liberalisation negotiations, particularly within the global trend of bilateral and regional free-trade agreements (FTAs), and especially with regards to cross-border data flows, data localisation, e-commerce and the overall aim of liberalising trade in information technology products and services. For the countries part of these efforts, there is a stringent necessity to address how these trade and investment agreements may affect their own set of national cybersecurity policies.

As far as the Asia-Pacific region is concerned, one of the most significant current issues has to do with the implementation of the Trans-Pacific Partnership (TPP), which contains explicit provisions regarding information security and privacy. Cybersecurity concerns have indeed been clearly integrated into the TPP trade agreement, which gathers so far 12 Pacific Rim countries, i.e. Australia, Brunei, Canada, Chile, Japan, Malaysia, Mexico, New Zealand, Peru, Singapore, the United States and Vietnam.[17] The final draft of the TPP was released on November 5, 2015.[18] Among the 30 chapters of the document, three of them are deemed to have the most compelling impact on national cybersecurity policies: "Telecommunications" (Chapter 13), "Electronic Commerce" (Chapter 14) and "Intellectual Property" (Chapter 18). Without seeking to be exhaustive regarding the various relevant TPP provisions impacting cybersecurity policy making, we may highlight a few noteworthy Articles from the Chapter 14 on "Electronic Commerce". For instance, Article 14.7, "Online Consumer Protection", and Article 14.8, "Personal Information Protection", address the concern of the security and privacy of Internet users by requiring commitment to and enforcement of adequate subsequent regulations. It is worth noticing that the diversity of national legal frameworks pertaining to this issue is fully recognised but that interoperability between the diverse legal regimes is fostered. Furthermore, Article 14.13, "Location of Computing Facilities", and Article 14.17, "Source Code", make a push against velleity towards in-country data localisation and software source code sharing requirements for foreign vendors and suppliers entering a TPP market. Regarding software source code sharing requirements, the TPP document adds the following precisions as stated in the relevant Article's paragraphs 1 and 2: "No Party shall require the transfer of, or access to, source code of software owned by a person of another Party, as a condition for the import, distribution, sale or use of such software, or of products containing such software, in its territory"; "For the purposes of this Article, software subject to paragraph 1 is limited to mass-market software or products containing such software and does not include software used for critical infrastructure". It is also worth mentioning the existence of the Article 14.16, "Cooperation on Cybersecurity Matters", which emphasises the need for and importance of close cooperation among TPP Parties regarding cybersecurity policy, capability, as well as information-sharing relating to cyberthreats and using of dedicated cooperative mechanisms. To sum up, we may consider that the TPP, if successfully implemented, would be potentially paving the way for an evolution of cybersecurity governance cooperation.

Besides the very specific example of the Trans-Pacific Partnership, which is first and foremost a mega-regional agreement, we may underline the existence of other broader international trade mechanisms and international agreements of global reach, which may have an impact on national cybersecurity and international trade. In this sense, the World Trade Organisation (WTO) entails for instance a very peculiar provision in the form of Article XXI on Security exception of the General Agreement on Tariffs and Trade (GATT).[19] This highly controversial provision is purely self-declaratory in nature and is based on the rationale of preserving State sovereignty. It would be somehow tempting to try to use it, as well as the Agreement on Trade-Related Aspects of Intellectual Property Rights (TRIPS), as it had been sometimes suggested regarding the case of the United States and Chinese economic cyberespionage accusations which would have been tentatively subject to countermeasures under the WTO. The WTO may to date nevertheless not provide an adequate dispute resolution framework regarding national security and cyber espionage. However, it is still relevant regarding issues such as procurement, market access, and foreign investment, as per to Article XXIII of the Government Procurement Agreement.[20] For instance, the issue of China's new regulations restricting the use of foreign information technology equipment by the banking sector has been brought at the WTO by the United States.[21] It shall be noted that the WTO Revised Agreement on Government Procurement entails a specific provision on "Security and General Exceptions" which may still be used in the name of national security and national defence. Meanwhile, one shall bear in mind that the WTO is pursuing its liberalisation of IT goods and services agenda, as demonstrated by the expansion of the Information Technology Agreement (ITA) agreed in December 2015. Besides the ITA expansion which will further eliminate tariffs on 201 IT products, it is worth noticing that the agreement contains a commitment to work to tackle non-tariff barriers in the IT sector. Among the WTO members who made up the expanded ITA, there are several Asia-Pacific parties: Australia, China, Hong Kong, Japan, Korea, Malaysia, New Zealand, the Philippines, Singapore, Taiwan, and Thailand.

Further to the examples of the TPP and the WTO, we shall consider the current evolution of the Wassenaar Arrangement, as an export control regime on conventional arms and dual use goods and technologies, which is a voluntary and multilateral arrangement, built on consensus-based decision-making. To date, Japan, South Korea, Australia and New Zealand are the only Asia-Pacific participating States. What clearly constitutes a new dedicated cyber amendment has been announced in December 2013 and stated in the form of additional items to the Wassenaar Arrangement's control lists.[22] For instance, addition of Category 5.A.1.j mandates export controls on certain forms of software and associated goods, specifically "IP network communications surveillance systems or equipment, and specially designed components therefor" and addition of Category 4.A.5 mandates export controls on "Systems, equipment, and components therefor, specially designed or modified for the generation, operation or delivery of, or communication with, 'intrusion software'." Under the Wassenaar Amendment, intrusion malwares, intrusion exploits and IP surveillance items shall be subject to export control. This cyber amendment has sparked harsh international criticism from the technical cybersecurity experts and security researchers' community as these changes would notably include research into 0-days.[23] After proposing a first translation of the Wassenaar Arrangement amendment into national legislation, the US Department of Commerce's Bureau of Industry and Security (BIS) announced in September 2015 that it was scrapping its initial proposal and that it would be drafting a second proposal. The European Union (EU) has for its part adopted a delegated regulation updating the EU list of dual use items subject to EU export controls, which came into force on December 31, 2014.[24] It is important to note that the Wassenaar Arrangement does not entail enforcement provisions and that it is based on voluntary compliance. Still, if effectively enforced, the Wassenaar Arrangement's cyber amendment shall have an impact on how businesses are able to legitimately share cybersecurity information and technology across borders. As a matter of fact, the question bounces back not only to the constant issue of international trade but also to the strategic issue of access for third-parties countries to cybersecurity technologies from the Wassenaar Arrangement participating States. On the other hand, private sector businesses may for their part find it hard to conform to restricting sales rules on foreign markets, which are inherently contrary to their commercial interest and their overall quest for profit. For participating States in the Wassenaar Arrangement, and when complying with subsequent provisions enforcement, there is a need to balance their cybersecurity export ambitions with their formal commitments, whereas for third-party States, there is an incentive for indigenous cybersecurity capability and technology, i.e. self-reliance, although as mentioned previously the key question remains their truly effective capacity to achieve tangible results in this domain. One fundamental issue revolves thus around what we would call the cybersecurity gap that is potentially growing between the have and the have-nots in terms of cybersecurity capability and technology, as it is the case regarding what is called the digital divide among countries.

## Conclusion

The Asia-Pacific region is no exception to the growing trend of inclusion of economic considerations into national cybersecurity policies. This current drive clearly stands at the nexus of national security, innovation capability and commercial interests. It is a challenging move as national economies are not decorrelated from the global economy which means that national cybersecurity industries are equally not disassociated from the global cyber supply chain. As a matter of fact, cybersecurity is obviously subject to trade-offs between national security imperatives and international trade requirements. As rightly stated by Microsoft Corporation in a White Paper entitled, "Cyber Supply Chain Risk Management: Toward a Global Vision of Transparency and Trust", "the question becomes, therefore, how do countries protect national security interests without inappropriately undermining the value produced by a global supply chain?"

The concomitant drive towards national sovereignty considerations pertaining to cybersecurity is clearly adding difficulty. The national/international dilemma relating to cybersecurity considered from the angle of policy making shall well be put in perspective with regards to the very transnational nature of the cyberthreats landscape. In a supposedly borderless cyberspace, we may wonder to what extent the increasing concern for national sovereignty tends to somehow reinstate what we could call cyber frontiers, thus leading to fragmented cyber markets, and may actually hinder the crucial need for international information and technology sharing and access in the global quest for cybersecurity enhancement as a needed collective action. Here again, there is a balance to be reached between what is non-negotiable from a national security point of view and what is negotiable from an international cooperation point of view.

Another challenging aspect may be considered. As cybersecurity has evolved into a key issue in global economic relations, we may wonder to what extent it tends to become more than an objective in itself, going beyond securing cyberspace, i.e. also possibly a tool for broader objectives in the political, economic and technological realms. There are a few hints of nascent velleities geared at nurturing what is called a cyber-industrial complex, which could possibly broadly materialise into a cyber military-industrial complex within the framework of the current digitalisation process of the military. As a matter of fact, among the main world arms producers and exporters, defence industries have already started entering the cybersecurity market. This would be somewhat the result of the exacerbation of the economic dimensions of cybersecurity viewed first and foremost as an industry and coupled with increasing national security concerns.

## NOTES

1. For instance, according to a report by the Organisation for Economic Cooperation and Development (OECD), China overtook the United States in 2004 to become the world's leading exporter of Information and Communications Technology (ICT) goods such as mobile phones, laptop computers and digital cameras.
2. "Asia-Pacific B2C E-commerce Report 2015", E-Commerce Foundation.
3. The cybersecurity market is estimated to grow to US$ 170 billion by 2020, at a Compound Annual Growth Rate (CAGR) of 9.8 per cent from 2015 to 2020, according to a Markets and Markets report. According to figures published by MicroMarketMonitor, the Asia-Pacific Cybersecurity Market is expected to grow to $32.95 billion by 2019, with an expected CAGR of 14.1 per cent for the period 2013-2019; this market contributes 17.21 per cent of the global market and will slightly grow to 21.16 per cent by 2019 (.
4. Ministry of Communication and Information Technology (IT), Department of Electronics and Information Technology, National Cyber Security Strategy-2013 (NCSP-2013), July 2, 2013.
5. Ministry of Communication & IT, Department of Telecommunications, National Telecom Policy-2012 (NTP-2012), June 13, 2012.
6. The US terminology "cybersecurity" is primarily technological, while China prefers the terms "information security "and "network security".
7. "Document 27" was released in 2003 by the National Network and Information Security Coordination Group.
8. National Security Law (July 1, 2015), Anti-terrorism Law (December 27, 2015), Draft Cybersecurity Law (July 6, 2015).
9. "Cybersecurity Strategy", National Centre of Incident Readiness and Strategy for Cybersecurity (NISC), Cabinet Decision, Government of Japan, September 4, 2015, at http://www.nisc.go.jp/eng/pdf/cs-strategy-en.pdf.
10. "Cyber Exports Korean Gov't to Triple Exports from Cyber Security Industry by 2019", at http://www.businesskorea.co.kr/english/news/ict/10604-cyber-exports-korean-govt-tripleexports-cyber-security-industry-2019.
11. "Internet agency vows to double domestic security industry", at http://www.koreatimes.co.kr/www/news/tech/2014/12/133_169685.html.
12. National Cybersecurity R&D Programme, National Research Foundation, Prime Minister's Office, Singapore.
13. National Cyber Security Masterplan 2018, Infocomm Development Authority of Singapore, at https://www.ida.gov.sg/Programmes-Partnership/Store/National-Cyber-Security-Masterplan-2018.
14. National Cyber-Security Policy, National IT Council Malaysia, at http://nitc.kkmm.gov.my/index.php/national-ict-policies/national-cyber-security-policy-ncsp.
15. The Compliance Gap, BSA Global Software Survey, June 2014, at http://globalstudy.bsa.org/2013/downloads/studies/2013GlobalSurvey_Study_en.pdf.
16. For an analysis of the correlation between unlicensed software use and malware encounter,

see, for example: "Unlicensed Software and Cybersecurity Threats", White Paper, International Data Corporation (IDC), January 2015.

17. Other countries have expressed interest in joining the TPP: besides Colombia, five Asian countries have announced interest, namely Indonesia, the Philippines, South Korea, Taiwan and Thailand. India has not yet indicated whether it has interest in pursuing TPP discussions, whereas China is not involved so far in the negotiations but acting proactively in advancing the Regional Comprehensive Economic Partnership (RCEP) negotiations.
18. Full text of the TPP can be accessed at https://ustr.gov/tpp/.
19. Full text of Article XXI can be accessed at https://www.wto.org/english/res_e/booksp_e/gatt_ai_e/art21_e.pdf.
20. Not all WTO members are parties to the Agreement, some members are observers (this is notably the case for India, Indonesia, Malaysia, Thailand, Vietnam), among which some of them are in the process of acceding to the Agreement (this is notably the case for China).
21. "China – Local Content Requirements for Purchases of Technology by the Banking Sector: Questions from the United States", Committee on Trade-Related Investment Measures, WTO, G/TRIMS/W/150, March 26, 2015.
22. Fully updated control list can be accessed at http://www.wassenaar.org/wp-content/uploads/2015/08/WA-LIST-15-1-2015-List-of-DU-Goods-and-Technologies-and-Munitions-List.pdf.
23. Typically, security researchers wishing to report security vulnerabilities to foreign companies would indeed need to register with the relevant countries and obtain an authorisation before disclosing their information.
24. For further details see: http://ec.europa.eu/transparency/regdoc/rep/3/2014/EN/3-2014-7567-EN-F1-1.Pdf; http://trade.ec.europa.eu/doclib/docs/2015/october/tradoc_153907.pdf.

# 18

# International and Regional Responses to Cybersecurity Challenges

*Nandkumar Saravade*

## The Cybersecurity Problem

Technological development and approaches are converging the physical, digital, and biological worlds in a manner that can potentially transform humankind, and will possibly guide the fourth industrial revolution. Technology has become the lifeline of critical infrastructures such as energy, telecommunication, financial services and healthcare. Emerging breakthroughs in fields such as artificial intelligence, robotics, the Internet of Things, autonomous vehicles, 3-D printing, sensors, nanotechnology, biotechnology, materials science, energy storage and quantum computing have outpaced Moore's law. There is a tremendous opportunity to generate, collect and utilise colossal amount of data which was earlier non-existent. Information is now generated, transacted and made available easily in real time. Organisations are capitalising on the unprecedented opportunities presented by this wave of digitisation and are transforming their business models.

Gartner forecasts that 6.4 billion connected things will be in use worldwide in 2016,[1] up 30 per cent from 2015, and will reach 20.8 billion by 2020. In 2016, 5.5 million new things will get connected every day. A Gartner survey states that organisations are investing in location-based services (31 per cent), wearable IT (31 per cent) and behavioural advertising/digital engagement (29 per cent), clearly hinting at technology becoming invasive. It also states that by 2018, 50 per cent of business ethics violations will occur through improper use of big data analytics. The volume, variety and velocity of data, on one hand has enabled data-driven development but also increased the risk exposure on the other. World Economic Forum developed a framework,[2] called "cyber value-at-risk" for organisations to calculate the impact of cyberthreats. As per its study, 90 per cent of companies worldwide recognise they are insufficiently prepared to protect themselves against cyber risks. The Ponemon Institute has estimated that criminal data breaches now cost companies an average of $ 174 per record. Globally, cyber incidents are growing by more than 50 per cent annually.

Attacks on critical infrastructure, cyber espionage, ransomware, etc. have seen a constant rise. The attackers are local, and global – driven by passion for crimes such as financial frauds or terrorism; crime syndicates; and nation-states attacking directly or using non-state actors for economic and political espionage. Attacks on critical infrastructure can have crippling effects on civilians, with outcomes similar to those achieved by traditional war. Unique characteristics of the Internet, namely offence dominance, difficulty in attribution of attacks, development of cyberweapons by states and the use of non-state actors to camouflage their actions are making cyberspace more and more vulnerable. Tracking cybercriminals and bringing them to justice in sovereign countries is increasingly difficult with challenges in collection of appropriate cyber forensics data, applicability of laws and acceptance by courts. Moreover, the applicability of international laws is not known, since the act of war by a state is difficult to establish – when started, whether ended.

Cybersecurity is indeed a multi-dimensional concept, a complex issue straddling many disciplines and fields. It is a global problem that needs a global solution. No government can fight cybercrime or secure its cyberspace in isolation. Cybersecurity is not just a technology problem that can be 'solved'; it is a risk to be managed by a combination of defensive technology, astute analysis and information warfare, and traditional diplomacy. Nations have to take appropriate steps in their respective jurisdictions to create necessary laws, promote the implementation of reasonable security practices, incident management, and information sharing mechanisms, and continuously educate both corporate and home users about cybersecurity. International cooperation is essential to securing cyberspace. When it comes to tracking cybercriminals, it is not only the laws dealing with cybercrimes that must exist in various countries, but the collection of appropriate cyber forensics data in various jurisdictions and their presentation in courts of law, which are essential to bring criminals to justice in sovereign countries.

Highlighted briefly are some recent and important challenges:

- Cyberattacks targeted at critical information infrastructures (such as energy, telecom, financial services, defence and transportation) have the potential of adversely impacting a nation's economy and public safety, and citizens' lives. Attack on Iran's nuclear facility (Stuxnet) has been an eye opener for every nation. Cyberattacks on Estonia in 2007 crippled an entire nation that was heavily dependent on the Internet for delivery of government and business services. These growing instances of cyber espionage for stealing critical information and intellectual property – e.g. OPM, NASA, Pentagon and PMO in India.

- Given the increased dependence on Information and Communications Technology (ICT), especially in operating critical sectors, and growing realisation of cyber risks, countries are doubting the integrity of these products, fearing that adversaries may introduce malicious codes/functions to do surreptitious surveillance, disrupt services, or at worst paralyse a nation. Alleviating such doubts and fears to continue benefitting from global ICT supply chain is one of the biggest challenges the world faces in cyber security today.
- Transitioning to 'Cloud' offers immense benefits especially in terms of cost, flexibility, scalability, and agility. According to a Gartner report,[3] 50 percent of enterprises will use Hybrid Cloud by 2017. Cloud technology is a great opportunity for users as well as service/technology provider organisations. However, there are major challenges – non-tariff barriers in the form restrictions on the global data flows, jurisdictional issues because of location of servers in the Cloud, stringent investigation/surveillance regulations in countries that give legal rights to Law Enforcement Agencies for accessing data, concerns over security and privacy of sensitive business and personal information hosted in the Cloud and sharing of ownership and accountability between the user and provider organisations. Full potential of the Cloud can only be realised if these challenges are addressed through appropriate policy measures, standards, contracts, legislations and diplomacy.
- Increasing vulnerabilities and exploitables in new technologies and platforms due to complex architecture coupled with interference in ICT supply chain by nation states and sponsored actors by insertion of backdoors, demeaning open standards development. Also, mass surveillance by nation states aggravates trust deficit in society.
- The cyberspace is being used by terrorists to proselytise, recruit new members, send encrypted communications, mount surreptitious surveillance and launch cyberattacks on government infrastructure. Sophisticated use of technology was made by 26/11 Mumbai attackers – Global Positioning System equipment, satellite phones, encrypted phones (BlackBerrys), CDs holding high-resolution satellite images, multiple cellphones with switchable SIM cards and e-mails routed through servers in different locations.

There is a deep lack of appreciation and comprehension of the enormity of the problem posed by cybersecurity challenges. These threats are a matter of concern for the national security, and for the financial, economic, social and political environment stability of a nation. However, unlike physical world threats, cyber risks are not fully understood by the decision-makers, resulting in sub-optimal strategising and investments.

## Importance of Cybersecurity to a Nation State

Cyberspace is borderless, but nations consider their parts of cyberspace infrastructure under their sovereignty. Which laws operate in cyberspace, and whose laws? Clearly, Cybersecurity is linked to National Security. Cybersecurity has become an important element of foreign policy due to relevance to national security, public safety and economic development. A new form of cyber diplomacy has taken centre stage and these issues are getting discussed at all levels – in forums such as the United Nations (UN), World Economic Forum, Organisation for Economic Cooperation and Development (OECD) and Asia Pacific Economic Cooperation (APEC), in multilateral, plurilateral, bilateral talks, conventions and treaties. Cyberspace becomes a component of foreign policy on the issues of the applicability of existing public international law to cyberattacks, setting the rules of acceptable behaviour in the cyber world, curbing terrorist activities, enhancing the respect of human rights in cyberspace, or bringing cybercriminals to justice.

Political decisions regarding the cyberspace have strong international implications that require international commitment and collaboration. Therefore, the diplomatic activity in the cyber domain has an important dimension of cooperation, of concluding diplomatic engagements and multi-level agreements, including with the private sector stakeholders. Concluding bilateral and multilateral agreements on cybersecurity, aims at coordinating policy and harmonising the legal framework at national level.[4]

Over the last two years, India has initiated dialogues on cybersecurity with the US, UK, Germany, the EU, France, South Korea, Russia, Japan, Australia, the UAE, Malaysia, Singapore[5] and Mongolia where exchanges have happened on cyber security, sharing of critical information, capacity building, research and development (R&D) and other issues.

In particular, cybersecurity is important for India, as its thriving Information Technology and Business Process Management industry is resting on its well-earned reputation for world-class data security practices of its service companies.

### *International and Regional Efforts*

- US strategy paper for cyberspace[6] derives strength of the Universal

Declaration of Human Rights (UDHR) and International Covenant on Civil and Political Rights (ICCPR) resolutions to advocate fundamental freedom in the Internet space. In line with the basic principles behind the World Trade Organisation (WTO), it recommends that the states should respect intellectual property rights. It recognises the importance of valuing privacy in the cyberspace, and necessitates need of protection from cybercrimes. The strategy also posits the international community to respect the right of self-defence. The strategy cautions about international implications of technical decisions taken by the nation states. In a bid to protect the common interest in advancement, it recommends a global structure and technical standards and recommends to keep the Internet away from national prestige and political control.

- Russia and China along with Tajikistan and Uzbekistan submitted a code of conduct[7] and convention on international information security. The draft does provide a reflection of the challenges that the countries facings managing the cyber space. It comprehends the contemporary challenges and provides due attention to the possible considerations that the global cyberspace needs to deal with.

  ***The Code of Conduct*** lays down the objectives of the cooperation, and incorporates all three dimensions, viz. politico-military, cybercrime and content. It defines principles for the cooperation under the framework of the UN. The code of conduct revolves around the following issues:

  (a) Use of ICT for hostile acts.
  (b) Proliferation of cyberweapons.
  (c) Cybercrime and cyberterrorism activities.
  (d) Respecting the freedom of speech, but complying to national laws.
  (e) Multilateral and democratic Internet governance.
  (f) Supply chain integrity issues.
  (g) Respecting sovereignty, territorial integrity and political independence.
  (h) Rights and responsibility of the state.

The draft of the Convention of International Security goes further in:

(a) Spelling clear objectives and aims.
(b) Defining threats to cyberspace.
(c) Defining terms.
(d) Identifying principles of cooperation.
(e) Suggesting measures for averting military conflicts.
(f) Prescribing restrictions on use of cyberspace.
(g) Suggesting means for criminalising illegal acts.
(h) Identifying criminal procedures.

(i) Advocating Confidence Building Measures for military use of cyberspace.

(j) Prescribing mechanism for conflict resolution.

- The Association of Southeast Asian Nations (ASEAN) has been pro-active in the region's efforts to tackle cybersecurity challenges and has undertaken various cyber confidence building measures. In 2003, ASEAN adopted the Singapore Declaration which emphasised the efforts to establish an ASEAN Information Infrastructure with a view to promoting interoperability, interconnectivity, security and integrity of cyber systems. The Master Plan on ASEAN Connectivity also dwells on the importance of cybersecurity and ASEAN has further set-up a Network Security Action Council (ANSAC). Similar efforts at integrating regional cybersecurity initiatives have been carried out under the aegis of the ASEAN Regional Forum (ARF), which conducts regular Inter-Sessional Meetings on Counter Terrorism and Transnational Crime where, inter alia, cybersecurity is discussed. The ARF has issued statements on cooperation in cybersecurity and organised workshops aimed at intensifying regional cooperation in the use of information and communication technologies.[8]
- International Telecommunication Union (ITU), a UN body, in 2007 launched Global Cybersecurity Agenda (GCA) and is built upon following five strategic pillars: Legal Measures, Technical and Procedural Measures, Organisational Structures, Capacity Building and International Cooperation. Similarly, Council for Security Cooperation in the Asia Pacific (CSCAP) initiated work on cybersecurity, but not much progress has been achieved in either of these initiatives.
- With respect to international rulemaking and capacity building for cybersecurity, active discussions take place at various forums such as the UN, G8, ARF, OECD, APEC and North Atlantic Treaty Organization (NATO). NATO has taken strides to develop cyberwarfare strategy. With respect to policies for critical infrastructure protection and rapid incident response, global initiatives have also been undertaken such as at the International Watch and Warning Network (IWWN), which is for government agencies, as well as at such meetings as the Forum of Incident Response and Security Teams (FIRST), Asia Pacific Computer Emergency Response Team (APCERT), which is a community of CERTs from the Asia-Pacific region.[9]
- World Summit on the Information Society (WSIS) Forum, co-organised by the ITU, United Nations Educational, Scientific, and Cultural Organisation (UNESCO), United Nations Development Programme (UNDP) and United Nations Conference on Trade and Development (UNCTAD), in close collaboration with all WSIS Action Line facilitators/

co-facilitators and other UN organisations, such as the UNCTAD, United Nations Conference on Science and Technology for Development (UNCSTD), United Nations Department of Economic and Social Affairs (UNDESA), Food and Agriculture Organisation (FAO), United Nations Environment Programme (UNEP), World Health Organisation (WHO), International Labour Organisation (ILO), World Meteorological Organization (WMO), International Trade Centre (ITC), Universal Postal Union (UPU), World Intellectual Property Organisation (WIPO), United Nations Office on Drugs and Crime (UNODC), United Nations Children's [originally International Children's Emergency] Fund (UNICEF) and UN Regional Commissions), is a mechanism for coordination of multi-stakeholder implementation activities, information exchange, knowledge creation and sharing of best practices, and continues to provide assistance in developing multi-stakeholder and public-private partnerships to advance development goals. Strong calls for multilateral action on cybersecurity were made by many governments during the WSIS +10 review conference.[10] While a multilateral instrument on cybersecurity has long been on the list of several governments, including China and Russia, the topic also seems to be getting more traction at the UN, a result of recent terror attacks around the globe.

- The UN Group of Governmental Experts (GGE) in 2004/05 failed to reach any common position, while in 2010 it reached a level of understanding on a relatively minimalist agenda as compared to the code of conduct and the convention submitted by Russia and China. This suggests that the international norms for cybersecurity are still in the initial phase, where it will require multiple deliberations and discussions to reach the next level from the current position accepted by the GGE understanding in 2010. The code of conduct and the convention are attempts to jump many stages of norm life cycle to impose stricter norms. The 2015 GGE on Developments in the Field of Information and Telecommunications in the Context of International Security examined existing and potential threats arising from the use of ICTs by States and considered actions to address them, including norms, rules, principles and confidence-building measures.[11] The GGE report discusses how international law applies to the use of ICTs by States.
    - o States have jurisdiction over the ICT infrastructure located within their territory.
    - o In their use of ICTs, States must observe, among other principles of international law, State sovereignty, sovereign equality, the settlement of disputes by peaceful means and non-intervention in the internal affairs of other States, as well as existing obligations under international

law, including those to respect and protect human rights and fundamental freedoms, based on the principles of humanity, necessity, proportionality and distinction.
  - o States must not use proxies to commit internationally wrongful acts using ICTs, and should seek to ensure that their territory is not used by non-State actors to commit such acts. However, the indication that an ICT activity was launched or otherwise originates from the territory or the ICT infrastructure of a State may be insufficient in itself to attribute the activity to that State. The Group noted that the accusations of organising and implementing wrongful acts brought against States should be substantiated.
- In addition, with respect to cybercrimes, efforts are being undertaken to deepen international cooperation in criminal investigations through frameworks and arrangements such as the Budapest Convention, International Criminal Police Organisation (ICPO) or INTERPOL. France advocated advancing Budapest Convention to cover cybersecurity-related aspects as well.[12] The UNODC,[13] as called upon by General Assembly resolution 65/230 requesting the Commission on Crime Prevention and Criminal Justice, identified the following elements of the framework essential for effectively dealing with cybercrime.
  - o The development of international model provisions on criminalisation of core cybercrime acts, with a view to supporting States in eliminating safe havens through the adoption of common offence elements
  - o The development of international model provisions on investigative powers for electronic evidence, with a view to supporting States in ensuring the necessary procedural tools for investigation of crimes involving electronic evidence
  - o The development of model provisions on jurisdiction, in order to provide for common effective bases for jurisdiction in cybercrime matters
  - o The development of a multilateral instrument on international cooperation regarding electronic evidence in criminal matters, with a view to providing an international mechanism for timely cooperation to preserve and obtain electronic evidence
  - o The strengthening of international, regional and national partnerships, including with the private sector and academic institutions, with a view to delivering enhanced technical assistance for the prevention and combating of cybercrime in developing countries
- Global Trade arrangements like the Trans-Pacific Partnership (TPP), Transatlantic Trade and Investment Partnership (TTIP), WTO and also

regional arrangements such as Regional Comprehensive Economic Partnership (RCEP) focus on cybersecurity as the common ground.

- Export control regimes like Wassenaar Arrangement on Export Controls for Conventional Arms and Dual-Use Goods and Technologies, is one of multilateral export control regimes that now seeks to cover cyber security tools and technology. The National Association of Software and Services Companies (NASSCOM)-Data Security Council of India (DSCI) had earlier opposed attempt by the Bureau of Industry and Security (BIS), US, to include intrusion software as part of the regime.[14]
- Formulation of international standards has become important as ICT systems are increasingly traded internationally. Maintaining technological standards of such systems is imperative to ensure their interoperability and security level. While various initiatives are underway for international standardisation, it is important to formulate and disseminate international standards of cybersecurity technology and to create interoperable regime. Voluntary or recommended adoption of international standards, in absence of global regulations, find new recognition. The International Organisation for Standardisation (ISO) continues to develop important standards on information security, cryptography, network security, supply chain security etc., while new standards on cyber-resilience, cyber insurance are getting formed. Comparatively, Privacy specific standards are difficult to forge due to lack of consensus and varying expectation of Privacy. In the last ISO SC27 Working Group meetings, hosted in India by NASSCOM-DSCI,[15] Indian delegation proposed formulation of two new standards – 'Privacy in Smart Cities' and 'Privacy in Smartphone Applications,' which was well received by the global delegates. Indian experts are also playing important role in development of other privacy related standards at ISO including User friendly Privacy Notices, Privacy in Internet of Things (IoT), Privacy in Identity and Access management, among others.

## *Way Forward for India*

On the background of these initiatives, India should enhance its outreach, partnership, and engagement at International and regional level to develop practical and implementable frameworks to address global cybersecurity challenges. It should contribute to, and take leadership in, global forums to protect its strategic interests. As cybersecurity is a global issue, the international community is continuously engaged through various forums, such as the UN, Internet Governance Forum (IGF), Internet Corporation for Assigned Names and Numbers (ICANN), ISO, ITU and Internet Engineering Task Force (IETF), to discuss issues such as Internet governance, cybercrimes, cyberwarfare, security standards, supply chain risks, information sharing and surveillance. At present, India's participation in such forums is not commensurate with its cyber economy. The government should create necessary structures and capabilities within the country to understand complex issues, do consultations with the key stakeholders and develop country's position on such issues. India should collaborate with regional and global players to enhance International cooperation. There is an urgent requirement of cooperation amongst countries for recognising and developing interoperable arrangements. Bringing up implementable proposals and taking countries in confidence to develop harmonised regimes is the need of the hour. Following would be the areas of focus to move towards a cyber secure world:

- Work with 'like-minded' nations and global institutions to develop cyber norms and acceptable behaviour for operating in cyberspace.
- Ensure its strategic and economic interests are addressed by ascertaining implication of security aspects in bilateral and multilateral trade dialogues.
- Strengthen participation in existing mechanisms and enhance cooperation amongst LEAs at international level to improve attribution and for solving cybercrime cases to bring criminals to justice.
- Focus on developing cybersecurity capabilities for improving assurance of information systems within the country, through adoption of standards, frameworks, skill building and research and development.
- Promote development of communities for trust building through cooperation and information exchange among government and private sector entities and security professionals members at regional and international level.

## NOTES

1. "Gartner Says 6.4 Billion Connected 'Things' Will Be in Use in 2016, Up 30 Percent From 2015", Gartner, November 10, 2015, at http://www.gartner.com/newsroom/id/3165317.
2. World Economic Forum, "Partnering for Cyber Resilience", at http://www.weforum.org/projects/advancing-cyber-resilience.
3. Charles Babcock, "Gartner: 50% Of Enterprises Use Hybrid Cloud By 2017", networkcomputing.com, January 10,2013, at http://www.networkcomputing.com/cloud/gartner-50-enterprises-use-hybrid-cloud-2017/2055920703.
4. Dana Danca, "Cyber Diplomacy – A New Component of Foreign Policy", *Journal of Law and Administrative Sciences*, March 2015, at http://jolas.ro/wp-content/uploads/2015/03/jolas3a7.pdf.
5. Press Trust of India, "India, Malaysia, Singapore And Japan Sign Pacts For Cyber Security", ndtv.com, January 27, 2016, at http://www.ndtv.com/india-news/india-malaysia-singapore-and-japan-sign-pacts-for-cyber-security-1270707.
6. "International Strategy for Cyberspace", May 2011, at https://www.whitehouse.gov/sites/default/files/rss_viewer/international_strategy_for_cyberspace.pdf.
7. Incyder news, "An Updated Draft of the Code of Conduct Distributed in the United Nations – What's New?", ccdoe.com, February 10, 2015, at https://ccdcoe.org/updated-draft-code-conduct-distributed-united-nations-whats-new.html.

8. "Inaugural Address by Secretary (East) at the ASEAN-India Cyber Security Conference in New Delhi (January 19, 2015)", Ministry of External Affairs India, January 19,2015, at http://mea.gov.in/aseanindia/Speeches-Statements.htm?dtl/22570/Inaugural+Address+by+Secretary+East+at+the+ASEANIndia+Cyber+Security+Conference+in+New+Delhi+January+19+2015.
9. "International Strategy on Cybersecurity Cooperation", Information Security Policy Council Japan, October 2, 2015, at http://www.nisc.go.jp/eng/pdf/InternationalStrategyonCybersecurityCooperation_e.pdf.
10. Monika Ermert, "WSIS+10: Roles, Responsibilities Remain Hot; Cybersecurity Treaty Demanded By Many States", Intellectual Property Watch, December 16, 2015, at http://www.ip-watch.org/2015/12/16/wsis10-roles-responsibilities-remain-hot-cybersecurity-treaty-demanded-by-many-states/.
11. "UN Group of Governmental Experts: Developments in the Field of Information and Telecommunications in the Context of International Security", Council on Foreign Relations, July 22, 2015, at http://www.cfr.org/internet-policy/un-group-governmental-experts-developments-field-information-telecommunications-context-international-security/p36949.
12. No. 10.
13. United Nations Office on Drugs and Crime, "Comprehensive Study on Cybercrime", February 2013, at https://www.unodc.org/documents/organized-crime/UNODC_CCPCJ_EG.4_2013/CYBERCRIME_STUDY_210213.pdf.
14. "NASSCOM-DSCI Comments on BIS proposal to include cyber security items in the export control regime", at https://www.dsci.in/sites/default/files/NASSCOM-DSCI%20comments%20on%20the%20%E2%80%98Proposed%20Rule%E2%80%99%20%20by%20BIS%20(US)%20to %20the%20Wassenaar%20Arrangement.pdf.
15. Data Security Council of India, "ISO/IEC/ JTC 1/SC 27 WG Meeting", at https://www.dsci.in/events/about/2298.

# 19

# A South Asian Regional Cybersecurity Cooperation (SARCC) Forum: Prospects and Challenges

*Munish Sharma and Cherian Samuel*

## Introduction

South Asia has experienced a long haul of robust economic growth over the last one decade. A significant factor in this economic development has been the digitisation of public services and taking advantage of vast business opportunities in Information and Communications Technology (ICT). Despite the accelerated growth of this sector, countries in the South Asian region have not concomitantly internalised cybersecurity as vital to their economic and national well-being, making the entire region vulnerable to terrorism and crime in cyberspace, both from state and non-state actors.

As an economic and geopolitical organisation, the South Asian Association for Regional Cooperation (SAARC) can play a pivotal role in building capacity and capability as well coordinate the cybersecurity efforts of all the members facing non-traditional security threats to both their populace and businesses.

Representing one-fifth of the global population and the fastest growing economic bloc, South Asia is home to vast opportunities in terms of businesses, jobs, technology start-ups and capital investments. Therefore, the region has both numbers and volume, culminating into growth momentum, which needs to be sustained. The region houses English-speaking skilled workforce, which has proved its enormous potential at the global technology market with both technical and managerial skills. In order to secure the services industry and ensure employment opportunities for a significant portion of skilled youth, South Asia needs to initiate a regional cybersecurity forum. Given the future plans, the integration of electricity grids and economies and interlaced banking systems, South Asian governments must bring cybersecurity as a key agenda item for discussion. Concerted efforts of all the governments, who have their own individual plans and models in place, could be leveraged to leapfrog from e-governance to m-governance, especially when the mobile telephony market in the region is performing extraordinarily well. However, any non-traditional security issue cannot pave its way into the discussions without overcoming the challenges.

## SAARC: Objectives and ICT

The very foundation of SAARC rests on seven key objectives. In a nutshell, SAARC aspires to promote the welfare of the peoples of South Asia and improve their quality of life through accelerated economic growth, social progress and cultural development in the region. The SAARC charter denotes active collaboration and mutual assistance in the economic, social, cultural, technical and scientific fields, and envisages cooperation among the members in international forums on matters of common interests.

As a technology, the ICT has emerged as an important vector for transforming social and economic parameters of development, especially in the developing and underdeveloped countries. The declining cost of hardware, computing resources, improved mobility and usability, deeper penetration and lower subscription costs, and a host of other factors have given thrust to new business initiatives and models throwing open employment opportunities for both skilled and semi-skilled workforce. On the other hand, threats to the security of information systems hosting volumes of data have increased the security costs. In order to secure the genuine interests of the society, business and themselves, governments need to ensure that stringent cybersecurity measures and policies are adopted and effectively implemented.

### *The Growth Story of South Asia*

According to World Bank data (2015), South Asia is the fastest-growing region in the world in terms of Gross Domestic Product (GDP) (see Figure 1). The South Asia Economic Focus report projects GDP growth rate to steadily increase from 7 per cent in 2015 to 7.6 per cent by 2017.[1]

**Figure 1: GDP Growth (annual %) Rate of South Asia and the World**

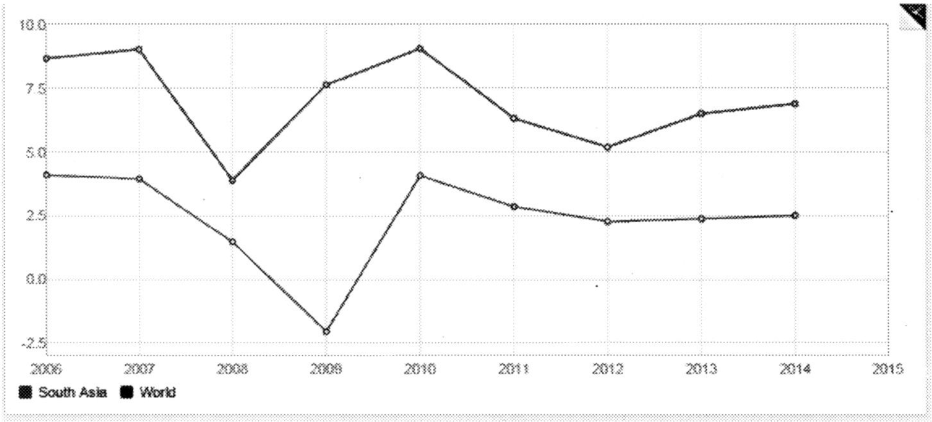


*Source:* World Bank.

Figure 2 indicates an upward trend in the absolute number of Internet users across the region. In developing countries, Internet penetration in rural areas is enabling governments in efficient delivery of services through their e-governance initiatives.

**Figure 2: Internet Users (per 100 people): South Asia and the World**

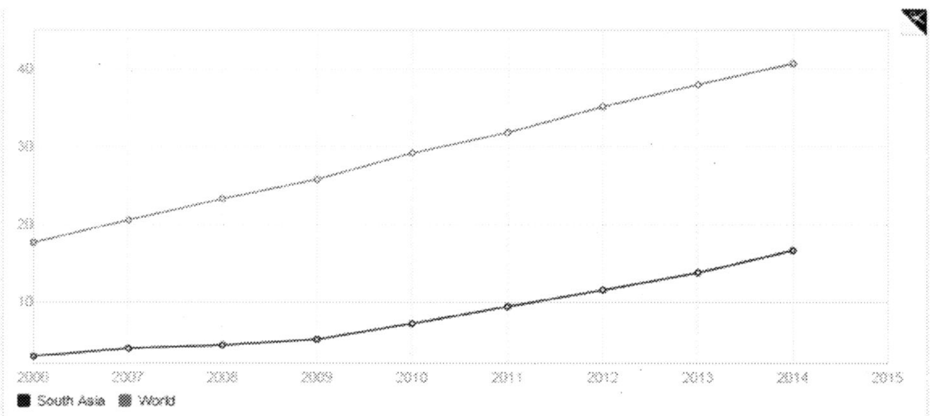


*Source:* World Bank.

On similar lines, the growth in the number of mobile cellular subscriptions suggests that the Internet reach is facilitated by mobile devices (see Figure 3). Very recently, India with more than 220 million unique smartphone users has overtaken the US as the world's second-largest smartphone market.[2] The healthy growth of mobile subscriptions and the smartphone market is compelling governments to move their governance practices to mobile platforms, and these include initiatives such as financial inclusion.

**Figure 3: Mobile Cellular Subscriptions (per 100 people): South Asia and the World**

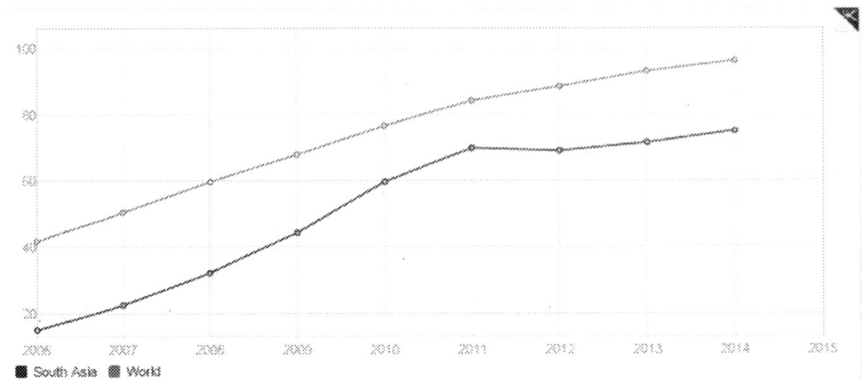


*Source:* World Bank.

All the countries in South Asia are faring better in terms of developing own e-governance solutions and setting up industries leveraging the ICT advantage. However, despite the geographical proximity, the economic interactions and dialogues related to key security matters have not been up to the mark.

## South Asia: Regional Integration

### *Economy and Trade*

South Asia is one of the most dynamic regions in the world, but it is also one of the least economically integrated. Intra-regional trade accounts for just 5 per cent of total trade, compared with 25 per cent in the Association of Southeast Asian Nations (ASEAN).[3] In order to maintain the impressive growth rate, as a regional entity, South Asia requires market integration to ensure the free flow of goods, services, and capital across borders.

Since the early 1990s, several attempts have been made at various bilateral, sub-regional, and multilateral levels to boost the economic integration among the members of South Asia, through a number of trade pacts. SAARC has taken initiatives to enhance integration, namely the South Asian Preferential Trading Arrangement (SAPTA), South Asian Free Trade Area (SAFTA), and more recently, SAARC Agreement on Trade in Services (SATIS), which was signed in 2010.[4]

SAFTA was envisaged to lead the trade among the members of SAARC towards a Customs Union, Common Market and Economic Union. The SAFTA Agreement was signed on January 6, 2004 during the 12th SAARC Summit held in Islamabad, and entered into force on January 1, 2006. Still, intra-SAARC trade flows under SAFTA are far below the potential.[5]

**Table 1: South Asia's Total Trade within the subregion and with the World**

| *Indicator* | *South Asia* | | | *World* | |
|---|---|---|---|---|---|
| | *Total Trade in billion US$* | *As percent of South Asia's Total Trade with the World* | *Total Trade Growth (%)* | *Total Trade in billion US$* | *Total Trade Growth (%)* |
| 1990 | 1.8 | 2.7 | N/A | 66.2 | N/A |
| 1995 | 4.4 | 4.2 | 43.3 | 104.4 | 26.9 |
| 2000 | 6.2 | 4.3 | 21.7 | 142.8 | 10.6 |
| 2005 | 17.3 | 5.3 | 30.6 | 324.1 | 32.5 |
| 2010 | 33.3 | 4.6 | 45.9 | 719.9 | 33.8 |
| 2011 | 40.5 | 4.3 | 22.1 | 951.1 | 32.1 |

Source: www.asiapathways-adbi.org.

However, statistical evidence suggests that intra-subregional trade among SAFTA members is rising slowly and steadily. As indicated in Table 1, South Asia's intra-subregional trade share has increased from 2.7 per cent in 1990 to 4.3 per cent in 2011.[6]

## *Services Sector*

The services sector has emerged as the growth engine in the SAARC economies over the past decade. Services sector is also emerging an important source of employment. Some SAARC countries have also begun to exploit the potential of services in trade.[7] Regional cooperation and integration of the service markets across the SAARC region may also assist in specialisation and strengthening the competitiveness of services industries in the South Asian countries.[8] Rapid economic growth and programmes to address issues of poverty and public distribution could be led by services. This approach has changed the development patterns because necessities like banking and healthcare could be carried online.

A wide spectrum of business processes are already globalised and digitised: the processing of insurance claims; e-publishing; remote infrastructure management and maintenance of IT networks; electronic medical records; etc. As the South Asian economies embark on the plans for services sector development, there are many initiatives on the ground as well as in the policy circles which are going to shape the regional dynamics, and simultaneously have a bearing on cybersecurity of individual states.

## *Cybersecurity Element in SAARC*

### *SAARC Agreement on Trade in Services*

In order to expand cooperation in trade and further deepen the integration of the regional economies, the SAARC Agreement on Trade in Services was signed at the 16th SAARC Summit held in Thimphu in April 2010. The Agreement entered into force on November 29, 2012, and thereafter the Expert Group on the SAARC Agreement on Trade in Services has been engaged in negotiating Schedules of Specific Commitments. The SAARC Commerce Ministers have been monitoring the progress in this regard and giving requisite directives aimed at time-bound actions for finalisation of Schedules of Specific Commitments.[9] If the Agreement meets its expectation in the near future, the flow of services across South Asia would surge, not only bringing benefits to the economies of the SAARC region, but also escalating the risks of cyber-led crime and terrorism.

### *SAARC Electricity Grid*

The idea of a SAARC grid was originally floated in 2009 as part of an Asian Development Bank report on SAARC energy trade.[10] The Foreign Ministers of all the eight countries signed the framework agreement on cooperation in power sector at Kathmandu Summit in 2014. The agreement is anticipated to facilitate electricity trading through grid connectivity.[11] As per the agreement, the SAARC grid would lead to optimal utilisation of regional electricity generating resources, enhanced grid security, and electricity trade arising from diversity in peak demand and seasonal variations. Article 10 of the agreement (on Electricity Grid Protection System) states: "Member States shall enable joint development of coordinated network protection systems incidental to the cross-border interconnection to ensure reliability and security of the grids of the Member States."[12]

Ensuring a resilient electric grid is paramount, since the entire critical infrastructure sectors depend upon electricity grids to deliver essential services.[13] While the threats to electricity grids have always existed in theory, recent incidents such as the cyberattack at the electricity distribution centres in Ukraine, which took 30 electricity substations offline and left 230,000 residents in darkness in December 2015, underscore that such threats no longer exist only in fiction. The blackouts in North India in 2012 were also initially speculated to be a consequence of cyberattacks, though subsequent investigations indicated otherwise. Going forward, the integrated electricity grids of South Asian countries would be dispersed across different territories and authorities, single point failure may cascade to other states, such as the European Blackout of 2006. Learning from these instances, SAARC needs unified cybersecurity architecture for entities responsible for generation and transmission of electricity in their national boundaries.

### *Banking and Financial Services: SAARC Payment Platform*

Banking and finance is another core focus area for the development of the South Asian region, poverty alleviation and efficient public distribution of government schemes or subsidies. The sector is a key critical infrastructure as well and an enticing target for cyberattacks.

An integrated banking system warrants comprehensive cybersecurity to mitigate any possibility of a cyber heist. For example, the US$ 81 million heist at the Bangladesh Bank in March 2016 covered many jurisdictions and areas of responsibility of various entities, spread across the globe. The success of banking and financial systems' integration and the financial inclusion initiatives would depend on the integrity and security of the technology behind these systems.

The idea of SAARC Payments Initiative (SPI) originated at the SAARC FINANCE conference on "Towards a Regional Payments Group", held at the Central Bank of Sri Lanka (CBSL) in July 2007. At the 16th SAARC FINANCE Governors' Meeting, the agreement to establish the SPI was reached, and therefore, SPI was established to facilitate the development of payments systems in the SAARC Region. The landmark achievements of SPI include standardised payment and settlement systems and establishment of Real Time Gross Settlement (RTGS) systems in member countries. As a focus area for the future, the SPI is working on internationally recognised benchmarks for efficient and safe electronic payment and settlement system infrastructure to cater to the emerging needs of the region. The basic aim is to promote electronic and other modes of funds transfers to facilitate financial inclusion. While working on the SPI, it is imperative that the cybersecurity angle should also be covered.

### *Telecommunications*

South Asian Telecommunication Regulator's Council (SATRC) was formed in 1997 by an initiative of Asia Pacific Telecommunity (APT) and the International Telecommunication Union Regulatory Forum for South Asia. The SATRC holds discussions and coordinates issues of common interest relating to regulations in telecommunication and ICT in South Asian countries, such as radio frequency coordination, standards, regulatory trends and issues, strategies for telecommunication development and telecommunication related international affairs.[14] Platforms like SATRC hold salience, given the exorbitant growth of mobile subscribers in SAARC countries. According to the telecom market forecasts by Dataxis, the SAARC region will have 1.3 billion active mobile subscribers by 2018, which will constitute 98 per cent of the total number of telecommunication access lines.[15]

This gives the impetus to leapfrog from e-governance to mobile-dominated governance model. The growing base of smartphone users would compel the governments to deliver the e-services on mobile platforms, as users are more likely to access Internet over mobile platforms. With an existing forum to discuss and cooperate over issues related to regulations in telecommunication and ICT, SATRC could be leveraged for cybersecurity issues prevalent in the telecommunication sector and the rapidly growing usage of mobile phone operating systems or applications, which are the key enablers of growth and economic development.

*E-governance Applications*

As a geographical entity, South Asia inhabits one-fifth of the global population. The governments across all the eight countries are developing ICT-based applications to deliver governance and services to the populace. This encompasses services such as taxation, healthcare, education, public distribution systems, banking services, etc.

In this context, ICTs are deemed to play an important role in stimulating economic and social growth in the developing countries, primarily by enabling novel and efficient methods for the delivery of government services. A number of e-governance initiatives and practices are prevalent in South Asia. The National e-Governance Plan of India is a single platform for all e-governance services provided by centre and state governments across India.[16] A similar portal is maintained by the Government of Nepal.[17] The Government of Bhutan[18] has a master plan in place, while Maldives[19] also practises a set of e-government initiatives. The e-Divisional Secretariat of Sri Lanka[20] handles e-governance-related activities. Pakistan is also developing a digital ecosystem infrastructure and institutional frameworks for the rapid delivery of next generation digital services, applications and relevant content to ensure efficiency, transparency and accountability.[21]

In the present scenario, ICT assets for digitisation of services are an organisational imperative. Technology, comprising of both hardware and software, enables the governments to undertake and execute these initiatives. The technology generates, stores, processes and disseminates quantum of information, in form of personal details and transactions. Threats to ICT assets, including the residing information, may arise from different sources. These could be internal, where information is misused or liable to unauthorised access, or the threats could be external, such as hackers, criminal organisations, terrorist organisations and foreign intelligence agencies. Therefore, a number of security measures for data, network and applications are put in place.

A platform for cooperation among South Asian countries to replicate and implement successful e-governance projects for mutual learning will improve the ability of the respective governments.[22] As the governments of South Asian countries have already embarked on the path to digitise the governance processes and functions, going forward the data residing in their information systems and the financial transactions or banking details would be at a greater risk. At present, the interactions between the governments are very limited in the context of e-governance initiatives and data security. Moreover, information is exposed to the expanding threats in the cyberspace. Over the years, these cyberthreats have augmented their capability to damage national interests, disrupt industry functions and harm individual or end users.

## Cybercrime: Threat to South Asian Integration, Economy and Services Sector

As South Asia is poised to emerge as an economic bloc, its telecommunications and services sectors have a central role to play. Telecommunications is the driver of infrastructure necessary to build the services sector. With the increased economic activity, integration of markets, exchange of goods and technologies, a new ecosystem will emerge where governments and business would need a secure cyber environment. An interconnected market or economy renders all the stakeholders prone to the risk from cybercrimes and the instances of cyberattacks.

According to McAfee and Centre for Strategic and International Studies analysis, cybercrime costs more than US$ 400 billion to the global economy annually.[23] Moreover, cybercrime damages the performance of companies and the respective national economies, in addition to trade, competitiveness and innovation.[24] An organisation can recover after the loss of operational data. But, if core intellectual property is compromised, the marketing plans, product design, and research and development, practically the whole company is at risk.[25]

The McAfee report observation regarding cybercrime states that:

- The cost of cybercrime will continue to increase as more business functions move online and as more companies and consumers around the world connect to the Internet.
- Losses from the theft of intellectual property will also increase as acquiring countries improve their ability to make use of it to manufacture competing goods.
- Cybercrime is a tax on innovation and slows the pace of global innovation by reducing the rate of return to innovators and investors.
- Governments need to begin serious, systematic effort to collect and publish data on cybercrime to help countries and companies make better choices about risk and policy.

Cybercrime is on the rise across all the countries of South Asia. In India, registered cases of cybercrime went up by 69 percent in 2014 as compared to 2013 (from 5,693 cases in 2013 to 9,622 cases in 2014), according to statistics from the National Crime Records Bureau (NCRB). The cases of cybercrime are increasing rapidly each year along with the rise in the number of Internet and social media users in Nepal. As per the Nepal Police statistics, the number of cybercrimes reported surged by 105 per cent in the last fiscal, 2014-15.[26] As per a report from Microsoft Corporation, Pakistan, Bangladesh and Nepal along with Indonesia, and Palestinian territories, attract the highest rates of attempted malware attacks.[27] All the eight countries are aware about the implications of cybercrime and the respective governments have already initiated measures to curb the growth of cybercrime through law enforcement apparatus. However, SAARC members have agreed to establish a cybercrime monitoring desk,[28] but there is absence of dialogue over cybersecurity or cybercrime among the SAARC members.

Nevertheless, cybercrime is poised to be severe in the future, despite the best efforts of governments, their law enforcement or cybersecurity agencies, corporations and their experts in this domain. Fundamentally, the growth of cybercrime is driven by the expanding number of services available online and the increasing sophistication of cybercriminals. The numbers of services offered online will eventually grow tremendously over the next couple of years, both for the social and economic benefit of the citizens. All the South Asian countries are almost at the same pedestal in terms of their risk exposure, preparedness, legal frameworks and institutional responses to the perils of cybercrime.

## Cybersecurity in South Asia: Prospects and Challenges

While there is a thrust on ensuring a peaceful and secure cyberspace for the global community, and many initiatives designed to implement the measures, there is a wide variation in the "cyber maturity" of different countries, the assistance provided by these organisations notwithstanding. In fact, as the example of South Asia shows, countries in the region are yet to internalise cybersecurity as essential to their economic, political and national well-being. This has many implications: a) It impacts the security of the citizen and makes the country vulnerable in ways unknown. b) It also affects the ability of countries in the region to contribute to the global conversation on cybersecurity.

South Asia is particularly affected since a regional organisation such as SAARC is finding it difficult to forge cooperation and consensus on even traditional security issues, and it has not yet taken up cybersecurity as an agenda item. This contrasts with the situation in South East Asia where despite differences on political issues, South East Asian nations have made significant movement towards cybersecurity, emphasising on their business and economic interests. Cybersecurity is a key discussion point under the ASEAN Regional Forum (ARF), holding regular workshops and seminars to operationalise cyber confidence building measures.[29] Traditional rivals like Australia, China and Japan are making efforts to lead cybersecurity efforts in the region.[30]

A South Asian Regional Cybersecurity Cooperation (SARCC) Forum under the institutional auspices of SAARC could pave the way for a resilient and comprehensive cybersecurity regime across South Asia. This would open up the window for providing inputs into both regional institutions dealing with security issues, such as the ARF and Shanghai Cooperation Organisation (SCO) and various international fora.

There are both challenges and opportunities on the road towards this goal; despite abundant human resources, the developing countries of South Asia are unable to tap the full potential and there is lack of skill-set development, especially in high-end technology research, which includes cybersecurity. Moreover, all the SAARC countries are heavily dependent on technology imports, including ICT and network security products. Moreover, governments have not been able to develop the desirable ICT infrastructure, which has widened the digital divide. The focus at present is to bridge the digital divide, without concomitantly addressing the issues of cybersecurity. There are also divergences, differences and legacy issues at the geo-political level, leading to anomalies such as the respective Computer Emergency Response Teams (CERTs) of India and Sri Lanka being active in Asia Pacific CERT (APCERT), while Pakistan CERT prefers to be active in the Organisation of Islamic Cooperation CERT (OIC-CERT). The goal of a South Asia CERT might be too ambitious, given the obstacles; still, a SARCC Forum would appear more within reach and would go a long way towards ensuring cybersecurity in the region, and to have effective contribution from South Asia to the ongoing global discourse on cybersecurity.

## NOTES

1. "South Asia, Now the Fastest-Growing Region in the World, Could Take Greater Advantage of Cheap Oil to Reform Energy Pricing", The World Bank, April 13, 2015, at http://www.worldbank.org/en/news/press-release/2015/04/13/south-asia-cheap-oil-reform-energy-pricing (Accessed February 22, 2016).
2. Rohit KVN, "India Replaces the US as World's Second-Biggest Smartphone Market: Counterpoint Research", *International Business Times*, February 03, 2016, at http://www.ibtimes.co.in/india-replaces-us-worlds-second-biggest-smartphone-market-counterpoint-research-665503 (Accessed February 22, 2016).
3. "Regional Integration in South Asia", The World Bank, October 03, 2015, at http://www.worldbank.org/en/region/sar/brief/south-asia-regional-integration (Accessed February 24, 2015).
4. Mustafa Moinuddin, "Economic Integration and Trade Liberalization in South Asia", *Asia Pathways*, August 27, 2013, at http://www.asiapathways-adbi.org/2013/08/economic-integration-and-trade-liberalization-in-south-asia/#sthash.gyeJXfno.dpuf (Accessed February 24, 2015).
5. This pales in comparison with other trading blocs. ASEAN intra-regional trade is about 25 per cent. See: "South Asian Free Trade Area (SAFTA)", at http://saarc-sec.org/areaofcooperation/detail.php?activity_id=5 (Accessed February 24, 2015).

6. Ibid.
7. Ram Upendra Das, "Regional Economic Integration in South Asia: Prospects and Challenges", Research and Information System for Developing Countries, Discussion Paper number 157, September 2009, p. 33 at http://www.eaber.org/sites/default/files/documents/RIS_Das_2009.pdf.
8. Ibid.
9. "SAARC Agreement on Trade in Services", at http://saarc-sec.org/areaofcooperation/detail.php?activity_id=46 (Accessed February 28, 2016).
10. Mahim Pratap Singh, "Union Power Minister Moots SAARC Power Grid", *The Hindu*, October 17, 2014, at http://www.thehindu.com/news/national/union-power-minister-piyush-goyal-moots-saarc-power-grid/article6512190.ece (Accessed February 28, 2016).
11. "SAARC Nations Sign Seamless Electricity Grid Deal", *The Economic Times*, November 27, 2014, at http://articles.economictimes.indiatimes.com/2014-11-27/news/56515642_1_saarc-summit-kathmandu-declaration-saarc-regional-convention (Accessed February 28, 2016).
12. "SAARC Framework Agreement for Energy Cooperation (Electricity)", at http://www.saarc-sec.org/userfiles/SAARC-FRAMEWORK-AGREEMENT-FOR-ENERGY-COOPERATION-ELECTRICITY.pdf (Accessed February 28, 2016).
13. "Cybersecurity", Office of Electricity Delivery & Energy Reliability (U.S. Department of Energy), at http://energy.gov/oe/services/cybersecurity (Accessed March 02, 2016).
14. Asia Pacific Telecommunity, "South Asian Telecommunications Regulators' Council", at http://www.apt.int/APTSATRC (Accessed April 28, 2016).
15. Telecom Press Release, Dataxis, January 18, 2016, at http://dataxis.com/pressrelease/fixed-broadband-subscribers-will-evolve-to-40-million-lines-by-2018-in-saarc-region/ (Accessed May 05, 2016).
16. "National e-Governance Plan", Government of India, at https://negp.gov.in/index.php (Accessed March 02, 2016).
17. "The Official Portal of Government of Nepal", at http://nepal.gov.np/portal/npgea/home?l=en&rn=1457451753599 (Accessed March 02, 2016).
18. Ministry of Information & Communications, "Bhutan e-Government Master Plan", Department of Information Technology & Telecom, January 2014, at http://www.moic.gov.bt/daden/uploads/2014/04/Bhutan-eGov-Master-Plan.pdf (Accessed March 02, 2016).
19. "National Centre for Information Technology", Republic of Maldives, at http://www.ncit.gov.mv/index.php/en/e-government (Accessed March 02, 2016).
20. Shoban Rainford, "e-Sri Lanka: An Integrated Approach to e-Government Case Study", http://unpan1.un.org/intradoc/groups/public/documents/un-dpadm/unpan040596.pdf (Accessed March 08, 2016).
21. Ministry of Information Technology, "National Information Technology Board", Government of Pakistan, at http://www.e-government.gov.pk/ (Accessed March 08, 2016).
22. "South Asia E-governance Summit", DQ Channels, November 19, 2006, at http://www.dqchannels.com/south-asia-e-governance-summit/ (Accessed March 08, 2016).
23. "Net Losses: Estimating the Global Cost of Cybercrime", McAfee and Center for Strategic and International Studies, June 2014, p. 2, at http://www.mcafee.com/in/resources/reports/rp-economic-impact-cybercrime2.pdf.
24. Ibid, p. 3.
25. Sam Visner, "Cybersecurity's Impact on Business", CSC, at http://www.csc.com/cscworld/publications/56901/56918-cybersecurity_s_impact_on_business (Accessed March 08, 2016).
26. A total of 39 cybercrime-related cases were filed last year compared to 19 cases in fiscal 2013-14. See: "Online Attacks on the Rise", *The Himalayan Times*, September 03, 2015, at https://thehimalayantimes.com/business/cyber-attacks-on-the-rise/ (Accessed March 11, 2016).

27. "Pakistan, Indonesia Lead in Malware Attacks - Microsoft Report", Reuters, May 05, 2016, at http://in.reuters.com/article/microsoft-cybersecurity-idINKCN0XW1L6 (Accessed May 06, 2016).
28. 18th SAARC Summit Declaration, November 27, 2014, at http://www.saarc-sec.org/press-releases/18th-SAARC-Summit-Declaration/121/ (Accessed March 11, 2016).
29. Embassy of the United States - Singapore, "Welcome Remarks by Amb. Wagar for the ARF Seminar on Operationalizing Cyber Confidence-Building Measures", October 21, 2015, at http://singapore.usembassy.gov/arf-seminar102115.html (Accessed May 06, 2016).
30. Simon Hansen, *Australia–China Cyber Relations in the Next Internet Era*, Australian Strategic Policy Institute, December 2015, at https://www.aspi.org.au/publications/australiachina-cyber-relations-in-the-next-internet-era/SR85_Australia_China_cyber_relations.pdf.

# 20

# REGIONAL SECURITY ARCHITECTURE IN ASIA: ENHANCING TRANSPARENCY AND CONFIDENCE AMONG MILITARIES ON CYBERSECURITY

*Caitríona Heinl*

## The Issue

Much attention has focused on the increase in the acquisition or development of advanced cyber technologies by militaries in the Asia-Pacific region, yet at the time of writing fewer concrete efforts have been made to address a lack in military-to-military dialogue on cyber-related matters. The international community is therefore encouraging states to be both 1) more transparent about the roles and responsibilities of their defence forces and security services in the cyber domain, and 2) pursue dialogue and other measures related to cyber issues to build confidence and ensure international stability.[1] This article focuses on the approaches, including existing institutional mechanisms that might be considered to foster such dialogue or other measures in the Asia-Pacific region. It does so because it is likely that these types of forums are important as a means to find areas of common interests where there may be like-minded interests.

Ultimately, states need to ensure an international environment of stability. While militaries might aim to be prepared to win in conflict, there should be an obligation to avoid escalation.[2] The military is after all an important stakeholder with an interest in a safe and secure cyberspace.[3] This article posits that although the cyber domain presents some unique challenges, cyber conflict or pure "cyber war" is not preordained for the same reasons that conflict is not inevitable. It thus considers the mechanisms that might help to mitigate such misunderstanding and escalation. In a recent address, USCYBERCOM Commander, Admiral Rogers highlighted that one of his key takeaways from the past two years is that cybersecurity is all about partnerships. He explains this phenomenon in the following words: "There is no single group, there is no single nation, there is no single segment ... there is no single entity that has all the answers ... This is a challenge that will require us to work together in collaborative and innovative ways."[4] A priority for 2016 will therefore include international partnerships, particularly since the international dynamic becomes critical as nations begin working on norms of behaviour and deterrence.[5]

## Challenges Facing Militaries in the Asia-Pacific: Geopolitical and Cyber-specific Factors Behind the Concerns Over Instability

### *Geopolitical Factors*

Cyber-related issues do not exist in a vacuum – if mistakes happen and conflict occurs, it is generally accepted that a cyber component will ensue. This extends to the regional level where within current regional discussions on cyber issues, concerns over the increasing environment of mistrust that is especially prevalent in the Asia-Pacific, are being flagged.[6] The existing geopolitical environment is naturally significant when analysing the development of cyber policies in the Asia-Pacific region. In short, the region is well known for its high national security sensitivities, exceptional levels of military modernisation and defence spending, on-going maritime and territorial disputes, and increasingly malevolent non-state actors who further complicate matters. Furthermore, there is a palpable uncertainty both about China's intentions as a regional military power and the United States' "pivot". There are high levels of strategic distrust and competition among the larger nations which must be carefully controlled in order to avoid an unnecessary escalation. And although it is clear that it is not in any state's interest that conflict occur, in such a tense environment mistakes might occur more easily.

### *General and Region-specific Cyber-related Challenges*

The very characteristics of cyber tools can in turn add even more layers of ambiguity and obfuscation, thus multiplying the likelihood of mistakes or miscalculations occurring. Government practitioners not only question the efficiency of existing diplomatic channels given that cross-border cyber incidents can happen at great speed, but there can also be difficulty in delineating what comes within the political or military realms.[7] These threats can challenge traditional state responses. Moreover, while most states or non-state actors may not have advanced capabilities for devastating attacks yet, there is a concern that unexpected outcomes could be caused by mistake if, for instance, less sophisticated cyber attacks are used against poorly protected networks that control core functions.[8] It is particularly challenging that these are dual-use technologies with both military and commercial or civilian applications.

Stronger regional cooperative mechanisms to deal with the challenging nature of these tools (the lessening significance of distance and land borders or the speed with which borders can be crossed, for example) have not been extensively implemented. Several factors could still hinder possible international military cooperation. For instance, given that many states recognise that cyberspace is a domain for military operations, it is most likely that they will try to obtain capabilities as best they can. In this region, cyber defence can be a sensitive subject, and both state intentions as well as their acquisition of capabilities can be ambiguous and hard to accurately measure. There are limited levels of transparency and this type of ambiguity can be a destabilising factor. This is especially the case as it makes it more difficult to establish common understanding and trust between parties.

What is clear is that these states are at different stages in terms of capabilities (for a closer outline of the nature of national level capabilities and evolving structures across the Asia-Pacific region, the Australian Strategic Policy Institute [ASPI] Cyber Maturity reports for the years 2014 and 2015 provide some useful indicators). Defence reports envisage that although countries like China will most likely be the most sophisticated players in the region over the next 20 years, other countries will develop cyberspace activities so as to project influence which they might otherwise find limited with conventional instruments.[9] Smaller countries may even believe this is a way to avoid open confrontation. So-called qualitative and quantitative differences in information and communications technologies (ICTs), and the ways in which they are used have already widened capability and performance gaps not only between developed and developing countries, but also among allies.[10] Consequently, these types of differences have implications when pursuing enhanced international military cooperation.[11]

In terms of strategy development and implementation, this field is still evolving. Many countries are determining the conceptual and doctrinal underpinnings of the role of military and armed forces in defending cyberspace, although to different degrees.[12] The Asia-Pacific region is no different and it is highly diverse meaning that states are at very different stages of both developing and implementing cybersecurity policies. Nonetheless, the growing understanding of cyber threats within regional militaries means that there have been many developments in organisation, cyber capability and doctrine.[13]

National level priorities for cybersecurity and cyber defence may not always align well with other states' priorities and a so-called lack of common understanding ensues. In fact, common understanding is cited as one of the most important factors for cooperation but unfortunately it is lacking in the region.[14] This can then make dialogue and progress at the regional level somewhat more difficult. Not only do these countries have significantly different levels of ICT development and adoption, but they can also face different threats (and have varying perceptions as to what constitutes a threat to the nation such as concerns about preserving social harmony or control). Domestic challenges differ from one another, and countries are at markedly diverse stages of economic development. Culture and ideologies can also vary significantly. The tendency of some countries in the region to view cyber operations through the lens of information warfare or information security, perhaps with tendencies more closely aligned to Sun Tzu's philosophy, means that there can be a further disconnect with "Western" doctrine (which is now beginning to use the term hybrid warfare to account for such developments).[15] A recent International Institute for Strategic Studies (IISS) strategic dossier highlights how significant this might be when it explains that, if implemented in full, Chinese thinking on information operations will make fundamental changes to the character of warfare.[16] This is especially the case for missions directed at soft targets, information systems, decision-making processes, and cognitive perceptions on the battlefield, as well as critical infrastructure like financial, energy and transportation systems.[17] The use of non-traditional war-fighting methods in such operations would it seems only exacerbate the change.[18]

Cognisant of the numerous plans to increase connectivity of critical infrastructure across the region, including for transport systems, ICT infrastructure, and energy or regional grids, collaboration is vital. Markets in the region are becoming increasingly interdependent and security is however essential to ensure continuing economic and social development. Moreover, such economic interdependence can in itself be a stabilising factor, although some defence strategists would remain sceptical of over-reliance on economic interdependence as a factor to prevent conflict given the experiences of Europe in the 20th Century.[19]

Another much discussed challenge in the cyber domain is that the use or expressed interest in using advanced cyber technologies by criminals, terrorists, hackers (including hacker activists) and proxy actors engaged or supported by government adds more layers of ambiguity to an already terse environment. This ambiguity is caused because it can sometimes be difficult to accurately attribute responsibility for an incident. Such growing levels of increasingly advanced as well as malicious crime raise the probabilities of misperception or mistake given the vagueness surrounding espionage and military activities. Resolving this problem of difficult attribution is a key way to be sure of intentions and to hold perpetrators responsible (whether criminal groups or malicious state actors). However, strong legal and technical attribution can still be both a difficult as well as slow process for the majority of states. Thus, state actors may be more tempted to act in the belief that this will be without penalty given their perceptions of high-level plausible deniability.

These higher levels of cybercrime and cyber-enabled crimes have already begun to arise across the region, specifically in Southeast Asia. In less digitally developed countries that did not previously have saturated Internet penetration rates, mobile smartphones have caused a major change in how quickly they have allowed for the increase in ICT use in the region over a short period of time. However, these countries do not all have the capacity to tackle cross-border crime effectively. Ongoing international initiatives in Southeast Asia can hopefully help to reduce this environment of ambiguity caused by crime. Such efforts include projects of INTERPOL Global Complex for Innovation (IGCI) and the multi-stakeholder Global Forum on Cyber Expertise (GFCE). GFCE, of which India is a founding member, aims to address rule-building, for a secure and open Internet, confidence and trust building, and capacity building by focusing on practical recommendations and best practices. Moreover, establishing cooperation or consensus on common areas like crime is identified as a first step toward stability and the widespread use of confidence-building measures (CBMs).[20] Nevertheless, another challenge in this field is how to control the production or even export of dual-use technologies. In collaborating, could it be possible that digital forensics tools used for law enforcement purposes may end up utilised by state parties for ill-intended purposes such as censorship or surveillance that does not respect universal rights.

## Regional Security Architecture: International Engagement on Cyber to Foster Transparency, Predictability, and Trust among Militaries in Asia

Even while this subject is still viewed as sensitive, creating an environment of international stability could be assisted with realistic and pragmatic measures that are taken by military stakeholders in conjunction with the political security community. In fact, several discussions held at the Observer Research Foundation, New Delhi, in 2014 reiterated this need for pragmatism. The highest levels of government generally accept that states need to cooperate on the key challenges raised by cross-border cyber threats. However, there can sometimes be less clarity on the specific details, mutually agreeable action points and deliverables between state parties so as to facilitate implementation. A challenge that now arises is how to ensure that dialogues or other measures can translate agreement into tangible action points that are mutually agreeable and will in fact be implemented.

The Chair's Statement at the Global Conference on Cyber Space (GCCS) in 2015 recognises this need for international cooperation to reduce risks, and discussions are focused on reaffirming the applicability of international law to state behaviour in cyberspace, as well as the development of voluntary, non-legally binding norms for responsible state behaviour in cyberspace during peacetime.[21] The need to both develop and implement CBMs to increase stability and prevent the risk of conflict as a result of misperceptions and miscalculations arising from the malicious use of ICTs is also recognised.[22] Given the nature of developments in this field, it is clear that more mechanisms to foster transparency, predictability and trust are needed so as to mitigate the risks of misperception and support stability. Notably, regional government practitioners now also recognise that this increasing of the level of transparency is vital to ensure an environment of stability and the creating of common concepts.[23] Trust is another key element of cyber stability as it supports confidence among parties that they will each adhere to rules of the road in accordance with international standards, conventions, law or consensus best practice, and that all can have reasonable confidence that the Internet will function as expected.[24] This need for trust is emphasised time and again by government practitioners through roundtables, panel discussions and policy documents.[25]

The question that now arises is how such transparency, trust and confidence can be created. Regional efforts are especially important as a means to help address this deficit in common understanding, and they are often key to international discussions. This is particularly the case when operationalising international agreements and practical measures. There are often, for instance, fewer ideological tensions between states with a focus on CBMs. And this can be helpful in this region where stark differences over culture and ideology remain. In fact, a U.S. International Security Advisory Board report on a potential architecture for enhanced international cooperation in promoting a peaceful, secure and open cyberspace environment advocates building on areas of consensus while exploring norms (that relate to core U.S. values), using a two-tier approach of both bilateral dialogues and discussions at the multilateral level.[26] This approach very closely reflects the nature of current developments in the field across the region. The Board argues that bilateral norms could be integrated into broader alliances, treaties, and agreements, while global or multilateral discussions through the United Nations Group of Governmental Experts (UN GGE) and other venues for instance should proceed in parallel.[27] It concludes that increasing consensus on the need for stability, norms as guides for best practice, real-time sharing of malware data, pursuit of greater capabilities for attribution of attacks, support of deterrence by denial and conventional means, and sharing expertise with allies and friends offer important opportunities for enhancing stability.[28]

It is clear that India is emerging as a regional leader in cyber policy discussions.[29] As are Japan, Singapore and Malaysia. However, if India is to continue its progress and become a leader in this field, it must ensure that its plans and institutional structures are implemented in as timely a fashion as possible. Japan, for example, has been very efficient in developing and implementing its policies and institutional reforms over the past year to 18 months. Singapore, a member of the Association of Southeast Asian Nations (ASEAN) which is central to the regional architecture, has also been strong on implementation.

### *Bilateral Developments*

Many bilateral discussions and strategic dialogues on cyber-related matters are already ongoing across the region. These bilateral efforts might then be later extended to larger groups of states. In this instance, having the defence community represented as one of the stakeholders can hopefully further assist in enhancing transparency and confidence building. For cyber defence cooperation itself, defence practitioners argue that while these are sovereign decisions, sovereignty is not in fact the decisive factor. Rather trust and shared interests are the more powerful drivers when deciding the degree of cooperation.[30] Consequently, bilateral cooperation may be easier to foster in terms of both establishing trust and identifying shared or common interests. In addition, traditional defence cooperation efforts that include dialogues or working groups could consider including cyber within the agendas. India has been particularly active in such international engagements, engaging in active cyber diplomacy both at the bilateral and international level.[31] It has, for instance, been dealing with several states like Japan and the U.S. at bilateral level, as well as the European Union (EU) and ASEAN.

### *Like-minded Groupings*

Cooperation efforts at the sub-regional level, in other words between like-minded groupings, could foster practices that might then extend to the regional level too.[32] This is the case in other security fields in the region.[33] In fact, several states already run bilateral or small cyber exercises with like-minded countries in other regions.[34]

Government practitioners in the region have specifically highlighted the utility of multilateral Memorandums of Understanding (MOUs) as well as international security and defence forums like the Shangri-La and Seoul Defense dialogues.[35] In fact a number of MOUs were agreed recently, both bilateral as well as multilateral. For example, the January 2016 Japan/India/Singapore/Malaysia MOU for sharing knowledge in detection, resolution and prevention of cyber incidents among their computer emergency response teams (CERTs). It is possible that such efforts could then be extended to a larger group at a later stage. Alternatively, collaboration could even be expanded from these agreements on CERTs to other areas of cooperation.

Such multilateral efforts may sometimes allow states to come to agreements outside other regional mechanisms or forums like the ASEAN Regional Forum (ARF), given that none are ideal platforms and each has its own strengths and limitations. Criticism is often levelled at these forums for numerous reasons including local politics and different priorities.[36]

The Seoul Defense Dialogue has previously discussed the military's role in cyberspace. A working group was established to promote pragmatic dialogue and enhance common understanding in order to assist in establishing structures for cooperation. In early 2015, Singapore Defence Minister Ng Eng Hen reiterated these recommendations and urged enhanced collaboration through multilateral platforms like the Shangri-La Dialogue.[37] While cybersecurity was not on the main agenda of the 2015 Shangri-La Dialogue, it has been within the agenda on prior occasions. There would be utility in continuing these discussions from previous Shangri-La Dialogues at the 2016 meeting. Even where previous public discussions on cybersecurity may not have been highly extensive at the dialogue, this is more likely a reflection of the forum's nature and the sensitivity of the subject. The closed-door side-meetings may still present another good opportunity to discuss steps ahead.

### *Parallel International and Regional-level CBMs*

While there are varying national interests and regional particularities to consider, it would be best if initiatives avoid duplication. They should complement as much as possible the consensus that has been reached at either international level or good practices in other regional platforms. This includes the work undertaken over recent years by the UN GGE on cyber norms, CBMs and capacity building as well as the ARF. India was on the 2013/2014 GGE and is a member of the ARF. Having the defence community engage in military-to-military dialogues and other practical measures could further complement the aims of building transparency and confidence to improve international stability through international political agreement.[38]

Caveats that arise however include the following: 1) There may sometimes be scepticism from members of the defence community about the utility of the norms and CBM process (although defence diplomacy has an important role in ensuring international stability); 2) It can be easier to create such mechanisms with like-minded countries or communities as opposed to potential adversaries;[39] 3) There must be the requisite political willingness to build transparency and confidence in the field; and 4) These processes can be slow and do not always keep up with the speed of developments in the field.

## *Patchwork of Regional Institutional Mechanisms: The "spaghetti bowl"*

In terms of regional institutional mechanisms and the regional security architecture, there is a patchwork of different forums, often dubbed a "spaghetti bowl". The ASEAN is central in this regional architecture that includes the ARF, ASEAN+3, East Asia Summit, and both the ASEAN Defence Ministers' Meeting Process (ADMM) and ADMM-Plus. At the time of writing, there does not always seem to be extensive coordination between the dialogues at the ARF and ADMM/ ADMM-Plus. This lack of coordination between the groupings is often visible in other fields too and the ASEAN post-2015 agenda vision aims to address this gap by increasing cooperation between the three community pillars, namely socio-cultural, political-security and economic. The vision further aims to enhance information sharing among ASEAN Plus One, ASEAN Plus Three, the East Asia Summit, the ARF, the ADMM, and ADMM-Plus. The nature of "cyber" means that questions relating to cybersecurity tend to have implications across many different policy portfolios which can be a challenge for cohesive policy development. In this case, these ambitions of the post-2015 agenda should hopefully assist in the development of more comprehensive cyber policies which already fall under each of the three community pillars. It will hopefully also reduce unnecessary duplication across the regional mechanisms, most particularly (and for the purposes of this article) between the ARF and ADMM/ADMM-Plus. Some recommendations in the past include, for instance, mutual participation in each forum or that the ASEAN Secretariat share the information with updates.

In these discussions related to cybersecurity policies for the region, there has sometimes been a tendency to make reference to Asia without clearly delineating what this is understood to mean. Consequently, some analyses on cybersecurity may sometimes be at cross-purposes when discussing the "region" given that it is extremely large and highly diverse. Institutional structures like ASEAN, ARF, ADMM-Plus and Asia-Pacific Economic Cooperation (APEC), among others, vary significantly in terms of their membership and how they operate. What is happening at the ARF, for instance, is not necessarily reflective of the nature of discussions at ASEAN level even though they might sometimes be spoken of as one and the same. Policy analysis and negotiations on cybersecurity policies might need some refining to take these nuances into account so as to achieve better outcomes. This similarly applies to the use of cyber terminology itself since it is not unusual that the understanding of a specific term might differ significantly between individuals or governments. This is not a minor point since there is still a need to understand how another government might view and utilise these terms. Building such common understanding over the near term helps facilitate policy development. Better defining and agreeing of the meaning of terms among stakeholders at national level (where this has not already been agreed) can especially assist regional and international level discussions.

The well-known default ambiguity in this region, mentioned earlier in the article, also needs to be reduced, particularly since cyber analysts consider it to be critical so as to set the foundation for further dialogue and cooperation on cyber issues.[40] In India's case, the government is clearly aware of these issues, yet recent reports assert that sometimes there can still be ambiguity on policy issues.[41] Implementation can also be a problem – the country has issued several policies and plans but their stage of implementation is sometimes uncertain. For example, the status of India's cyber command remains unclear.[42]

The "ARF Work Plan on Security of and in the Use of ICTs" was adopted in September 2015, following a call by the foreign ministers in 2012 to ensure security for ICT. Several workshops have been held in the meantime with an emphasis on CBMs and capacity building measures. Most recently, China and Malaysia co-hosted the ARF Workshop on Cyber Security Capacity Building, and a joint seminar was hosted by the U.S. Department of State and Cyber Security Agency of Singapore to discuss operationalising cyber CBMs. Such workshops are in themselves a CBM in that they bring a network of contact points together on a regular basis to discuss the field and conduct table-top exercises.[43] Where members of the national delegations can include the defence forces as one of the stakeholders (that often includes law enforcement, foreign affairs, justice/home affairs among others), this may better assist such dialogues. In some cases over recent years, delegations have been encouraged to include a representative from the defence community. Even where they do not in fact participate, national level coordination that informs different stakeholders about developments may be beneficial. This could be as simple as regular calls between the coordinator or civilian agencies and defence representatives responsible for this portfolio, as is the case in some countries like the United States.

The ADMM and ADMM-Plus are key defence forums within ASEAN that focus deliberately on practical cooperation, and they can perhaps alleviate this type of ambiguity among the defence community. However, although there have been Defence Minister calls at ADMM meetings to work together to restrain cyber threats, there does not seem to be much progress on how this might actually be achieved. Nonetheless, while the ADMM is sometimes criticised for achieving few concrete outcomes, meeting to discuss the common challenges of cyber threats could be a CBM in itself. This would hopefully be then supportive of the ongoing work in the ARF and other regional forums. Military-to-military relations might even be easier to establish given common hierarchies, terminologies and structures that can transcend national differences.[44] In particular, there can be a shared focus on concrete implementation of policies that may sometimes rival parallel negotiations between civilian ministries. This, too, can sometimes be seen at the ADMM where there is already a strong network of government practitioners who regularly attend meetings together. As a community, they tend to like to focus on action points and implementation. Nevertheless, it would be best if these efforts complement and support the work of parallel forums.

The Philippines recently proposed to include these discussions in the ADMM-Plus process by establishing an expert working group.[45] Singapore's Defence Minister, Ng Eng Hen, also called for enhanced collaboration through multilateral platforms like ADMM-Plus, with a view to strengthening the regional security architecture.[46] The Network of ASEAN Defence and Security Institutions (NADI) in its 2013 cybersecurity workshop equally suggested that the ADMM-Plus meet to discuss these challenges.

The question that remains is whether there is sufficient political willingness across the region to pursue these suggestions. Given that over recent years there has been an increase in government awareness across the region as well as better clarity at national level on positions (which includes many national level developments like the establishment of cybersecurity agencies, coordinators, and national policies or frameworks), it would seem that there might now be more willingness to engage internationally.

These Track 1 meetings can provide a venue for exchanging information on institutional structures and good crisis management practices thus enhancing communication and common understanding. By way of example, exchanging information on best practices in how to attract, train and retain experts given the much-discussed common shortage of cybersecurity experts from both technical and policy backgrounds. This is a problem that faces India too.[47] Discussions could therefore even consider how to best approach interdisciplinary education or the joint training of military and civilians to enhance mutual understanding and establish networks of trust.[48]

In many countries, the private sector supports the military with capacity, products or expertise and so exchange of best practices or lessons learnt in retaining experts could be considered not only between the public and private sectors but also between international partners.[49] The UN GGE 2015 report specifies in fact that while states have a primary responsibility to maintain a secure and peaceful ICT environment, international cooperation would benefit from the appropriate participation of the private sector, academia, and civil society. Private sector involvement is especially important in this field in order to ensure that discussions are properly informed. For instance, not only are critical infrastructures often owned or operated by the private sector, but since advanced technologies also often come from it, this means that military may sometimes feel challenged by how closely it might need to collaborate. This issue has been flagged in India where the cyber field is described as presenting different challenges to other theatres of war since the security services need to work with the private sector.[50]

Member states of ADMM and ADMM-Plus could explore how to progress further with traditional military CBMs like hotlines; regular exchanges of defence officials; military-to-military engagements and dialogue; information sharing (on institutional structures for example); joint exercises; official military-to-military contact points; and crisis communication procedures. Such exchange of information and good practices helps build more trust.[51] It would be best if such military CBMs for cyberspace were highly tailored in order to be effective. Strategies can also have a strong declaratory function which means that this might present an opportunity to reduce the risk of conflict.[52] In fact, while several countries have in the past criticised the development of national cybersecurity strategies or military cyber doctrine and capabilities as destabilising (an arms race), this is not necessarily a threat to international peace and security.[53] Focusing on common challenges like the terrorism threat is another avenue for enhanced collaboration.

Existing hotlines could be expanded to include cyber issues and so present a further opportunity for enhancing stability (although such hotlines can naturally have their own limitations). For example, if the ADMM maritime hotline that was proposed by the Brunei chairmanship were to come to fruition, this could possibly be extended to include cyber matters, as could the hotline between India and Pakistan.

Formal Track 1.5/Track 2 and informal initiatives like workshops, seminars or roundtables can further facilitate this type of progress. For instance, while NADI met in 2013, it could possibly host a second workshop since the landscape has changed in the interim. NADI is a Track 2 forum that complements the ADMM and provides recommendations from defence officials and analysts who discuss security matters that might be considered too sensitive at official Track 1 meetings. In addition, the Council for Security Cooperation in the Asia Pacific (CSCAP) and ASEAN-Institutes of Strategic and International Studies (ASEAN-ISIS) could continue to look at these issues.

Informal closed-door roundtables like that co-organised by the S. Rajaratnam School of International Studies (RSIS)/Leiden University in late 2014 to explore how to improve domestic civil-military coordination in cybersecurity and international military cooperation in cyberspace can also be highly effective. The 2014 roundtable brought civil-military stakeholders (including civilian agencies, the defence forces and academia) from across Asia and Europe together in an informal setting. Such initiatives can facilitate discussion where issues might be deemed too sensitive for other forums. In which case such gatherings can be informative for participants still forming their own positions. They can also help to create an informal network of experts and foster debate.

Larger international conferences could also be held with a view to bringing experts from across the government stakeholder groups together as well as academia, research institutes, and the private sector. The global political and normative climate can be shaped by states either arranging or supporting international conferences or other processes as a way of promoting their preferred approach to employing ICTs.[54] GCCS-2015 for instance, as part of the London Process, included the defence community as one of the stakeholder groups participating in the conference. India is particularly good at engaging in such international forums and conferences.

Regional academic and research institutes could provide more extensive analyses on these issues. They would bring a deep understanding of the nuances of the region to discussions on cyber where the literature has sometimes been criticised as thin. In addition, the UN GGE 2015 report outlines the possible exchange of personnel between research and academic institutions. This is a recommendation that the government-affiliated defence think tanks in the region, like the Institute for Defence Studies and Analyses (IDSA), might consider.

### *Other Good Practices*

Regional good practices and expertise can inform each other (such as the Organisation for Security and Cooperation in Europe (OSCE), Organisation of American States (OAS), North Atlantic Treaty Organisation (NATO), and EU, among others) as well as inform evolving international discussions like those held by the UN GGE. The OSCE Decision on regional cyber CBMs of 2013 recommends, for instance, that states should consider discussions in other relevant international organisations working on the same issues. This helps to avoid duplication and to build on knowledge already available. The EU Cyber Defence Policy Framework of 2014 mentions the importance of international cooperation too and the need to ensure a dialogue with international partners, specifically NATO and other international organisations.[55] It confirms continued support for the development of CBMs in cybersecurity, to increase transparency and reduce the risk of misperceptions in state behaviour by promoting the ongoing establishment of international norms in this field.

The EU Cybersecurity Strategy of 2013 specifically highlights collaboration on improving cyber defence training and exercise opportunities for the military in the European and multinational context. There may also be good practices to consider based on how the European Defence Agency established its framework for "achieving more without losing sovereignty over assets and resources" with projects in cyber defence training and exercise ranges, among other options.[56] Such training sessions can provide another opportunity to increase communication and trust by bringing experts together. Given that at the ADMM-Plus, a "Plus" country like India, is expected to be able to work with the ADMM to build capacity so as to enhance regional security and promote capacity building in the fields of defence and security in the region, such training and capacity building cyber defence exercises may be well-suited to the ADMM-Plus process.

The Multinational Capability Development Campaign (MCDC) has received attention too as it is a neutral unclassified level platform with less political constraints. Both Japan and South Korea are observers and there may be opportunity for other states from the region to discuss good practices through this platform. Moreover, given that the Asia-Pacific region is particularly prone to natural disasters, there may be opportunity for law enforcement to work with armed forces post-crises (perhaps even post-conflict crises) on cyber capacity building where law enforcement has traditionally played a role in peacekeeping, capacity building and reconstruction efforts.[57] The expertise of regional and international law enforcement bodies that assist in building cyber capacity and capabilities might even be leveraged.[58] The ARF has spent quite some time looking at Humanitarian Assistance and Disaster Relief (HADR), including how to enhance military practical cooperation. Although HADR is primarily a task for the civilian agencies, there is often a need in the region for a response from military forces if they have unique attributes for critical response. This could possibly be extended to include cyber capacity building or post-crisis reconstruction.

## NOTES

1. Chair's Statement, Global Conference on Cyberspace (GCCS) 2015, No. 38, April 17, 2015.
2. Author's observations, Civil-Military Relations Panel, GCCS 2015, April 17, 2015.
3. Sergei Boeke and Caitríona Heinl, "Civil-Military Relations & International Military Cooperation in Cyberspace", Research project supported by the Netherlands Ministry of Defence, April 2015.
4. Cheryl Pellerin, "Cybercom Chief Details Strategic Priorities for 2016", DoD News, US Department of Defense, January 21, 2016 at http://www.defense.gov/News-Article-View/Article/643954/cybercom-chief-details-strategic-priorities-for-2016?source=GovDelivery.
5. Ibid.
6. Author's observations, ASEAN Regional Forum Workshop on Cyber Security Capacity Building, Hosted by the People's Republic of China and Malaysia, Beijing, July 29-30, 2015.
7. Author's observations, S. Rajaratnam School of International Studies (RSIS) - Leiden University CTC Roundtable on Civil-Military Relations in Cyberspace, organised with the support of the Netherlands Ministry of Defence, Singapore, November 18-19, 2014.
8. James Clapper, "Worldwide Threat Assessment", 2013.
9. UK Ministry of Defence, "Strategic Trends Programme, Regional Survey - South Asia out to 2040", October 2012.
10. International Institute for Strategic Studies (IISS), "Evolution of the Cyber Domain: The Implications for National and Global Security", November 2015.
11. Ibid.
12. Neil Robinson, "EU Cyber-Defence: A Work in Progress", European Union Institute for Security Studies, Brief Issue 10, March 2014, p. 2.

13. Australian Strategic Policy Institute (ASPI) and International Cyber Policy Centre (ICPC), *Cyber Maturity Report 2015*, February 2016.
14. Author's observations, No. 7.
15. The United Nations Institute for Disarmament Research (UNIDIR) Report, *The Cyber Index: International Security Trends and Realities 2013* further explains: "For example, most Western states believe that freedom of access to the cybersphere is a basic human right, which must be protected by law and regulations. Other states, in contrast, favour the concept of 'information security', and thus are seeking the right to limit the access of their citizens to the public cybersphere if the stability or survival of the regime is deemed to be at stake. Obviously, these two ideological approaches are in direct conflict, and complicate the effort to find multilateral solutions to the problems that face all states."
16. IISS, No. 10.
17. Ibid.
18. Ibid.
19. For further analysis on these points, see: Joseph Nye, *"Can China Be Deterred in Cyber Space?"*, *The Diplomat*, February 2016.
20. International Security Advisory Board (ISAB), *Report on A Framework for International Cyber Stability*, July 2, 2014.
21. Chair's Statement, No. 1.
22. Ibid.
23. Author's observations, No. 6.
24. ISAB, No. 20.
25. The importance of building trust has been emphasised on numerous occasions, including the GCCS discussions held on this topic, in roundtables on civil-military relations, and the Institute for Defence Studies and Analyses (IDSA) Asian Security Conference held in New Delhi in February 2016.
26. ISAB, No. 20.
27. Ibid.
28. Ibid.
29. ASPI ICPC, No. 13.
30. W. Roehrig and R. Smeaton, "Cyber Security and Cyber Defence in the European Union: Opportunities, Synergies and Challenges" European Defence Agency, 2013.
31. Author's observations, IDSA, 18th Asian Security Conference, New Delhi, February 2016.
32. Richard Youngs, "Keeping EU-Asia Reengagement on Track", Carnegie Europe, January 2015, p. 4.
33. Ibid.
34. W. Roehrig and R. Smeaton, No. 30.
35. Author's observations, No. 7.
36. IDSA, Asian Security Conference, New Delhi, February 2016.
37. Jermyn Chow, "Ng Eng Hen: Deeper Issues beyond the ISIS Threat", *Straits Times*, January 27, 2015.
38. Sergei Boeke and Caitríona Heinl, No. 3 .
39. Author's observations, No. 7.
40. ASPI ICPC, "Asia-Pacific Cyber Insights", Results of a multi-stakeholder roundtable held in Kuala Lumpur in February 2015.
41. ASPI ICPC, No. 13.
42. Ibid.

43. The ARF has been active in hosting several table-top and discussion exercises including the ARF Workshop on CBMs co-hosted by Australia and Malaysia in 2014.
44. Author's observations, No. 7.
45. The ADMM-Plus countries include the 10 ASEAN Member States and eight Plus countries, namely Australia, China, India, Japan, New Zealand, Republic of Korea, the Russian Federation, and the United States.
46. Jermyn Chow, No. 37.
47. Sergei Boeke and Caitríona Heinl, No. 3.
48. Ibid.
49. Ibid.
50. IDSA, No. 36.
51. Author's observations, No. 2.
52. Sergei Boeke and Caitríona Heinl, No. 3.
53. IISS, No. 10.
54. Ibid.
55. Council of the European Union, *EU Cyber Defence Policy Framework*, 15585/14, November 18, 2014, p. 2.
56. W. Roehrig and R. Smeaton, No. 30.
57. Sergei Boeke and Caitríona Heinl, No. 3.
58. Ibid.

# 21

# The Role of Military in Cyberspace: Case of Republic of China (Taiwan)

*Li-Chung Yuan*

## Strategic Value of Cyberspace

Since the advent of the digital era, the dramatic rise in Internet usage has presented a major challenge which states must unavoidably address. States have built up their information infrastructure to cope with both domestic and international affairs, including civil, government and military sectors. The continuing growth of network systems, devices and platforms means that "cyberspace is embedded into an increasing number of capabilities".[1] However, as a Chinese idiom states: "While water can carry a boat, it may also capsize it." That is to say, while a state may benefit enormously from heavy reliance on cyberspace, in doing so, it becomes more vulnerable to being capsized by attacks, crimes and terrorist acts generated through this medium. The task of securing cyberspace thus rates as one of the most serious challenges for national security, public safety and the economy in modern times.

Furthermore, the rapid growth of Internet has transformed cyberspace from a social platform into a potential battlefield in which states can contest for power. Historically, state power has been directly associated with its territory, which could be expanded to a limited extent through competition.[2] The intangible information network platform of cyberspace is now presenting a new potential military arena which challenges traditional, geographic doctrines of military strategy. As such, research into the effects of cyberspace on existing military doctrine is becoming essential. The pervasiveness and importance of cyberspace lend it a global strategic value. It is likely that this value also makes cyberspace a contested space for states.

## How Cyberspace becomes a Potential Virtual Battleground

The actors in the potential battleground of cyberspace are not necessarily only states, but also non-state actors such as non-government organisations (NGOs) and even private companies. Unlike the modern state system, in which a war is usually waged by states, it is possible that an attack in cyberspace may be conducted by just one individual person. If traditionally a war could be identified based on whether a state's territoriality is intruded, the same proposition can be reflected onto cyberspace. In other words, war can be defined based on encroachment onto virtual territory in cyberspace.

The digital age is manifested not only in civilian circumstances, but also in the military field. Research indicates that "the sophisticated information technologies on which the best armed forces, and all modern societies rest, have now become an attractive target for potential adversaries".[3] Consequently, it is likely that cyberspace is developing into a potential battleground in the digital age. The Internet, rapidly expanding in scope, constitutes a new theatre for belligerent politics, further transforming cyberspace from a social and political platform into a potential battleground. This will affect national security as traditional borders cannot protect a state's territory from cyberthreats, and government bodies no longer dominate communication systems and relevant techniques in cyberspace.[4] So why exactly is cyberspace considered a potential battleground? And what differences are there with traditional battlegrounds in terms of warfare? To begin with, it is contestable whether military strategy based on geographical battlegrounds can fit the battleground of cyberspace. After the invention of Internet, states began to build up their information infrastructure. States have now started to compete with one another for dominance in this potential domain, and it can be expected that a parallel virtual landscape will be drawn in cyberspace which will represent state territoriality differently to that of the physical world. Traditional thinking about the existing doctrine of military strategy will consequently be challenged based on this new proposition. In addition, state policy regarding warfare and security will need to be re-formulated, to respond to new military strategies which may well be generated to match this new battleground.

## Three Sectors Sharing Cyberspace: Civil Society, Government and Military

Historically, civil networks were derived from the Advanced Research Projects Agency Network (ARPANET), which was a research network for military purposes.

Cyberspace, which technically comprises transport and physical layers and was also derived from the ARPANET, potentially binds together the military, government and societal sectors, as they are all constructed on the same virtual information platform. This has caused the boundary between the battlefront and the rear to disappear. In addition, though attacks in cyberspace are waged by actors in a state's territory, these actors are not necessarily acting for that particular regime or state. In sum, cyberspace manifests itself both in the military field and in civilian circumstances, as both share the same information platform. This feature of cyberspace means that as a battlefield, it falls into the category of irregular warfare, as Lawrence Freedman stresses. Freedman also points out that the battlefield is transformed from one which is separated from civil society to one which is combined with civil society.[5] In terms of cyberattacks, the division between the battlefront and the rear is no longer clear; as such, existing military doctrines based on this division must be challenged.

Though the design of information infrastructure may vary across states, the so-called Open System Interconnection (OSI) model is a world-wide standard which defines a networking framework for implementing protocols in seven layers, namely, from top to bottom, the Application, Presentation, Session, Transport, Network, Data Link and Physical Layers. In addition, the Transmission Control Protocol/Internet Protocol (TCP/IP) is the radical protocol for the Transport and Network Layers of the OSI in all sectors. Therefore, for instance, both the US Network-Centric Warfare and Global Information Grid, which are the most advanced information architecture for military operations in the world, operate on a so-called 'Convergence Layer', which is established upon the same Transport Layer and Network Layer shared by civil infrastructure, in order to connect various networks and telecommunications. In this way, dominant capability in this battlefield can be increased whilst the ability of securing the civil information infrastructure can be reinforced. The relations among civil society, government and military are shown in Figure 1.

**Figure 1: The Relations between Civil Society, Government and Military**

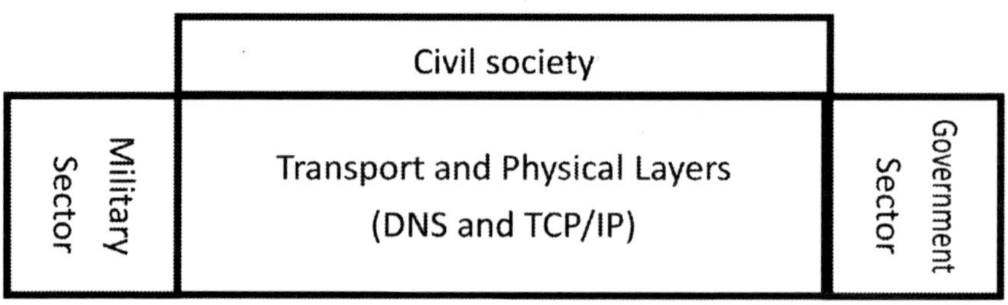

In Figure 1, the bold lines between different areas can be assumed as servers, such as proxy servers and web servers with firewall or encryption/decryption, acting as an inspection mechanism in order to monitor communication and information exchange between the three sectors. However, all of them share the same infrastructure: the Transport and Physical Layers constructed by the Domain Name System (DNS) and TCP/IP.

## Asymmetry and Vulnerabilities of Cyberspace

Cyberspace has become indispensable for modern human lifestyles across the world, but the features of cyberspace mean conditions of asymmetry and vulnerability are created. On traditional battlegrounds, the scale of warfare, identified by the size of the battlefield, manpower involved and causal destruction, may be limited geographically, whereas the exponential growth of cyberspace presents a very different scenario. Due to the rise in Internet usage, people are already becoming accustomed to deal with their affairs in virtual cyberspace instead of in a physical environment. The range of functions relying on cyberspace is expanding day by day. For instance, the fundamental sectors of state such as agriculture, electricity and water supply, defence, government administration, information and telecommunication, public transportation, banking and finance and mail systems and goods supply chains all rely on cyberspace.

However, state reliance on cyberspace offers an attractive target for adversaries, as many attacks, crimes and terrorist acts can be generated through the medium of cyberspace. Due to the reach of information technology in the digital age, any individual may be able to conduct cyberattacks or simply spam junk emails to cripple computer servers. This makes the virtual field of cyberspace much more vulnerable than a traditional territory. Cyberattacks can leap over state border inspections to create damage to information infrastructure. Put simply, cyberspace offers a feature of asymmetry beneficial for cyberattackers. As this asymmetry and vulnerability are caused by the technical features of cyberspace, it is essential to understand the physical construction of cyberspace in order to prevent potential threats.

## A Non-State Space: Anonymity of Actors

As recent research indicates, the speed and anonymity of cyberattacks make the actors indiscernible, whether they are terrorists, criminals, nation states or even just individuals carrying out random malicious attacks.[6] In traditional warfare, an enemy can be recognised by its visible hostile military action, and the target of attack is usually a regime or a state. According to Schmitt's definition, in human history enemies could generally be identified as unknown people or strangers, and existentially deemed to be something different.[7] Moreover, in a conventional war, the enemy becomes the public enemy, because everything that has a relationship to a collectivity of humans, particularly to a whole nation, becomes public by virtue of such a relationship. This perspective posits that hostility and the public enemy are crucial prerequisites in the concept of warfare. However, in cyberspace, for any individual or group, the existence of the enemy is indiscernible. Thus, identifying hostility or the enemy is a crucial factor when attempting to secure cyberspace. In addition to the existence of an enemy, recognising a conflict of armed forces also makes it possible to discern what can be classed as a war. However, the actors who produce threats or even launch attacks in the battleground of cyberspace do not necessarily identify with a regime or a state. The features of cyberspace increase the ambiguity of hostility in such a battlefield.

## The Role of Military in Cyberspace

Due to the features of cyberspace, discussed above, it is arguable that the role of military in cyberspace, unlike in the geographical world, has become ambiguous. In other words, the actors in cyberspace are not only states but also non-state actors. It is thus very difficult to identify an attack waged in cyberspace, in order to carry out defensive count-attacks. Besides, it is also just as difficult to define it according to the international legal environment. In terms of an attack in cyberspace, a problem thus arises as there could be no immediate physical casualties in a cyberattack. Existing definitions, based on the features of conventional war, may not easily match the battlefield of cyberspace in the digital age. The world is therefore faced with a "legal vacuum":[8] national or international laws cannot impose justice or prevent transnational hostile actions in cyberspace; a further problematic issue is how it will be possible to define a true war based on existing law in order to justify legal defensive action in cyberspace.

In terms of an organisation in charge of the cyber domain, taking the US as an example, there are two kinds of argument on who should lead the cybersecurity efforts for the country. Supporters of the Department of Homeland Security (DHS) argue that a credible civilian government cybersecurity capability cannot originate from a military intelligence organisation. On the other hand, advocates of a greater role played by the Department of Defense (DoD) argue that it is the only organisation with the capability and monitoring infrastructure to protect computer networks. An interesting question thus arises: What role does DoD play in cyberspace? Due to DoD's sufficient experience and manpower, it has been asked to take on the leadership responsibility for cybersecurity.[9] The DoD has been forward leaning in establishing policy and making organisational changes for cybersecurity. Therefore, the DHS signed a memorandum of understanding with DoD that allows the National Security Agency to support DHS cybersecurity efforts and established a personnel exchange between the two agencies.[10]

However, there are numerous drawbacks of making DoD in-charge of the country's overall cybersecurity. The legal restrictions of conducting domestic activities by the military are the primary concern. The DoD lacks regulatory authority and law enforcement powers over domestic affairs. Furthermore, DoD has less experience with regulatory matters and procedures. To this end, the design of the military to lead cybersecurity for civilian systems would be politically difficult.[11] On the other hand, placing DHS to play a leading role in terms of cybersecurity efforts is the most likely alternative because it has the basic regulatory functions and significant experience in cybersecurity issues. Similar situations arise in other countries as well, including Taiwan.

## Case of the Republic of China (Taiwan)

In order to provide a comprehensive picture of the relations between military and cyberspace, let us consider the case of the Republic of China (ROC; hereafter referred to as Taiwan).

### *External Cyberthreats to Taiwan*

Owing to its developed IT industries and infrastructure, hackers are likely to choose Taiwan as a potential target to launch cyberattacks. Especially, the political and military stalemate between Taiwan and China has caused Taiwan to become one of the likely targets. The cyberattacks can potentially jeopardise Taiwan's critical information infrastructure and compromise the government's daily operations, which makes cyberwarfare an ideal choice, to intimidate Taiwan with non-conventional forces. According to internal statistics, Taiwan's National Security Bureau (NSB) website encountered 3.4 million attempts of intrusions in 2012, the majority for reconnaissance purposes. Among them, more than 70,000 were malicious intrusions, an average of 209 daily attacks on the NSB website.[12] However, within the following two years, this figure doubled in 2014 as a total of 7.2 million hacking attempts were detected, among them 238,764 were malicious.[13]

There are at least two bureau-level People's Liberation Army (PLA) units conducting cyberespionage on Taiwan.[14] The PLA's 61398 Unit located in Shanghai is widely believed to be the central element.[15] In the case of China's launch of military operation against Taiwan, the PLA is likely to paralyse the military's cyberspace and communication capability. Through cyber operations, once conflict breaks out, China could interfere or sabotage Taiwan's military, government and important civilian targets. Nevertheless, the borderless characteristic of cyberspace also allows other countries to carry out cyberattacks against Taiwan. The targets of these threats are scattered at all levels in the society. According to some domestic experts, the cyberwarfare capabilities of Taiwan's military are insufficient, making the information and communication systems vulnerable.[16] Similar opinions were reflected by American experts.[17] It is imperative for Taiwan to prioritise development of asymmetric capabilities in the cyber domain to deter the increasing coercive cyberthreats.

### *Taiwan's Internal Mechanism to Respond to Cyberattacks*

Taiwan has been endeavouring to enhance its government's cyber defence and protection capability. To implement information and communication security plans in order to ensure cybersecurity, the National Strategy for Cybersecurity Development Programme (2013-2016) was proposed in 2013. Another measure is the creation of a taskforce for tackling the issues of information and communication security policy, notification and response mechanisms, review and consultation on major programmes, and coordination and supervision of inter-ministerial affairs concerning information and communication security.

From the central government's perspective, the National Information & Communication Security Taskforce (NICST) under Taiwan's Executive, Yuan (the Cabinet) was established in 2000 in charge of the inter-ministerial coordination and to develop a public and private partnership for defending Taiwan's information and communication security. It must be emphasised that pursuing cyber offence or espionage capabilities is not the task of NICST, rather, it is designed for a purely defensive capability with the aim to protect Taiwan's information infrastructure. The core missions of the NICST are: cyberspace protection, cybercrime investigation and critical infrastructure protection.[18] Like other countries, Taiwan has two Computer Emergency Response Teams (CERTs) in place, and collectively they cover cybersecurity incidents across the computer network. Government responsibility for network information and security is fulfilled by the Ministry of National Defense (MND). Structurally speaking, the Information and Electronic Warfare Command was created in 2004 directly under the MND, making it a legitimate military cyber force.

### *Role of Taiwan's Military in Cybersecurity*

According to NICST, tasks are assigned to the MND to ensure the safety of military information and communication systems; that is to say, the MND is only in charge of its own IT system and does not play a leading role in the cyber domain of the overall government. However, the precautious measures of the MND have so far blocked the potential threats that may damage the computer and information system of Taiwan's armed forces. These measures have effectively protected all valuable assets and intelligence from adversary forces.

There are three major institutional actors in Taiwan's cyber-defence infrastructure: the NSB, MND, and Criminal Investigation Bureau. Guided by the NSB and NICST, the MND has established a joint operations data security and protection mechanism, and taken advantage of various exercises and training to materialise the concept of Critical Information Infrastructure Protection (CIIP) in order to improve and enhance the protection capability of information security.[19] For example, the Office of the Deputy Chief of the General Staff for Communication, Electronics and Information (J-6 of MND) plans and conducts annual cyber exercises.

With regard to the bona fide cyber force, the Information and Electronic Warfare Command of the MND currently has three units with 3,000 personnel to carry out cyber protection, Internet intelligence gathering and electronic warfare missions. The fourth cyberwarfare unit is to be developed as part of the government's overall efforts to beef up its cyber capability.[20] Taiwan's Democratic Progressive Party (DPP) stressed in its Defense Policy Blue Paper, that the role of military in cyberspace will dominate the military development in cyberwarfare capability, in that a new service will be built to protect the cyberspace that is different from the physical territory protected by other Services. In details, the following priorities are addressed in force development:

> Immediately reinforce the armed forces' information warfare capabilities, with the establishment of a world-class cyberwarfare capability as the ultimate objective, while also gradually expanding the comprehensive defense of the national digital territory, to make Taiwan into not just a world leader in information technology, but also one in information security. Simultaneously, Taiwan should take the initiative to share its experience and skills in this area to form a system for cybersecurity cooperation among democratic nations. An essential first task toward accomplishing this goal would be combining all competences relating to information, communications, and electronics currently present across all military units, to establish an independent Fourth Service alongside the Army, Navy, and Air Force.[21]

The fourth service, or the so-called Taiwan's Cyber Command, according to DPP's Defense Policy Blue Paper, will be established in 2019 with 6,000 personnel to be fully filled by 2024, military and civilian personnel each accounting for half. This plan will change Taiwan's strategic cyberwarfare inclination from reactive to proactive. Simply speaking, the role of the military cyber force will be enhanced. Numerous resources and personnel, as mentioned above, are expected to put into this service. Going forward, Taiwanese military's cyberwarfare capability may be promoted from protective and reactive to be proactive. In light of Taiwan's defence goal, "Effective Deterrence", the enhanced cyberwarfare capabilities can be useful in deterring potential threats from adversaries.

### *Public-Private Partnerships (PPPs)*

Cybersecurity is a national priority, and public-private partnerships (PPPs) are considered to be an ideal tool in securing cyberspace. The past decade has seen the creation of many PPPs seeking to answer the call for greater cybersecurity.[22] To date, PPPs have successfully built trust in small pockets and reduced barriers to cross-sector sharing. This has been possible, partly owing to the improved intra-governmental coordination.

While so far there are no defined PPPs in Taiwan for cybersecurity, the Taiwan CERT/Coordination Centre (TWCERT/CC) plays an important role by liaising with the private sector in order to implement and coordinate incident response procedures. The NSB recommends that it is essential to integrate and cooperate with civilian telecommunication carriers to plan and execute pre-emptive protection measures in order to build multi-layered defence mechanism against cyberattacks.[23] Private sector experts participate as special committee members to contribute and provide their expertise to both NICST and NSB. The MND coordinates with NICST and NSB to pilot a public-private sector relationship intended to demonstrate the feasibility and benefits of voluntarily increasing the sharing of information about malicious or unauthorised cyber activity and protective cybersecurity measures. By adopting the common IT industry standards for certification, the MND is in line with the standards adopted in the private sector and other government agencies. Sharing these processes across the rest of the government can ensure and enhance the capability of the military's cyber force.

## Conclusion

While Taiwan has made important progress in its cybersecurity, it is also facing a new conundrum in this domain. A major challenge is the integration and cooperation among countries and different systems established by the major stakeholders in its cybersecurity infrastructure. Cooperation with other parties is needed and areas of cooperation could include the protection of critical infrastructure such as telecommunication networks, financial systems and electricity supplies, and establishing international rules on cyber issues. For the military, careful prioritisation on defence resources is essential to allow for the greater growth of military's cyber capability.

Vulnerabilities in a country's cybersecurity infrastructure could potentially affect the viability of a collective cyber defence. Even though a report on "Taiwan Strait Posture Status" described that Taiwan was leading the world in the areas of anti-virus techniques,[24] the military balance over the Taiwan Strait has been tilting to China's favour, which is a serious concern for Taiwan and also for the world. In order to gain superiority, both sides on the Taiwan Strait have tremendously invested in the development of cyberwarfare capabilities. Taiwan's military did not play the most prominent role in the nation's overall cybersecurity efforts; rather, the military's cyber efforts are integrated into the overall cybersecurity structure of the country. In light of the recent pronouncements by the MND's defense report and by the new ruling party, DPP's Defense Policy Blue Paper, it is clear that the development of both offensive and defensive capabilities in the cyber domain will become a major goal in the modernisation of Taiwan's military. Taiwan must integrate and leverage all of its intellectual, financial and scientific resources in order to successfully defend itself in the cyber domain.

## NOTES

1. US Department of Defense, *The DoD Cyber Strategy2015*, Office of the Secretary of Defense, Washington D.C., April 17, 2015.
2. T.W. Luke, "Simulated Sovereignty, Telematic Territoriality: The Political Economy of Cyberspace", The Second Theory, Culture & Society Conference, August 10-14, 1995.
3. Daniel Moran, "Geography and Strategy", in J. Baylis, J. J. Wirtz and C. S. Gray (eds.), *Strategy in the Contemporary World: An Introduction to Strategic Studies*, 3rd ed., Oxford University Press, New York, 2010.
4. John Schwartz, "A Nation Challenged: The Computer Networks; Cyberspace Seen as Potential Battleground." *The New York Times*, November 23, 2001.
5. Lawrence Freeman, "Does Strategic Studies Have Future?, in J. Baylis, J. J. Wirtz and C. S. Gray (eds.), *Strategy in the Contemporary World: An Introduction to Strategic Studies*, 3rd ed., Oxford University Press, New York, 2010.
6. D. Kelly, R. Raines, R. Baldwin, B. Mullins and M. Grimaila, "A Framework for Classifying Anonymous Networks in Cyberspace", Third International Conference on Information Warfare and Security, Omaha, April 24-25, 2008.
7. C. Schmitt, *The Concept of the Political*, University of Chicago Press, Chicago, 1996.
8. Since the onset of cyberterrorism, the question of who is responsible for the content of policy in cyberspace has often been debated.
9. Paul Rosenzweig, "10 Conservative Principles for Cybersecurity Policy", Backgrounder No. 2513, The Heritage Foundation, Washington DC, 2011, at www.heritage.org/research/reports/2011/01/10-conservative-principles-for-cybersecurity-policy.
10. "Memorandum of Agreement between the Department of Homeland Security and the Department of Defense Regarding Cybersecurity", September 2010, at www.dhs.gov/xlibrary/assets/20101013-doddhs-cyber-moa.pdf.
11. Kevin P. Newmeyer, "Who Should Lead U.S. Cybersecurity Efforts?", *Prism* 3 (2), NDU Press, Washington D.C., March 2012, p. 123.
12. National Security Bureau (NSB), "How Does Roc Deal with Attacks from Cyberforce and Hackers and How to Strengthen the Mechanism for Information Security", *National Security Report 2013*, Legislature Yuan, April 29, 2013.
13. Jason Pan, "NSB Warns of Rising China Cyberattacks", November 21, 2014, at http://www.taipeitimes.com/News/front/archives/2014/11/21/2003604932.
14. Mark A. Stokes, Jenny Lin and L. C. Russell Hsiao, "The Chinese People's Liberation Army Signals Intelligence and Cyber Reconnaissance Infrastructure", Project 2049 Institute, 2011.
15. See http://www.bbc.com/zhongwen/trad/world/2013/02/130219_china_hacking.shtml.
16. New Frontier Foundation Defense Policy Advisory Committee, "Taiwan's Military Capacities in 2025", Defense Policy Blue Paper No. 9, May 2015, p. 31.

17. Mark Stokes, "Revolutionizing Taiwan's Security, Leveraging C4ISR for Traditional and Non-Traditional Challenges", Project 2049 Institute, 2013.
18. National Information & Communication Security Taskforce (NICST), at http://www.nicst.ey.gov.tw/en/Default.aspx.
19. *ROC National Defense Report*, 2015, p. 133.
20. Russell Hsiao, "Critical Node: Taiwan's Cyber Defense and Chinese Cyber Espionage", *China Brief*, XIII (24), December 5, 2013.
21. New Frontier Foundation Defense Policy Advisory Committee, No. 16, p. 17.
22. The White House, "Cyberspace Policy Review: Assuring a Trusted and Resilient Information and Communications Infrastructure", U.S. Government Printing Office, Washington DC, 2009.
23. NSB, No. 12.
24. Daniel Ventre, *Information Warfare*, Wiley-ISTE, London and Hoboken, 2009, p. 80.

# 22

# CYBERSECURITY POLICY IN JAPAN

*Yasuaki Hashimoto*

## Introduction

Japan is now one of the major powers in cyberspace. At present, the number of Internet users is over 100 million and accounts for 82.8 per cent of all citizens.[1] Cybersecurity, therefore, becomes one of the most important issues to consider. This chapter intends to show briefly the present cybersecurity policy structure of Japan.

## Increasing Cyberattacks in Japan

The number of cyberattacks was 3 billion in 2005, 5.8 billion in 2010, 7.8 billion in 2012, 12.8 billion in 2013, 25.6 billion in 2014 and reached 54.5 billion in 2015. From 2013 to 2015 the cyberattacks grew twice every year.[2] In 2015, Japan had over 1,700 cyberattacks every second. As Japan is going to host the XXXII Olympiad and the Tokyo 2020 Paralympic Games four years from now, the threat of cyberattacks is particularly pertinent. Cyberattack is a clear and present danger that is on the rise.

## Cyberattacks against Japan

In Japan, there were a number of Distributed Denial of Service (DDOS) attacks in the past. High-tech companies, the Diet and Japanese diplomatic offices are prone to cyberattacks. Japan faced high-level Advanced Persistent Threat (ATP) attacks as well. One real example of an ATP attack was the industrial espionage attempt at the Japanese aerospace defence industry in 2011. The cyberattack unfolded after an email for a business meeting was sent by the Society of Japanese Aerospace Companies (SJAC) to the member companies. Soon after, a reproduced email with malware that showed a new schedule, was sent again to the same addresses. Some PCs of the member companies were infected, and it is believed that sensitive information was stolen using backdoors.[3]

## National Security Strategy and Cybersecurity

Under such circumstances, Japan now recognises cyberattacks as a serious problem to the Japanese national security. The National Security Strategy,[4] published on December 17, 2013, emphasises the stability of cyberspace. Moreover, "the Strategy, as fundamental policies pertaining to national security, presents guidelines for policies in areas related to national security, including sea, outer space, cyberspace, official development assistance (ODA) and energy".[5] This Strategy shows that the cyberspace, ocean and outer space are the so-called Global Commons ("Cyberspace, a global domain comprised of information systems, telecommunications networks and others, provides a foundation for social, economic, military and other activities. Meanwhile, risks of cyber-attacks with the intent of steal classified information, disrupt critical infrastructure and obstruct military systems, are becoming more serious"), and says that "protecting cyberspace ... is vital to secure national security".[6]

## Basic Act on Cybersecurity

Japan adopted the Basic Act on Cybersecurity on November 12, 2014.[7] The Basic Act was proposed by the Liberal Democratic Party, New Komeito, Democratic Party and others on a bipartisan basis, as the importance of cybersecurity is understood across political parties. (Some important basic laws such as the Space Basic Act and Ocean Basic Act were also adopted in Japan almost at the same time as the Basic Act.) The main points of this Basic Act on Cybersecurity are as follows:

### *(a) The maintaining of national cybersecurity should be done by national leadership*

In this respect, Articles 10 and 11 of the Basic Act provide the following clauses:

- Article 10: The Government shall be required to take necessary measures for the implementation of cybersecurity policies including legislative, financial, or taxation measures.
- Article 11: In providing cybersecurity policies, the Government shall make an effect to develop administrative organizations and to improve administrative management.

### *(b) Cybersecurity is maintained while respecting the freedom of information*

According to Articles 1 and 3, the citizens' rights including freedom of information shall be respected while cybersecurity is maintained.

- Article 1: It is an urgent issue to ensure the free flow of information and protect cybersecurity simultaneously, the purpose of this Act is to comprehensively and effectively promote cybersecurity policy.
- Article 3 (6): The promotion on the cybersecurity policy shall be required to be carried out with intent to be careful not to wrongfully impinge upon citizens' rights.

### *(c) The national and local governments set a policy on cybersecurity*

The central Government as well as local governments shall prepare the adequate regulations for cybersecurity.

- Article 4: In accordance with the basic principles prescribed under the preceding article (hereinafter referred to as the "basic principles"), the Government shall bear the responsibility to formulate and implement comprehensively cybersecurity policies.
- Article 5: In accordance with the basic principles, local governments shall bear the responsibility to formulate and implement proactive cybersecurity policies in consideration of the appropriate division of roles with the Government.

### *(d) All infrastructure providers and the cyber-related companies shall try to cooperate with the national/local cybersecurity policy (Best effort obligation).*

The best effort clauses for Critical Information Infrastructure (CII) operators as well as cyberspace-related business entities are included in the Basic Act.

- Article 6: In accordance with the basic principles and for the purpose of stable and appropriate provision of their services, CII operators shall make an effort to: deepen their awareness and understanding of the critical value of cybersecurity; assure cybersecurity voluntarily and proactively; and cooperate with the measures on cybersecurity taken by the Government or local governments.
- Article 7: In accordance with the basic principles, cyberspace-related business entities (here and hereinafter, referring to those engaged in business regarding the maintenance of the Internet and other advanced information and telecommunications networks, the utilisation of information and telecommunications technologies, or involved in business related to cybersecurity) and other business entities shall make an effort to assure cybersecurity voluntarily and proactively in their businesses and to cooperate

with the measures on cybersecurity taken by the Government or local governments.

*(e) The educational/research organisations shall try for personnel training to contribute to cybersecurity (Best effort obligation). For cybersecurity at the national level, an adequate number of engineers is necessary.[8] Article 8 provides the best effort clauses for educational and research organisations:*

- Article 8: In accordance with the basic principles, universities and other educational and research organisations shall make an effort to assure cybersecurity voluntarily and proactively, develop human resources specialized for cybersecurity, disseminate research and the results of research on cybersecurity, and cooperate with measures taken by the Government or local governments.

## National Cybersecurity Strategy

The latest Cybersecurity Strategy, adopted on September 4, 2015, establishes policies for stabilising cyberspace. The main points of this strategy are as follows:

### *(a) Cyber threats as a critical challenge to national security*

The Japanese Government recognises that securing cyberspace is necessary. National Cybersecurity Strategy outlines:

> The increasing dependency of socio-economic activities on cyberspace and the evolution of organized and highly sophisticated methods, or modus operandi, of cyberattacks that might be state-sponsored have caused grave damages and exerted negative impacts on the people's daily lives and socio-economic activities, and consequently, threats against national security have become more serious year after year.[9]

### *(b) Creating secure – Internet of Things (IoT) systems for future*

The Japanese Government aims at creating the secure IoT systems capable of meeting market needs by 2020, and subsequently enhancing the international reputation of Japan's IoT systems.

> The Government will promote the idea of 'Security by Design', an approach to incorporate the assurance of security into the initial phase of the planning and design of the entire IoT systems including the existing systems connected to them. More specifically, as to IoT systems-related business, the Government will promote security measures for these systems in a cross-sectoral manner, based on the Security by Design approach, and will give its prioritised support to the growth of such new business.[10]

### *(c) Promoting information gathering on vulnerabilities and monitoring of cyberattacks.*

The Government will promote information gathering regarding vulnerabilities, e.g. software vulnerabilities; and the condition and enhancement of the systems to monitor the Internet and detect cyberattacks and other cyber events.

In order to protect the users of cyberspace from cyber risks which require urgent attention, such as a possible exploitation of a vulnerable device for being used as a springboard of a cyberattack, and to build a safe and beneficial Internet environment, the Government will elaborate necessary measures to prevent a damage possibly induced by malware infection, in addition to provide security alerts and tips for the user(s) of a compromised device.[11]

### *(d) Enhancing measures to advance cybercrime response and investigative capabilities using the public-private sector cooperation.*

The Government will promote the improvement of structural arrangements for: the enhancement of information gathering to obtain a better understanding of cyber threats; the advancement of cybercrime-related investigative capabilities; cybercrime control, international coordination, and more. In addition, since advanced technical knowledge and skills are essential to investigation, cybercrime control, and the prevention of damage and the spread of damage, the Government will promote the accumulation of technical know-how, technologies, and other skills necessary to this end, by improving the structural arrangement for digital forensics, such as the technological advancement for malware analysis and others as well as the sophistication of the systems to monitor the Internet and detect cyberattacks and other cyber events. Likewise, the government will promote human resources development and technological development soundly. Furthermore, for the purposes of cybercrime investigation and prevention, the Government will aim at the active use of knowledge and experiences of the private sector and the enhanced public-private partnerships including personnel exchange programmes between the public and private sectors.[12]

### *(e) Enhancing effective and prompt public-private information sharing to protect CII, such as information-communication, electricity and finance.*

Various kinds of social infrastructures have ensured people's living and economic activities, and a wide range of information systems has been used for the functions of these social infrastructures. In the circumstances, the public and private sectors must work together to protect CII, in particular, information and communications services, electric power supply services, and financial services, of which the functional failures or deterioration would risk enormous impacts to the people's living conditions and economic activities. Such task cannot be entirely designated to the Government as a sole stakeholder, leaving the private sector with no responsibility, or vice versa; rather, it calls for strong public-private partnerships.[13]

The Government already has set the "Basic Policy of Critical Information Infrastructure Protection"[14] to guard the Japanese CII: "Having achieved substantial results in protecting Japan's CII, these existing measures will be implemented as they have been."[15]

### *(f) Protecting governmental bodies by advancing cyber defence capabilities.*

> The Government will take government-wide, multi-layered measures based upon the assumption of cyberattacks. This must also include contingency plans for the possibility – a certain entity would be used as a springboard for the entity that is the original target of a cyberattack. In promoting these measures, the Government will ensure that they are based on the common standards for the governmental bodies, and will conduct risk analysis intending to perform its administrative responsibilities, for the optimisation of these measures as the entire governmental bodies.[16]

### *(g) Enhancing response capabilities of relevant governmental bodies, such as law enforcement agencies and the Japan Self-Defense Forces.*

> Japan will strengthen the capabilities of law enforcement agencies, the Self-Defense Forces, and other relevant organisations, both in quality and quantity...to encounter cyberattacks targeting the classified information owned by the governmental bodies, Japan will promote the efforts related to counter-cyber-intelligence in the Cabinet Intelligence and Research Office and other relevant entities.[17]

### *(h) Contributing actively to the development of international rules and norms regarding cyberspace at the United Nations and in other international settings.*

Japan is interested in the applicability of existing international law to cyberspace:

> Japan will actively engage in the discussions on the application of specific individual international laws, and subsequently contribute to the development of international rules and norms regarding cyberspace with the view that existing international law is applicable to cyberspace. Japan will cooperate with domestic and foreign stakeholders in these discussions and actively promote the development of international rules and norms with a view that ensuring openness, interoperability, autonomy and the free flow of information in cyberspace will make a significant contribution to the development of society, economy, and culture.[18]

### *(i) Promoting confidence building internationally and cooperation for cybersecurity capacity building of other countries*

> Japan will actively provide information on its fundamental viewpoint and share it with many countries at multilateral conferences, such as UN, as well as bilateral

cyber dialogues and conferences. Moreover, Japan will promote international confidence building by creating multi-layered contact mechanisms, e.g. points of contact among countries or the private sectors, for translational cybersecurity incidents during peacetime and conduct contact exercises and other measures.[19]

Japan will further actively cooperate on capacity building as a responsible member of the international community with a basic principle of the free flow of information.[20]

## National Centre of Incident Readiness and Strategy on Cybersecurity (NISC)

Under the Cybersecurity Strategy, the NISC in the Cabinet is regarded as the headquarters of the national cybersecurity. NISC's main roles are as follows:

(a) Drafting Strategy and relevant plans.
(b) Promoting international cooperation.[21]
(c) Operating Government Security Operation Coordination team (GSOC).
(d) Coordinating cooperation in order to protect CII through 18 Capabilities for Engineering Protection, Technical Operation, Analyses and Response (CEPTOARs) (1. Telecommunications, 2. Cable TV, 3. Broadcast, 4. Bank, 5. Securities, 6. Life insurance, 7. Non-life insurance, 8. Aviation, 9. Railroad, 10. Electricity supply service, 11. Gas supply service, 12. Water supply service, 13. Medical care, 14. Logistics, 15. Chemical, 16. Credit, 17. Oil, 18. Governmental administrative services)
These 18 CEPTOARs are established for sharing information about prevention and quick recovery of damage by cyberattacks. Also, the CEPTOAR Council is responsible for the information sharing among all CEPTOARs.
(e) Promoting the investigation and analysis of cyberattacks.

NISC has established a multilayered structure to ensure international and domestic cooperation for securing cyberspace at the national level. It includes public-private, interagency within a central and local government, and inter-industry cooperation.

## Japan Self-Defense Forces and Cybersecurity

Ministry of Defense and Japan Self-Defense Forces are interested in the stability of cyberspace. For this purpose, the Cyber Defense Unit was established on March 26, 2014. This joint unit for Cyber Defense is combined with three units of Ground, Maritime and Air Self-Defense Forces and is operating under the Command Control Communication Computer Systems Command (C4SC). Main roles of this Unit are information gathering and sharing information about cyberattacks against the Ministry of Defense and Self-Defense Forces, protection

**Figure 1: Japan Self-Defense Forces and Cybersecurity**

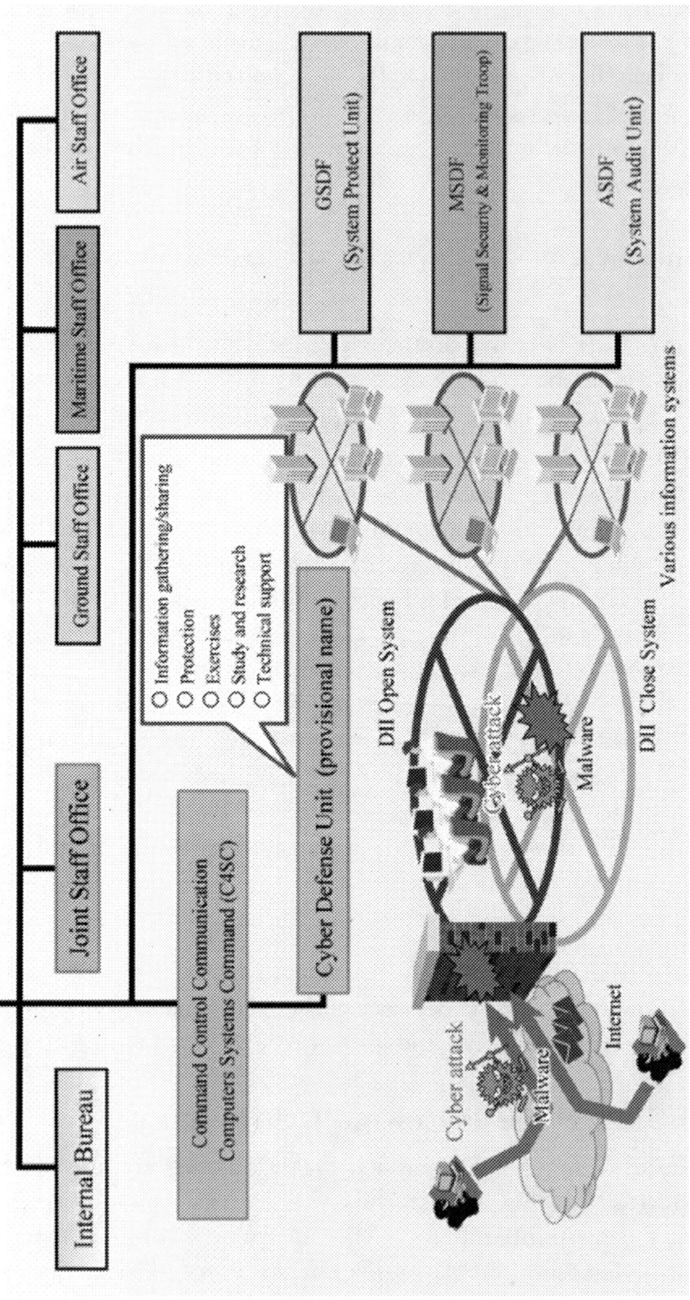


*Source:* http://www.mod.go.jp/e/about/answers/cyber/index.html

of Defense Information Infrastructure (DII), setting exercises, study and research of cybersecurity, and providing technical support. However, as pointed our above, this joint Cyber Defense Unit only defends the Ministry of Defense network system, not the whole cyber network in Japan, though the Self-Defense Forces defend the entire national territory (land, sea and airspace) in the real world.

## International Cooperation for Cybersecurity

Japan actively seeks international cooperation for the stability of cyberspace. As a consolidated response to tackle cybercrime, Japan joined the Convention on Cybercrime,[22] which was drafted mainly by the Council of Europe. Forty-seven countries including European countries, the US, Canada and Japan ratified this Convention and agreed on information sharing as well as cooperating on investigations. Japan signed it in 2001, and ratified it in 2012. Among Asian countries, only Sri Lanka and Japan have joined the Convention.

Moreover, Japan also has been pushing forward cooperation between US and Japan on cybersecurity. In the Guidelines for the Japan-U.S. Defense Cooperation concluded in April 2015, the bilateral cooperation on cybersecurity was specified as follows:

> To help ensure the safe and stable use of cyberspace, the two governments will share information on threats and vulnerabilities in cyberspace in a timely and routine manner, as appropriate.[23]

In addition, other cooperation channels including Japan-US Cyber Dialogue and Japan-US Cyber Defense Policy Working Group have been established, leading to a deepening in bilateral cooperation.

Besides Japan-U.S. bilateral cooperation, international collaboration is also being steadily carried out. The Forum of Incident Response and Security Team (FIRST) was already set up in 1990 for the purpose of a cybersecurity incident information sharing between each country's Computer Security Incident Response Team (CSIRT). An international forum called the Meridian Conference fosters international collaboration on information sharing, among other issues of concern, for the protection of the CII. Japan also seeks cooperation with other Asian countries for enabling a safer cyberspace, for example, through Japan-ASEAN Cybersecurity Policy Meetings.

## Conclusion

Japan is promoting cybersecurity as an IT advanced country. Japanese Government has established National Security Strategy, National Cybersecurity Strategy, Basic Act on Cybersecurity and various security-related organisations. Cybersecurity, however, cannot be achieved by only one nation's effort because a large number of cybercrimes and attacks are crossing national borders. Bilateral, multilateral, regional and global cooperation is essential for more secured cyberspace.

## NOTES

1. The number of Internet users has reached 100 million in Japan. Among them mobile device users were 82 million in March 2016 (the end of FY 2015). http://www.soumu.go.jp/johotsusintokei/field/tsuushin01.html (in Japanese).
2. The number of attacks was counted by the National Institute of Information and Communication Technology (NICT).
3. https://www.ipa.go.jp/files/000024541.pdf, p. 2.
4. This was Japan's second National Security Policy. The first was the Basic Policy on National Security in 1957. The earlier policy had only eight lines in Japanese, while the latest National Security Policy includes over 30 pages in Japanese (and English). The 1957 Basic Policy included the following aspects (translated by the author):
   1. Japan supports the UN activities, international cooperation and expects international peace.
   2. Japan establishes the basis for stabilizing national lives, promoting patriotism and ensuring national security.
   3. Japan prepares defense capability gradually for self-defense according to the national strength.
   4. Japan responds to the aggression on the basis of Japan-US security structure until the time when the UN prepares the effective responding ability.

   Japanese security policy was based on this Basic Policy for 56 years, from 1957 to 2013.
5. *National Security Strategy* (in English), "I. Purpose", p. 2.
6. *National Security Strategy* (in English), "III. Security Environment Surrounding Japan and National Security Challenges, 1. Global Security Environment and Challenges, (4) Risks to Global Commons", p. 9.
7. The Basic Act on Cybersecurity (in English), at http://www.japaneselawtranslation.go.jp/law/detail/?ft=1&re=01&dn=1&co=01&ia=03&x=0&y=0&ky=%E3%82%B5%E3%82%A4%E3% 83%90%E3%83%BC %E3%82%BB%E3%82%AD %E3%83%A5%E3%83%AA%E3%83%86%E3%82%A3&page=1
8. According to the Information-technology Promotion Agency (IPA), there are 230,000 cybersecurity engineers who work in the Japanese companies (over 100 workers) at present. However, 190,000 engineers do not have the skill required for their jobs. Moreover, Japan needs additional 22,000 engineers because the actual number of engineers needed is 252,000. See http://www.ipa.go.jp/security/fy23/reports/jinzai/ and http://www.ipa.go.jp/files/000014184.pdf (both in Japanese).
9. The Government of Japan, *Cybersecurity Strategy*, "2.2 Increasing Threats in Cyberspace", Provisional Translation, September 4, 2015.
10. Ibid., "5.1.1 (1) Promoting New Business Harnessing Secured IoT Systems".
11. Ibid., "5.2.1 (1) Building a Safe and Secure Cyber Environment for Users".
12. Ibid., "5.2.1 (3) Enhancing Measures against Cybercrimes".
13. Ibid,. "5.2.2 Measures for Critical Information Infrastructure Protection".
14. Established in 2014 by the Information Security Policy Council and revised on May 24, 2015 by the Cybersecurity Strategic Headquarters.

15. The Government of Japan, No. 13.
16. Ibid., "5.2.3 (1) Strengthening Defense Capabilities of Information Systems and Promoting Multi-layered Measures against Presumed Cyber Attacks".
17. Ibid., "5.3.1 (1) Enhancing Response Capabilities of Relevant Governmental Bodies".
18. Ibid., "5.3.2 (1) Establishing the International Rule of Law in Cyberspace, i. Developing international rules and norms".
19. Ibid., "5.3.2 (2) Building International Confidence Measures".
20. Ibid., "5.3.2 (4) Cooperating for Cybersecurity Capacity Building".
21. For example, on November 5, 2012, the Japan and India cyber discussion (the 1st Meeting of Japan-India Cyber Dialogue) was held. The 2nd meeting will be likely in early 2016. See "FACT SHEET: Japan and India, Working Together for Peace and Prosperity", December 5, 2015, at http://www.mofa.go.jp/files/000117791.pdf,.
22. http://www.coe.int/en/web/conventions/full-list/-/conventions/treaty/185/signatures
23. *The Guidelines for Japan-U.S. Defense Cooperation*, "VI. Space and Cyberspace Cooperation, B. Cooperation on Cyberspace", April 27, 2015.

## REFERENCES

Basic Law for Cyber Security, November 12, 2014, at http://law.e-gov.go.jp/htmldata/H26/H26HO104.html (in Japanese).

Cyber Security Strategy, September 4, 2015, at http://www.nisc.go.jp/active/kihon/pdf/cs-senryaku-kakugikettei.pdf (in Japanese).

Council of Europe, Convention on Cybercrime, European Treaty Series No. 185, Budapest, November 23, 2001, at http://www.europarl.europa.eu/meetdocs/2014_2019/documents/libe/dv/7_conv_budapest_/7_conv_budapest_en.pdf (in English).

Japan National Security Strategy, December 17, 2013, at http://www.cas.go.jp/jp/siryou/131217anzenhoshou/nss-e.pdf (in English).

National Institute of Information and Communication Technology (NICT), at http://www. nicter.jp/nw_public/scripts/index.php#nicter (in Japanese).

Japan Ministry of Defense, *The Guidelines for Japan-U.S. Defense Cooperation*, April 27, 2015, at http://www.mod.go.jp/e/d_act/anpo/shishin_20150427e.html (in English).

**Disclaimer:** The views expressed in this chapter are the author's own and do not represent any organisation with which the author is affiliated.

# 23

# South Korean Legal Initiatives to Combat Cybercrime and Enhance Digital Economy

*Il Seok, Oh*

## Introduction

Computer networks and information systems have governed almost every element of our daily lives since the early 21st century, and provided considerable advantages. However, as a result, cyberattacks and cybercrimes, too, have made inroads, and damaged critical computer systems and disturbed their functions. Cybercrimes threaten not only individual lives but also sustainability and the economy of a state. Therefore, a state needs to design appropriate legislations and national plans to respond against cybercrimes, and enhance digital economy. In this chapter, I will introduce and explore South Korean legislations and policies to combat cybercrimes and enhance digital economy.

## Recent Major Cybercrimes

### *Cyberattacks against South Korea*

According to Kang Dong-won, an independent lawmaker, the number of cybercrimes in South Korea, including hacking and distributed denial-of-service (DDoS) attacks, was 108,223 in 2012, 116,961 in 2011 and 122,902 in 2010. A domestic media reported that there were about 296 cyberattacks by unidentified hackers per day in 2012.[1]

Particularly, the January 25, 2003 cyber crisis caused nationwide malfunctions of the Internet services, which made people acknowledge the importance of cybersecurity. The DDoS attacks On July 7, 2009, affected 68 portal sites, including the South Korean governmental websites. Moreover, the South Korean governmental websites and financial institutions were damaged by the March 4, 2011 DDoS attacks, and 273 network systems in Nonghyup (NH) Bank were destroyed after the April 12, 2011 hacking. The March 20, 2013 cyberattacks paralysed computer networks in major broadcasting companies such as KBS, MBC and YTN. The June 25, 2013 DDoS attacks affected 155 servers in 69 institutions including governmental authorities and media companies.

In December 2014, South Korea's nuclear power plant operator, Korea Hydro and Nuclear Power (KHNP) reported that its computer system had been breached, and resulted in the leaking of personal details of 10,000 KHNP workers, designs and manuals for at least two reactors, electro flow charts and estimates of radiation exposure among local residents. The cyberattacks were made between December 9 and 12 by sending 5,986 phishing emails containing malicious codes to 3,571 employees of the nuclear plant operator, and the cyberattacks continued until March 2015.

The malicious codes used for the nuclear-operator hacking were the same in composition and working methods as the so-called 'kimsuky' that the North Korean hackers use. Fortunately, control systems at the nuclear plant were not harmed by the cyberattacks. The hacker could not have accessed classified information because KHNP's internal server was isolated from the Internet. Consequently, KHNP increased its security efforts to defend against possible additional attacks.

The South Korean Government has concluded that the hackers were intentionally trying to create confusion in the Korean society; South Korean prosecutors asserted that North Korean hackers were responsible for repeated disclosures of information, including blueprints of South Korean nuclear reactors gleaned from cyberattacks, as well as threats to extort money and destroy the nuclear facilities.[2]

## *North Korean Cyberattacks*

Suffering a shrinking economic situation, the North Korean Government has encouraged cyberattacks which have been cost-benefit asymmetric measures causing confusion in the international society including South Korea and the US. Furthermore, even if the cyberattacks turned out to be arising from North Korea, it could deny its involvement and argue that the South Korean Government must prove the allegation with clear and convincing evidence.[3]

To launch cyberattacks more systematically, the North Korean Government established Unit 121 comprising both intelligence and attack components. The Unit 121 has about 3,000 experts who have been systematically trained for cyberwarfare. North Korea has also developed logic bombs. North Korea has instigated cyberattacks using Internet game sites and hiding hacking tools.[4]

Major recent cyberattack cases, such as the July 7, 2009 and March 4, 2011 DDoS attacks, March 12, 2011 NH Bank hacking and March 20 and June 25, 2013 cyberattacks as well as the 2014-2015 cyberattacks against the nuclear power plant were traced to have North Korea connections.

### *Personal Data Disclosed*

In 2014, an employee from a personal credit ratings firm, Korea Credit Bureau (KCB) was arrested and accused of stealing data from customers of three credit card firms, Kukmin Bank Card, NH Bank Card and Lotte Card, while working for them as a temporary consultant – personal data of at least 20 million bank and credit card users in South Korea was leaked.[5] The stolen data included the customers' names, social security numbers, phone numbers and credit card details.[6]

## *Cybercrime Statistics*

#### *Cybercrime Arrests*

According to the Korean National Police Agency's "Status for Cyber Crime Arrest", cybercrimes have increased over the last 10 years: Total cybercrimes occurred in 2015 were 155,366 compared with 77,099 in 2004. However, the arrest cases decreased to 86,105 in 2013 from 147,069 in 2009 (see Table 1). It seems likely that the decreased arrest cases have resulted from the difficulty to trace newly developed cybercrime tactics and technologies.

**Table 1: Cybercrime Statistics**

| *Year* | *Total* | | |
|---|---|---|---|
| | *Occurred* | *Arrested* | |
| | | *Cases* | *People* |
| 2004 | 77,099 | 63,384 | 70,143 |
| 2005 | 88,731 | 72,421 | 81,338 |
| 2006 | 82,186 | 70,545 | 89,248 |
| 2007 | 88,847 | 78,890 | 88,549 |
| 2008 | 136,819 | 122,227 | 128,635 |
| 2009 | 164,536 | 147,069 | 160,656 |
| 2010 | 122,902 | 103,809 | 111,772 |
| 2011 | 116,961 | 91,496 | 95,795 |
| 2012 | 108,223 | 84,932 | 86,513 |
| 2013 | 155,366 | 86,105 | 92,621 |
| 2014 | 110,109 | 71,950 | 59,220 |
| 2015 | 144,679 | 104,888 | 75,250 |

*Source:* http://www.police.go.kr/eng/main/contents.do?menuNo=500109. Simple comparison with statistics from the previous two years is unavailable due to the adoption of a new classification system since July 2014. Cybercrime statistics of 2014 include data from January to June 2014.

Because cybercrimes are committed in the virtual world using advanced technical tools, it is very difficult to arrest cybercriminals and prove their liability. Further, lack of expert officers in cybercrimes at the National Police Agency makes it more difficult.[7]

*Types of Cybercrime*

The Korean National Police Agency has classified cybercrime into hacking and virus, internet fraud, cyber violence, illegal website operation, illegal copying and sales, and others (see Table 2).

**Table 2. Cybercrime Statistics (by type)**

| *Year* | *Total* | *Hacking and Virus* | *Internet Fraud* | *Cyber Violence* | *Illegal Website Operation* | *Illegal Copying and Sales* | *Others* |
|---|---|---|---|---|---|---|---|
| 2004 | 63,384 | 10,993 | 30,288 | 5,816 | 2,410 | 1,244 | 12,633 |
| 2005 | 72,421 | 15,874 | 33,112 | 9,227 | 1,850 | 1,233 | 11,125 |
| 2006 | 70,545 | 15,979 | 26,711 | 9,436 | 7,322 | 2,284 | 8,813 |
| 2007 | 78,890 | 14,037 | 28,081 | 12,905 | 5,505 | 8,167 | 10,195 |
| 2008 | 122,227 | 16,953 | 29,290 | 13,819 | 8,056 | 32,084 | 22,025 |
| 2009 | 147,069 | 13,152 | 31,814 | 10,936 | 31,101 | 34,575 | 25,491 |
| 2010 | 103,809 | 14,874 | 35,104 | 8,638 | 8,611 | 17,885 | 18,697 |
| 2011 | 91,496 | 10,299 | 32,803 | 10,354 | 6,678 | 15,087 | 16,275 |
| 2012 | 84,932 | 6,671 | 33,093 | 9,055 | 3,551 | 15,111 | 17,751 |
| 2013 | 86,105 | 4,532 | 39,282 | 7,873 | 2,953 | 13,567 | 17,898 |

*Source:* http://www.police.go.kr/eng/main/contents.do?menuNo=500075. Simple comparison with statistics from the previous two years is unavailable due to the adoption of a new classification system since July 2014. Cybercrime statistics of 2014 includes data from January to June 2014.

While most of these classifications are self-explanatory, internet fraud and cyber violence bear further elaboration. An Internet fraud is the use of Internet services or software with Internet access to defraud victims or to otherwise take advantage of them; for example, by stealing personal information, which can even lead to an identity theft.[8] A very common form of Internet fraud is the distribution of rogue security software; the Internet services can be used to present fraudulent solicitations to prospective victims, to conduct fraudulent transactions, or to transmit the proceeds of fraud to financial institutions or to others connected with the scheme.

Cyber violence means violence, or its equivalent, carried out in cyberspace or on the internet.[9] Cyberspace is also transforming the face of teenage violence to be cyber bullying which can come in various forms, but among the most widespread is through social media.[10]

Although there are no official statistics, the total loss resulted from cybercrimes in South Korea has been estimated to US$ 4.87 billion annually.[11]

*Cyber Financial Crimes*

In South Korea a number of cyber financial crimes, such as phishing, smishing, pharming and memory hacking, have been newly developed and have caused actual financial damages. Phishing is the attempt to acquire sensitive information such as usernames, passwords, and credit card details (and sometimes, indirectly, money), often for malicious reasons, by masquerading as a trustworthy entity in an electronic communication.[12] Victims typically receive a fake e-mail, resembling that of a legitimate organisation, which directs them to a bogus website link where they are asked to update personal information, such as passwords or bank account numbers. Any information the users enter on the page is thus stolen.

SMS phishing or smishing is a form of criminal activity using social engineering techniques for text messages on cell phones sending fake coupons or invitation letters. Clicking on the coupons or invitation letter, the victim installs a malicious code which then steals personal information stored on the smartphone.

Memory hacking inserts malicious codes into the data in the PC memory, malfunctioning the bank security programme, for example, and stealing an unsuspecting victim's money.

Pharming is a cyberattack that redirects a legitimate website's traffic to a fake site. Pharming can be conducted either by changing the host's file on a victim's computer or by exploiting a vulnerability in the Domain Name System (DNS) server software.[13] In one case, in South Korea, 184 victims' financial information was stolen by pharming, and the victims' "authorized certificates" were re-issued and $ 130 million were withdrawn by a hacker.

Because of the National Police Agency's aggressive investigation, the number of smishing cases decreased to 4,900 in 2014, compared to 29,700 in 2013, and the number of memory hacking cases also decreased to 155 in 2014, compared to 463 in 2013. In 2014, the total loss arising from cyber financial crimes increased sharply to ₩ 56,200,000,000 (about US$ 47 million), compared with that of ₩ 25,000,000,000 (about US$ 21 million) in 2013 and ₩ 5,700,000,000 (about US$ 4.75 million) in 2012.[14]

However, during this period, the number of pharming cases increased from 3,218 in 2013 to 7,101 in 2014. The number of occurrences of cyber financial crime decreased to 15,596 in 2014, compared to 33,442 in 2013.[15] Through these statistics, the loss arising from pharming has increased the total loss arising from cyber financial crimes.

## Recent Policies and Plans to Respond to Cyber Crimes[16]

The South Korean Government acknowledged cybersecurity as the one of main factors to enhance national security. In the *National Security Strategy 2014*, cybersecurity was emphasised as follows:

> In addition to provocation from NK [North Korea], threats and challenges against national security can manifest themselves in various forms: transnational threats such as terrorism, cyberattacks, climate change, epidemics, natural disasters and accidents, and as yet unknown future threats. In the Information Age, in which the world is connected through various networks, the possibility of cyberattacks is also a serious security issue. Being used in an increasing number of fields including the business, academic, military, and cultural areas, cyberspace offers a great number of benefits to humankind. The anonymity and trans-nationality of cyberspace, however, have brought various threats in such forms as cybercrimes and cyberattacks. Such cyber threats underline the need for the establishment of not only a domestic response system but also bilateral and multilateral cooperation mechanisms, some of which include fostering confidence-building measures between countries and establishing international norms.

After the March 20 and June 25, 2013 cyberattacks, the South Korean Government established the "National Cyber Security Comprehensive Measures" in 2013. Under the Measures, Critical Information Infrastructures were increased to 292 to strengthen the response and resilience of critical functions to cyberattacks and cybercrimes. Further, the Government has made a plan to educate and train about 5,000 persons as white hackers by 2017. Under these Measures a cybersecurity threat information sharing system shall be established for greater cooperation among the National Cyber Security Centre in National Intelligence Service, Ministry of Science, Information and Communications Technology and Future Planning (MSIP), Korea Communications Commission, Ministry of Security and Public Administration, Financial Service Commission and Personal Information Protection Commission.

The Government has also acknowledged the importance of responding to cyberattacks and cybercrimes, and designated a Special Secretary of Cyber Security in the Blue House, the office of the President of South Korea, in January 2015. Three months later, the Secretary of Cyber Security in the Blue House was appointed, who controls cybersecurity policies and initiatives.

After the cyberattacks on the nuclear power plant, the South Korean Government established "Comprehensive Measures to Enhance National Cyber Security" to create a total responding system in March 17, 2015. According to the Measures, the Government shall 1) develop Critical Cybersecurity Technologies, and educate and train Core Experts, 2) encourage Cybersecurity Response Team, recruit Workforce, and enhance Cybersecurity Industry, 3) promote International Cooperation, and 4) amend and revise laws and regulations related to cybersecurity.

## Major Legislations against Cybercrimes

### *Electronic Financial Transaction Act*

The South Korean Government enacted the Electronic Financial Transactions Act to ensure the security and reliability of electronic financial transactions and develop the national economy by advancing the electronic financial industry.

According to the Act, financial institutions and electronic financial business operators shall have a duty to ensure secured processing of electronic financial transactions.[17] They shall comply with the standards set by the Financial Services Commission (FSC) for the information technology fields of manpower, facilities and electronic apparatuses, necessary for electronic transmission or processing and electronic financial business to secure the safety and reliability of electronic financial transactions.[18]

Financial institutions and electronic financial business operators shall appoint a chief information security officer, responsible for managing electronic financial business and information technology security which forms the basis of electronic financial business.[19] They shall analyse and evaluate the vulnerabilities of electronic financial infrastructure and shall report the results to the FSC.[20]

According to the Act, no one shall access electronic financial infrastructure without access authority, or manipulate, destroy, conceal or leak stored data by exceeding his/her access authority. No one shall destroy the data of electronic financial infrastructure, or use programmes, such as computer viruses and logic bombs, with the intention of obstructing the operation of electronic financial infrastructure. No one shall abruptly send large amounts of signals with the intention of obstructing the operation of electronic financial infrastructure, or causing a fallacy in information processing by means, such as inducing the processing of a wrong order.[21] When the occurrence of intrusion incidents leads to disturbance, paralysis or destruction of electronic financial infrastructure, financial institutions and electronic financial business operators shall notify these to the FSC. When making notifications, they shall analyse the cause of the incidents, and take necessary measures to prevent the spread of damage caused by the intrusion incidents.[22]

According to the Act, when a person who uses electronic financial transactions suffers any damages as a result of an accident arising out of forgery or alteration of the means of access or in the course of electronically transmitting or processing the conclusion of a contract or a transaction request, the financial institution or electronic financial business operator concerned shall compensate the damages.[23]

## *Cyber Security Industry Enhancement Act*

In accordance with the 2015 "Comprehensive Measures to enhance National Cyber Security", South Korean policy-makers and legislators in consultation with individuals, vendors, and governmental agencies decided that cybersecurity industry should be encouraged to support robust cybersecurity activities with best technologies. Hence, Cyber Security Industry Enhancement Act was enacted by the National Assembly on June 22, 2015, and came into effect on December 23, 2015.

According to the Act, South Korean central and local government and municipals shall initiate policies to encourage cybersecurity industry and prepare measures to allocate budgets to fulfil the policies.[24] The MSIP shall prepare a Cyber Security Industry Enhancement Plan.[25] Public agencies shall submit to the MSIP "purchase demand information" that annually describes the cybersecurity products and services required for the agencies.[26] To develop cybersecurity industry and to guarantee the quality of cybersecurity products and services, public agencies shall pay reasonable consideration when they take contracts with the vendors.[27] For these contracts, MSIP shall draft a model contract in consultation with the Korean Fair Trade Commission and recommend it to the public agencies and vendors.[28] Moreover, when public agencies issue unreasonable orders or violate other laws and regulations, the MSIP shall investigate the orders by comprehensive public-private monitoring activities and ask the agencies to revise or correct them.[29]

Further, assessment agencies registered to the MSIP shall provide Assessment on Cyber Security Preparedness for Internet Service Providers (ISPs) or Online Intermediaries.[30] The South Korean Government shall provide technical or budget assistance required for the Assessment to the registered agencies.[31]

According to the Act, the MSIP shall develop cybersecurity technologies and standards,[32] educate and train cybersecurity experts[33] and enhance international cooperation.[34] The MSIP shall also designate outstanding cybersecurity technologies and vendors.[35]

## *Network Protection and Critical Information Infrastructure Protection*

Act on Promotion of Information and Communications Network Utilisation and Information Protection has been enacted for the improvement of citizens' lives and the enhancement of public welfare by facilitating utilisation of information and communications networks, protecting personal information who use information and communications services and developing an environment in which people can utilise information and communication networks in a sounder and safer way. Under this Act the MSIP or the Korea Communications Commission shall prepare a policy for establishing a solid foundation for an information society through the promotion of utilisation of information and communications networks, the stable management and operation of such networks, the protection of personal information of users, and other related activities.

Further, to protect critical information infrastructures from cybercrimes, the Critical Information Infrastructure Protection Act was enacted in 2001.

## *Personal Information Protection Act*

### *Disclosure of Personal Information and Tort Case*

Under the negligence tort cases, the plaintiff shall prove the duty, breach of duty, damages and causation. When personal information is negligently lost, stolen, hacked or disclosed, a plaintiff shall demonstrate these elements to recover the damages against the companies or organisations which collect, retain, maintain and control personal information.

The Supreme Court of Korea, on the major personal information breaching case, the so-called "Auction Case", decided that the ISP, Auction, had not breached its duty to protect personal information which it had collected and stored, even though the system was hacked and information disclosed, because the company had followed and complied with the regulations and guidelines. However, the regulations and guidelines were only the minimum standards which the ISPs must comply. The company, Auction, had not set up firewalls or used encryption on the personal information under its control.

### *Compensation for Damage in Disclosure of Personal Information according to Personal Information Protection Act*

The Personal Information Protection Act aims to protect the rights and interests of all citizens and to realise the dignity and value of each individual by protecting personal privacy from collection, leakage, misuse and abuse of personal information.

The Act, allows the person whose personal information has been disclosed to claim compensation against the personal information manager, who manages or controls the information directly or via another person. In such cases, the personal information manager cannot be exempted from responsibility unless it is proved that such an act has been performed neither intentionally nor by negligence.[36]

After several personal information disclosure or leak cases came to light, the National Assembly amended the Act on July 24, 2015 and inscribed a punitive damages clause. According to the clause, when a person's personal information is stolen, disclosed, changed, counterfeited or destroyed by intent or gross negligence of the personal information manager, courts shall award punitive damages within three times that of the normal damages. However, if the personal information manager proves that there was no gross negligence or intent, the punitive damage clause shall not be applied.

*What are the Damages in Personal Information Disclosure Case?*[37]

Many district courts have decided and approved, in the damage claim cases related to breach of personal information, a monetary compensation on pain and suffering or emotional distress of a person whose personal information under the ISPs or personal information manager's control was disclosed. Courts have concluded that it should be commonly acknowledged with practical social experience that when a person's personal information is disclosed, the person suffers pain and suffering or emotional distress. Therefore, courts have concluded that ISPs or information managers shall have to make monetary compensation for the pain and suffering or emotional distress.

On January 22, 2016, the Seoul Central District Court also reached the same conclusion in the KCB personal information disclosure case. The court ordered the three companies, Kukmin Bank Card, NH Bank Card and KCB to give each victim ₩ 100,000 (about US$ 83) as damages on pain and suffering.[38] However, in February another District Court dismissed the plaintiffs' damage claims against Lotte Card in another personal information disclosure case. This time the court determined that although the personal information was disclosed, there was no actual disclosure or damage because the information was not transferred to third party.[39]

Pain and suffering or emotional distress are unclear, imprecise and speculative terms; therefore, they need to be elaborated upon. The damages caused by the victims' suffering therefore need to be carefully reviewed. It must be ascertained whether they are just worried or anxious that their personal information might be used for unlawful behaviour or criminal conduct. Such kinds of anxieties cannot be compensated with damages because there was no actual or immediate damage. It may be claimed that such emotional distress is likely to cause future harm very similar to the harm caused when a person is exposed to harmful gases, acids or radioactive materials. These kinds of future harms shall be compensated because there is a certainty to them. However, the victim whose personal information has been disclosed has not actually been exposed to harmful materials or situations. Therefore, the pain and suffering or emotional distress, suffered by the victim, cannot be compensated as future harm.

Pain and suffering or emotional distress is an economic loss which cannot be compensated, because it is a consequential loss. However, when a tortfeasor or a contract breaching party has foreseen or should have foreseen the secondary harms due to Phishing and Pharming arising from the personal information disclosure, it can be compensated. Nonetheless, it is almost impossible for the victim to prove the foreseeability and causation.

Again, pain and suffering or emotional distress, resulting from breach of the duty to protect personal information, are not compensated under tort or contract claims, because they are too speculative, unclear, and imprecise. Expecting only ISPs and personal information managers to ensure the protection of personal information, Korean governmental authorities disregard their own role. They assume that personal information protection should only be one of the consumer protection activities. There is a case to be made for the Korean Fair Trade Commission to bolster its role in this field based on the clause of "unfair trade" of the Monopoly Regulation and Fair Trade Act.

## Conclusion

With advanced information and communication technologies and infrastructure, the South Korean economy has recovered from the 2008 financial crisis. However, the country has suffered several cyberattacks which have caused actual social costs and losses. To prevent cyberattacks and cybercrimes, the South Korean government has decreed several cybersecurity policies and laws and regulations. The Government has acknowledged cybersecurity as one of the main factors of national security, and developed National Cyber Security Comprehensive Measures, appointed Secretary of Cyber Security in the Blue House and established Comprehensive Measures to Enhance National Cyber Security.

The South Korean Government has enacted the Electronic Financial Transaction Act to safeguard the security and reliability of electronic financial transactions and develop economy by enhancing electronic commerce and decreasing costs related to financial transactions. To support cybersecurity activities with best technologies, the Government has also enacted the Cyber Security Industry Enhancement Act. To protect critical information infrastructure from cyberattacks, the Critical Information Infrastructure Protection Act has been enacted since 2001. Moreover, the Personal Information Protection Act has also been enacted. According to the Act, the person whose personal information is disclosed may claim for compensation against a person or an institute that manages or controls it. After several personal information disclosure cases emerged, Korean legislators amended the Act and inscribed a punitive damages clause as well.

Undoubtedly, the above-mentioned South Korean initiatives and activities would be of immense value for other states to develop and establish their own cybersecurity policies and legislations.

### NOTES

1. "More than 100,000 cases of cybercrime occur in S. Korea every year", N K News, October 4, 2013, at http://english.yonhapnews.co.kr/northkorea/2013/10/04/62/0401000000AEN20131004007400320F.html (Accessed March 1, 2016).
2. Bruce Klinger, "The U.S. Needs to Respond to North Korea's Latest Cyber Attack", March

20, 2015, at http://www.heritage.org/research/reports/2015/03/the-us-needs-to-respond-to-north-koreas-latest-cyber-attack (Accessed March 1, 2016).

3. Il Seok Oh, Seung Yeol Lee and So Jeong Kim, "Effective Legal and Political Measures against North Korea's Cyber Warfare", *Policy Studies*, 186, 2015, p. 34.
4. Ibid.
5. "20 million people fall victim to South Korea data leak; FSS calls on financial institutions to improve protections against insider leaks", January 19, 2014, at http://www.databreaches.net/20-million-people-fall-victim-to-south-korea-data-leak-fss-calls-on-financial-institutions-to-improve-protections-against-insider-leaks/ (Accessed March 1, 2016).
6. "20 Million People Fall Victim to South Korea Data Leak", *Security Week*, January 19, 2014, at http://www.securityweek.com/20-million-people-fall-victim-south-korea-data-leak (Accessed March 1, 2016).
7. I jump over the cyber crime police, January 5, 2015, at http://www.kookje.co.kr/news2011/asp/newsbody.asp?code=0300&key=20150106.22008193319 (Accessed March 1, 2016).
8. "Internet Fraud", at https://en.wikipedia.org/wiki/Internet_fraud (Accessed March 1, 2016).
9. "Cyberviolence", at https://en.wiktionary.org/wiki/cyberviolence (Accessed March 1, 2016).
10. "In digital age, teen violence takes new turn", *The Korea Herald*, December 8, 2015, at http://khnews.kheraldm.com/view.php?ud=20151208000959&md=20151211003712_BL (Accessed March 1, 2016).
11. Steven Levy, "Leaked confidential company ... Korea helpless hole is on the Web", June 26, 2015, at http://news.mk.co.kr/newsRead.php?year=2015&no=611797 (Accessed March 1, 2016).
12. "Phishing", at https://en.wikipedia.org/wiki/Phishing (Accessed March 1, 2016).
13. "Pharming", at https://en.wikipedia.org/wiki/Pharming (Accessed March 1, 2016).
14. "[24 journalists at] cyber crime economy culture to take "Report", January 1, 2014, at http://news.mk.co.kr/column/view.php?year=2014&no=3423 (Accessed March 1, 2016).
15. "Cyber crime financial damages one nyeonsae 25 billion '! 600 billion ... 'pharming' surge", *Sinhuieun News* January 29, 2015, at http://www.mt.co.kr/view/mtview.php?type= 1&no=2015012908211802132&outlink=1 (Accessed March 1, 2016).
16. Il Seok Oh, "Cyber Security Law and Policy", Presentation to Foreign Officer Training Course, Cyber Security Training and Exercise Center, National Security Research Institute of Korea, January 16, 2016.
17. Art. 21. para 1 of Electronic Financial Transaction Act.
18. Art. 21. para 2 of Electronic Financial Transaction Act.
19. Art. 21-2. para 1 of Electronic Financial Transaction Act.
20. Art. 21-3. para 1 of Electronic Financial Transaction Act.
21. Art. 21-4. of Electronic Financial Transaction Act.
22. Art. 21-5. of Electronic Financial Transaction Act.
23. Art. 9. para 1 of Electronic Financial Transaction Act.
24. Article 3. of Cyber Security Industry Enhancement Act.
25. Article 5. Para 1. of Cyber Security Industry Enhancement Act.
26. Article 6. Para 1. of Cyber Security Industry Enhancement Act.
27. Article 10. Para 1. of Cyber Security Industry Enhancement Act.
28. Article 10. Para 3. of Cyber Security Industry Enhancement Act.
29. Article 10. Para 2. of Cyber Security Industry Enhancement Act
30. Article 12. Para 2. of Cyber Security Industry Enhancement Act.
31. Article 12. Para 3. of Cyber Security Industry Enhancement Act.
32. Article 14. Para 1. of Cyber Security Industry Enhancement Act.

33. Article 15. Para 1. of Cyber Security Industry Enhancement Act.
34. Article 16. Para 1. of Cyber Security Industry Enhancement Act.
35. Article 18. Para 1. and Article 19. Para 1. of Cyber Security Industry Enhancement Act.
36. Article 39. para 1.of Personal Information Protection Act.
37. Il Seok Oh, "A Study on the Compensated 'Damages' arising from breach of Duty to Protect Personal Information", *Ewha Law Journal,* 19 (3), 2015, pp. 29-31.
38. "1 eokgeon information leakage surprise in the first sentence ... per person © 100,000 restitution?", May 25, 2016, at http://www.dt.co.kr/contents.html?article_no=2016012502100251800001 (Accessed March 1, 2016).
39. "Three cards four private information sonbae .. mixed court ruling", *Financial News,* February 5, 2016, at http://www.fnnews.com/news/201602051621400203 (Accessed March 1, 2016).

# 24

# GLOBAL CYBERSECURITY ENVIRONMENT: PERSPECTIVES OF THE US AND CHINA IN COMPARISON

*Cuihong Cai*

## Concept and Shaping of Cybersecurity Environment

The term security refers to the state of being free from danger or threats. The interconnective digital technique however certainly makes the connotation and extension of the term "security" extremely generalised. "Security" in such a case not only belongs to a state, region or certain social organisation, but refers to the living state of each group or individual who uses the network or is greatly affected by the network technique all over the world.

The cybersecurity environment is related to the cybersecurity threat which is perceived by a certain subject. The concept of "security environment" is still ambiguous, although it is very popular. The security environment as understood in the field of international security study mainly investigates the external environment, which does not involve the internal environment of a state. The global cybersecurity environment therefore refers to the current conditions and events related to cyberspace that affect the security, stability and development of the countries in the world. Generally, cybersecurity refers to at least two different images, namely which kinds of security threats are confronting cyberspace and who is being threatened in cyberspace, and therefore should be protected.[1] Several characteristics of the cybersecurity environment are discussed in the following sections.

### *Cybersecurity Environment Scope could be Large or Small*

As far as the components of "cybersecurity environment" are concerned, cybersecurity concepts, strategies, structures, institutions as well as relations among all states may be deemed as integral parts according to an early analysis. Some scholars tend to regard the cybersecurity environment as an overall comprehensive security environment including political, economic and social development. Nevertheless, a more intuitive understanding about cybersecurity environment summarises major cybersecurity threats.

The cybersecurity environment scope could be large or small. One of the reasons is related to the fact that it is difficult to absolutely define the boundary of cyberspace as the security object. For example, Martin Libicki deconstructs cyberspace into a three-layer model. The "physical layer" mainly refers to the physical infrastructures and equipment forming cyberspace; the "grammar layer" refers to the software and protocol which make the information transfer possible; and the "semantic layer" refers to the information flowing in cyberspace.[2] Such different structure layers contain different security risks, but it is difficult to conclude whether they cover all meanings of cybersecurity. Choucri and Clark even suggest that the "user layer" should also be incorporated into the framework of cyberspace and cybersecurity.[3] As far as the information layer (i.e. the above-mentioned semantic layer) is concerned, the information which represents the code and value could be further distinguished. If the latter is deemed as the natural information extension, the cyberspace boundary will be extended to the dimensions of subjective human activities, such as values, beliefs and thoughts.

### *Severity of Threats to Cybersecurity Environment Varies*

The threats to the cybersecurity environment have different severities, which extend from the non-intentional accidents (such as software or system errors) with the lowest severity to cyberwarfare with the highest severity. The malicious acts of different forms such as cybercrimes, cyber espionage and cyberterrorism lie somewhere between these two extremes. However, the boundary between different types of cybersecurity threats is very obscure. Although cyberwarfare is regarded as the most destructive activity, as far as the loss caused and overall scale are concerned, cybercrimes and spy activities seem to be much closer to the direct threats to international security. How to accurately judge cyberthreats is the key to evaluate the cybersecurity environment.

According to different behavioural subjects, the cybersecurity threats may be divided into four types: hacker attacks, organisational cybercrimes, cyberterrorism and cyberwarfare supported by states. Generally speaking, the threat severity to national security increases from the individual, non-state actor to the state actor.

The potential hazard of the cybersecurity threat to national security should not be underestimated, no matter which kind of threat it is.

### *Cybersecurity Environment is Essentially a Subjective State*

The cybersecurity environment reflects the cognitive differences in different subjects. A by-product of rapid development of Internet technology is, rising of cyber fear. Cyber fear refers to the psychological panic of the general public because of the fragility of Internet society, Internet dependency and development of Internet technology. It has a deep-rooted psychological origin, which is historically the reflection of technical pessimism, with danger exaggeration and overreaction as inherent psychological properties. Therefore, on the one hand, people tend to exaggerate the cybersecurity risks and potential threats and underestimate the social bearing capacity to respond to the cybersecurity problems; and on the other hand, they easily believe the cyber-threat rhetoric of politicians, media and enterprises, rather than making rational judgment according to objective facts and logical inference. Besides fear, the cybersecurity environment is also shaped by the interest pursuing behaviour of various groups and power games of states. It means that cybersecurity is shapeable to a great extent. The over-interpretation of cybersecurity risks further strengthens the threat cognition, which results in conflicts and control-oriented security practices. In fact, it creates a cybersecurity environment with less mutual trust and weakening rules, resulting in the self-fulfilling prophecy of heightened cyber conflicts.

### *Shaping of Cybersecurity Environment is Related to Discourse*

Cybersecurity environment is shaped by discourse construction, too. Through several approaches such as circumstance assumption, analogy and logical deduction, the securitisation subjects construct different cyberthreat images. Hansen and Nissenbaum divide the construction process of such images into three types:[4] The first is "hyper-securitisation", i.e., the cybersecurity discourse depends on the imaginary disaster scene to make the severity and urgency of the security image much higher than the real threat. The discourse of hyper-securitisation often emphasises on the destructive consequences which may be caused by a sudden cybersecurity event, especially the failure of the core information system of such fields as society, finance and military, which may lead to collapse of the whole social and political order. In such a context, other historical experience of weapons of mass destruction (such as nuclear weapons and biological weapons) is often used to speculate the potential effects of the cybersecurity threats,[5] although there is no real precedent for such experience in cyberspace. The second is the discourse of "everyday security practice", which connects the cybersecurity image with general experience of the audience and their daily knowledge, making the danger scene "close at hand". By using the parables such as "virus", "infection" and "loophole" which are close to the daily life, the discourse constructs the direct relation between cybersecurity and audience, which could be used to increase public support for the corresponding security strategies and propositions of certain measures. The third is "technification", in which the authoritative discourses about the cybersecurity problems are granted to the technical experts of cybersecurity. Such discourses often describe cybersecurity as the unavoidable problem which has radically resulted from technical development.

There are several examples here. Firstly, people generally use terms from biomedicine and military to exaggerate cybersecurity problems to impress the public. For example, terms such as "virus" and "worm" are widely used to describe the technical disturbances in the Internet. And terms such as "cyberwarfare", "cyberweapons of mass destruction", "cyber Pearl Harbor" and "cyber 9/11" are increasingly used by the public, especially in the US. Secondly, the normal cybersecurity competition among different states is exaggerated. Some politicians and scholars are often in a state of extreme nervousness and regard general cyber spy activity as cyberwarfare. For example, Mike McConnell, former director of the National Security Agency declared in *Washington Post*, "The US is fighting a cyber warfare. However, we are losing it ...."[6] Moreover, the media often regards the cyberattacks which occurred in Estonia in 2007 and Georgia in 2008 as models of modern cyberwar and demonstrate that cyberwar has become a disturbing reality, although it has been shown that those cyberattacks did not have a great effect on the two countries.

## Reconstitution of Security Environment in Cyberspace

### *All actors in the network environment have the capacity to launch attacks*

The most important characteristic of cybersecurity is diversification of actors. In the network environment, all actors such as governments and non-governmental organisations have the capacity to launch attacks, and the actors are more diversified. For a state, the threats of the traditional and modern worlds mainly come from equivalent actors, i.e. states, but the attacks in cyberspace may possibly come from non-governmental actors, such as master hackers or terrorist organisations. Moreover, the attacks by non-governmental actors do not mean that the destruction capacity is reduced in anyway. In contrast, the attacks using cheap tools and methods such as information bombs, virus programmes and Trojans are sufficient to disable the financial system, water conservancy and power supply system of a state. Anyone may be able to launch an attack, and the attack behaviour may be hidden completely.

### *The launching capacity in the network environment is not limited by geographical distance*

In the strategic game of the traditional world, the important method for checks and balances is power projection. The costs increase with the distance of the power projection. Meanwhile, the power efficiency shows a decline trend. As John Mearsheimer mentioned, the geographic limitation results in scarcity of global hegemony, because geographic factors such as a huge body of water restrain the effective launch of power. However, there is no geographic concept in the network world. The network capacity may be launched in any place and at any time or across the oceans or continents, without the effects of limitation of range capacity.

### *Threats to the network environment cannot be attributed quickly and effectively*

The decision can be made quickly only after effective attribution is made. Failure of attribution for the behaviour of the actors results in misjudgement. Technical factor is the primary reason for the difficulty of effective attribution in cyberspace. The network techniques change quickly, various network loopholes exist and attacking methods for various network loopholes are changing and unpredictable. In addition, there are lots of actors and considerable information in cyberspace, all of which make it more difficult to attribute cyberspace behaviours.[7] Moreover, as a result of the numerous actors in cyberspace, it is difficult to define and distinguish them through their behavioural characteristics. In the traditional security field, distinguishing combat personnel from non-combat personnel and vengeance principle of proportionate damage and punishment are explicitly stipulated in the Geneva Conventions.[8] However, it is not just difficult to timely detect the threats in the cyberspace environment but also to distinguish cyber espionage activity from cyberwar behaviour, individual behaviour from the national behaviour, military target from the civil target and combat personnel from the non-combat personnel even though they are detected. The network environment also makes it easier to orchestrate false flag attacks.[9] The actors with the advanced network techniques may launch attacks by hiding his or her identity with the technological means or using the IP address of others. As far as the technological logic is concerned, any computer may be attacked and used as an attack platform or attack channel. Therefore, the cybersecurity defender often does not know where the threats are from, who launches the attack and who will be in charge of such attack behaviour. Even when the state, organisation or individual declares responsibility for such network behaviour, we have reasons to doubt that they are lying.

### *Effective deterrence is not feasible in the network environment*

Deterrence means the capacity and will for war which must be actively shown by the state for the purpose of discouraging hostile attack. Glenn Snyder divides the deterrence methods into deterrence by punishment and deterrence by denial.[10] The former means to revenge the attacker in a manner of punishment, by forcing it to realise that launching an attack is unworthy. Such deterrence method is based on the (counterattack) striking capacity of the actor. The latter emphasises on effective defence capacity. It means to make the enemy realise that it cannot achieve the expected attack purpose, and therefore, gives up its attack intention. Because the "secondary striking capacity" cannot be ensured and the vengeance method, an eye for an eye, is based on the "strategy to ensure mutual destruction", the application of deterrence by punishment in cyberspace has several problems. What is different from the traditional military strike is that the network counterattack cannot make the counterpart definitely realise the revengeful consequences and does not have explicit expectation of great auxiliary casualties and the decision-makers cannot forecast the costs and benefits either. The pure method of deterrence by denial also has its limitations. The counterparts will not be deterred if they think that the costs of network attack are low, even though they realise that their attacks may not be successful. Meanwhile, once the cyberattack method is shown, the counterpart will quickly make a defence strategy and develop the method for a counterattack. Such "advantage hiding" feature which requests to hide the network defence measures inevitably reduces the deterrence effects.[11]

## Common Characteristics of Global Cybersecurity Environment

The global cybersecurity environment is confronted with a variety of problems, such as network information inundation, information pollution, information infringement, information monopoly and cybersecurity crisis. There is considerable false, redundant and junk information in cyberspace, which affects further development and utilisation of the information resources. The information junk such as obscenity, violence, terror and superstition seriously disturbs the social order. The cyberspace openness allows it to become a hotbed of information infringement, such as infringement against intellectual property rights and personal privacy. Meanwhile, cybercrimes, such as cyber thefts, spying for economic, political and military intelligence, emerge again and again. Besides all these above, the global cybersecurity environment also has the following characteristics:

### *Hysteretic Nature of Security Technique*

Cyberspace has open and anonymous characteristics, providing numerous chances for cybercrimes and cyberattacks. It is much easier and safer for the criminals or attackers to act in such a hidden space, i.e. cyberspace, than in the real world.

Compared with the methods of cyberattacks, development of the cybersecurity technique has a hysteretic nature. The better the security techniques and products, the higher the attack techniques and methods. Therefore, we can only consider the cybersecurity environment from the concept of "risk" and perspective of "process". In this regard, it seems that there is no "whole set of solutions" to solve the cybersecurity problems, nor solution without risks. What we can do is to strike a balance between the risks and benefits within the sustainable limit. Cybersecurity is only relative security, rather than absolute security. Different network systems have different requirements for cybersecurity; different networking development phases also have different security demands.

### *Imbalance of Power Structure*

There are big differences among various countries concerning cyber power. The US has obvious advantages. The Internet technology mainly originates from the US, and thus, the relevant technical standards are made for maintaining its interests. In order to achieve its dominant position, the US not only dominates the technical standards for cyberspace, but also controls the cyberspace norms. At present, there are 13 Domain Name System (DNS) root servers all over the world. Among the 13 root servers, one main root server is set in Dulles, Virginia, and there are 12 auxiliary root servers, with 9 in the US and one in the UK, Sweden and Japan, respectively. The US manages the cyberspace affairs with its roles in the Internet Corporation for Assigned Names and Numbers (ICANN), Internet Society (ISOC) and other organisations.[12]

The secret monitoring project, "Prism" launched in 2007 by the US, which was disclosed by Edward Snowden, proves the imbalance of the cyberspace power structure. Such monitoring plan exposes the attempt of the US to seek for the cyberspace hegemony position, because such project aims at not only its strategic competitors, but also its allies such as Western European countries. According to the information disclosed by *The Guardian*, as per the top secret administrative instruction signed by President Obama in October 2012, Offensive Cyber Effects Operations (OCEO) was a non-traditional way for the US to achieve its target.[13] Therefore, Obama ordered his senior national security and intelligence officials to draw up an overseas target list for cyberattacks.[14] Although many countries study cyberweapons more or less, only the US and Israel reportedly have used destructive cyberweapons such as Stuxnet, Duqu and Flame.[15]

### *Deficiency of Institutional Norms and Mechanism*

It is impossible for the global cyberspace to form a mature institutional framework system in a short time. At present, there are some rules and systems at the technology level, but there are still no institutional norms for the global cyberspace. The international treaties such as the Convention on Cybercrime are mainly about cooperation concerning cybercrime, with the purpose to address cybercrime. Its jurisdictional function is very limited, and it is not implemented by a wide range of member states – an insufficient response to global cybersecurity problems. In order to reach a unified consensus about the Internet rules and regulations, the International Telecommunication Union (ITU) has been actively promoting to conclude an international treaty about the Internet rules and regulations, but several big countries have different opinions about the nature and implementation of the treaty.

There are mainly two reasons for the deficiency of the institutional norms for the global cyberspace. One is the anarchic nature of cyberspace. There is no supranational authority in cyberspace, which could have the power to lead the institutional construction of the global cyberspace and manage global cyberspace affairs. The other is that different actors have different demands for cyberspace institutional norms, different proposals and conceptions about these norms, thus resulting in difficulty of constructing an effective institutional framework for global cyberspace in a short time span.

### *Insufficient Mutual Trust*

Failure of quick and effective attribution of threats in the network environment and cyberspace virtuality destroys the truth of communication intentions among the cyber actors, which results in insufficient trust in cyberspace. According to the level of trust, the strategic mutual trust mainly has three levels of manifestation, including high trust, medium trust and low trust.[16] For example, the mutual trust between the US and China belongs to the low trust level. The level of trust is different in different fields of Sino-American relations. Compared with the fields such as economics, politics and military, the level of strategic mutual trust in terms of the cyberspace between the US and China may be deemed to be the lowest. As indicated by Kenneth Lieberthal, cyber is a realm in which the most hostile images each side has of the other are being reinforced.[17] The US and China doubt and worry about each other's cyberspace behaviour and policies, which seriously affects their strategic mutual trust in cyberspace.

The Transparency and Confidence-Building Mechanisms (TCBMs) should also be strengthened for the cybersecurity environment. An important aspect of building TCBMs is to carry out the cyber arms control. Russia called upon the international community to carry out the cyber arms control early in 1998. However, the US thinks that it is impossible to apply the traditional weapon control policies to cyberspace. The differences between the US and Russia restrict the realisation of cyber arms control. In addition, building of TCBMs also requires the international community to coordinate the concept of cybersecurity. For example, no consensus on "cyberattack", "cybercrime" and "cyberwar" has been reached, which will result in unnecessary misunderstandings or – misjudgements in cyberspace.

## Cognitive Differences between the US and China about Cybersecurity Environment

As a subjective cognition, the security environment changes among different groups and countries. China and the US, too, have cognitive differences in this area. According to the SWOT analysis method, the environment analysis may be divided into four strategic factors including strengths, weaknesses, opportunities and threats. But threats are the most important aspect of the cybersecurity environment analysis, as cybersecurity interests remain a top priority. While the United States defines global cybersecurity environment from the perspective of "threats", China tends to define it from the perspective of "development". The threat-based approach defines it from the perspective of "others", and the development-based approach focuses more on the "self" needs, with the main purpose to enhance the development of cyberspace as well as guarantee the domestic stability of society. Therefore, the social-political stability is regarded as the core national interest of China, and it implements the network filtering and monitoring techniques accordingly to maintain its social and political stability.

The American core cybersecurity interests include the following three aspects:

> First, the security of key infrastructure. The American economic, political and military operations highly depend on the network. Therefore, preventing the finance, telecommunication, energy, transportation, water supply and emergency service facilities from cyberterrorism and other cyberattacks becomes its crucial national interest. The US had 15,000 networks and 7 million computers operated in dozens of other countries which support the US military operations such as training, intelligence and command control early in 2011, and the number might be much higher today.[18] This makes the US more susceptible to cybersecurity concerns than other countries.
>
> Second, freedom of actions in cyberspace, including access to the network systems of other countries. For example, the Prism project of the National Security Agency monitors and collects data of the telecommunication and network systems of many countries all over the world. The US argued that such activities comply with its national legislation, and its war on terror with the essence of maintaining its legal national interests.
>
> Third, security of the commercial and technical secrets. The anonymity and connectivity of the Internet provides convenience for data and information theft. It is said that the American enterprises suffer the losses of hundreds of billions of US dollars each year because of cyber information theft. Therefore, protecting the intellectual property rights, technical patents and commercial secrets becomes an important cybersecurity interest of the US.

Based on these core national interests in cyberspace, the US employs a pre-emptive cybersecurity strategy. It hopes to achieve cyber deterrence and seek the dominant position in cyberspace through the pre-emptive cyberspace strategy to support its leading position in the world. Five pillars were proposed in the first Cyberspace Operation Strategy which was issued by the US Department of Defense in 2011, and the first two especially reflected its pre-emptive cyberspace strategy.[19] First, cyberspace was listed as the "operation field" which is paratactic with land, sea, air and outer space. It was the first time that cyberspace was included into the military action category. Second, passive defence is changed to the active defence in order to more effectively prevent and attack invasion and other hostile acts upon the American network system. Several "No. 1s" of the US may show its pre-emptive characteristics in terms of the cyberspace strategy. The US is the first country to propose to take cyberspace as a battlefield; it is the first country to establish a Cyber Command and it is the first country to carry out the cyberwar practice.

China has the largest number of network users and has developed into the largest e-commerce market in the world. Therefore, keeping the stable connectivity of cyberspace is helpful for China to promote economic development and social progress, strengthen international competitiveness and increase new strategic opportunities. However, compared with the US and some other Western countries, the research and development (R&D) capacity of the network techniques, network product competitiveness and network application in China are still weak. The difference between China's demands for cybersecurity and actual capacity to ensure cybersecurity makes China take a defensive attitude toward cybersecurity.

The Chinese core cybersecurity interests have the following three aspects:

- First, social and political stability. The Internet is an information dissemination platform with low cost and high efficiency as well as a cyber public opinion field of great influence, especially with the rapid development of the social media. China is in the transitional phase of developing into an industrialised society. During this process, the new and old social contradictions are interwoven. Therefore, the Chinese Government needs to appropriately manage cyberspace from the perspective of avoiding negative impact of the cyber public opinions on its social and political stability.

  Thus, in China's case, the biggest cyber threat is any factor that affects its social and political stability. Any anti-government or anti-social activities through cyberspace; dissemination of words and deeds destabilising the society; cyberspace activities inciting ethnic hatred and terrorism; planning, organisation and implementation of any acts of subversion, division or sabotage; violent separatist terror attacks aimed at China's territorial integrity and political power consolidation through the network; and public opinion attacks on information network that could undermine the consolidation of the Chinese regime, the stability of the political system as well as the unity and harmony of all peoples, along with any other equivalent acts all fall into the primary category of threats to national security. With regard to the most serious political threat, China has a special term, the "three forces" (sangu shili), namely terrorists, separatists and extremists. The Chinese Government worries that unrestricted Internet access or uncontrolled information or dissent might become a tool of subversion and pose a significant threat to Chinese political security.

- Second, security of information infrastructure and network system. With the rapid development of the informatisation process, the economic and social operations in China have increasingly become dependent on cyberspace. Therefore, safeguarding the security of the information infrastructure and various important information systems relating to the national economy and the people's livelihood has become very important. Therefore, it is easy to understand two buzz words with regard to China's information infrastructure and network systems security in the past two years, which are "network security review" (wangluo anquan shencha) and "independent controllability" (zizhu kekong). China believes that, in order to maintain the security of critical infrastructure, import censorship must be established when it comes to servers, routers, switches, storage devices and other network facilities. Meanwhile, China believes that, in order to safeguard the network information security fundamentally, it must vigorously foster and support the domestic networking industries, increase the import substitution rates of network products and software in key areas with a targeted and step-by-step approach.
- Third, security of network information and data. A large number of network data of the Chinese Government, enterprises and users involves the important national interests, but there are still problems such as lack of legal basis and technical methods for China to maintain the data rights and interests. The network activities generate considerable data, whose ownership and management right are separated. For example, the foreign network enterprises in China manage a lot of data and information belonging to the Chinese network users. If such foreign enterprises share the data and information with the governments of their own countries, according to the their own laws, it will damage the national interests and national security of China.

The international cybersecurity pressures that China is confronted with have two main stimulating factors: the Prism project revealed by Snowden as well as the rapid development of new technology. The publication of Project Prism has triggered the pre-existing security anxiety of each country, including the European Union (EU), about the United States' ability to abuse its technical advantages. For

China, 2013 is the starting year when national cybersecurity threats have become clearer. During this year, in addition to Project Prism, subsequent in-depth reports about the Stuxnet virus, as well as the mass demonstrations spreading from East and North Africa to South America, were further evidence that the country was facing a huge challenge in cyberspace.

Of course, although the priority of national interests in cyberspace is different, China and the US have common interests in terms of safeguarding security of cyber infrastructure, maintaining the connection of international network and attacking cyberterrorism and cybercrime, which is an important foundation for China and the US to cooperate in the cybersecurity field. However, the global concern regarding the cybersecurity interests of the US and the priority domestic concern about the cybersecurity interests of China result in differences in opinions about global cybersecurity institutions between them, such as the arguments over whether to maintain or reform the current global cyberspace governance system; the governance model of "multi-stakeholder" or the UN-led model; and brand-new cyberspace institutions or the extension of traditional institutions.

## Conclusion

The global cybersecurity environment may be understood from several perspectives. The cybersecurity environment scope could be large or small and the severity of threats to the cybersecurity environment varies. In fact, the cybersecurity environment is a subjective state and related to discourse construction. The network environment has the reconstitution role for the security problems. All actors in cyberspace have the capacity to launch attacks. There is no geographic concept in the network environment, and the launching capacity therefore is not limited by geographical distance. The threats to the network environment cannot be attributed quickly and effectively and it is also impossible to perform effective deterrence in the network environment. The global cybersecurity environment is confronted with a number of problems, including information inundation, information pollution, information infringement, information monopoly and cybersecurity crisis. It also has some common characteristics, such as hysteretic nature of the security techniques, imbalance of strength structure, deficiency of institutions and norms and insufficient mutual trust.

As a subjective cognition, the security environment changes with different groups and countries. The US and China, too, define cybersecurity environment differently, with the US viewing it from the perspective of "threats" and China from the perspective of "development". The core cyberspace interests of the US include security of key infrastructure, freedom of action in cyberspace and security of the commercial and technical secrets, while China's primary national cyberspace

# About the Contributors

**Alexandra Kulikova** is the Global Stakeholder Engagement Manager for Eastern Europe and Central Asia at ICANN, and also acting as PIR Center Consultant (non-staff). Alexandra's research interests within the program and beyond include national and global internet governance, privacy and data protection online, state and corporate policies on ICT security, international cyber-strategies and policies.

**Amit Sharma** is currently serving as Additional Director in the Office of the Scientific Advisor of Defence Minister, Defence Research and Development Organization (D.R.D.O.), Ministry of Defence, Government of India. He has worked in the field of Information Security, Information warfare, Strategic Information Dissemination Systems, Net Centric Warfare, C4I2SR systems and Secure and survivable networks.

**Arvind Gupta** is the Deputy National Security Adviser of India and Secretary in the National Security Council Secretariat (NSCS). Previously, he was the Director General of Institute for Defence Studies and Analyses (IDSA). He is a former Indian Foreign Service Officer, and has served in diplomatic missions in Moscow, London and Ankara. He held the Lal Bahadur Shastri Chair on National Security at the IDSA from 2008 to 2011.

**A. Vinod Kumar** is an Associate Fellow at the Institute for Defence Studies and Analyses (IDSA), New Delhi and a visiting faculty at the Institute of Foreign Policy Studies (IFPS), University of Calcutta, Kolkata. His areas of expertise include nuclear policy issues, missile defence, foreign policy and strategy. He has written extensively in acclaimed publications like Bulletin of the Atomic Scientists, The National Interest, Strategic Analysis, Asia Times and Vayu Aerospace Review, among others.

**Caitríona Heinl** joined the Centre of Excellence for National Security (CENS) at S. Rajaratnam School of International Studies (RSIS) as a Research Fellow for cybersecurity issues in October 2012. She has published articles in peer-reviewed journals and policy advisory reports on topics that include international and regional cooperation, country case studies, and national security implications of emerging technologies.

**Candice Tran Dai** is Vice President and Cyberspace Program Manager at Asia Center, France. She has also been working as a consultant in international business development strategy since 2006, advising European companies regarding their market access and international development in China and Southeast Asia. She is focusing on issues relating to knowledge society, national ICT development strategy, as well as political and cybersecurity issues.

**Caroline Baylon** serves as the director of the cybersecurity research program at the Center for Strategic Decision Research in Paris, France and was previously the lead researcher on cyber security at Chatham House (The Royal Institute of International Affairs) in London, United Kingdom. Her research is focused on critical infrastructure protection, notably on cyber security challenges for nuclear facilities and on cyber security threats to satellites. She is currently carrying out two research projects, one on curbing the proliferation of cyber weapons and another on cyber proxy armies, funded by the UK government.

**Cherian Samuel** is an Associate Fellow at IDSA. He is an alumnus of Madras Christian College, and of Jawaharlal Nehru University, New Delhi. He has previously worked with India Abroad and was a Research Fellow in the US Studies Program at the Observer Research Foundation. He was co-ordinator of the IDSA Task Force on Cyber Security which published a report on "India's Cyber Security Challenges" in March 2012.

**Cuihong Cai** is associate professor of international relations at the Center for American Studies of Fudan University. Prior to the present job, she worked for the Foreign Affairs Office of Fudan University during 1996-2001. She has authored Political Development in the Cyber Age (2015), U.S. National Information Security Strategy (2009) and Internet and International Politics (2003), as well as several dozen of articles and papers on cyber-politics, cyberspace governance, cybersecurity strategy and Sino-US relations.

**Gillane Allam** is a career diplomat of the Ministry of Foreign Affairs of Egypt. During her service abroad, she has served as a diplomat in the Permanent Missions of Egypt to the UN in New York & Specialized Agencies in Vienna. She has held the posts of Ambassador consecutively to India, Australia, New Zealand & countries of the Pacific. Post retirement, she taught at the Graduate School of The Arab Academy in Cairo, and joined the Egyptian Council for Foreign Affairs (ECFA).

**Greg Austin** is a Professorial Fellow with the EastWest Institute in New York and a Professor at the Australian Centre for Cyber Security at the University of New South Wales, Canberra, at the Australian Defence Force Academy. He is the author of several highly reviewed books on international security, especially on Asia. His latest book is Cyber Policy in China (Cambridge: policy 2014).

**Il Seok, Oh** is Senior Researcher at Institute of Legal Studies, Korea University Law School, an expert in Contract, Tort, Oil and Gas Law, and Information Security Law. He has a Ph.D from Korea University and an LLM from the Northwestern University School of Law, Chicago.

**Jana Robinson** is currently Space Security Program Director at the Prague Security Studies Institute (PSSI). She previously served as Space Policy Officer at the European External Action Service (EEAS) in Brussels. She was also a Space Security Advisor to the Czech Foreign Ministry, seconded to the EEAS. From 2009 to 2013, she worked as Resident Fellow at the European Space Policy Institute (ESPI), seconded from the European Space Agency (ESA), leading the Institute's Space Security Research Programme.

**Kah-Kin Ho** is Senior Director for public sector at FireEye. Earlier, he was with Cisco for more than 18 years and Headed Strategic Security, where he played a key role in developing and shaping Cisco's strategic positioning in security that aligns with customer requirements. He also serves in the Advisory group of EUROPOL European Cyber Crime Center (EC3) and teaches Cyber Security Strategy and Policy at ETH Zürich.

**Liam Nevill** is currently working in the Australian Strategic Policy Institute's International Cyber Policy Centre, researching and writing on international and domestic cyber policy issues. Prior to joining ASPI Liam worked at the Australian Department of Defence on strategic and international defence policy issues. He has previously worked in policy roles in the Department of Health and Ageing, and the Northern Territory Treasury.

**Li-Chung Yuan** is currently teaching at the Graduate Institute of Strategic Studies in the Republic of China (ROC) National Defense University (NDU) as an Assistant Professor with the rank of Colonel. With 22 years of military service, he has served as teaching assistant and squadron commander in the Air Force Academy, translation officer at the Institute of National Strategic Studies (a defense think tank), staff officer at the Intelligence Division (J-2) of the Ministry of National Defense, Air Combat Command, and the Combined Logistics Command.

**Liina Areng** assumed the duties of Head of International Relations at Estonian Information System Authority in March 2014. Prior to her current position, she coordinated NATO Cooperative Cyber Defence Centre of Excellence's (NATO CCD COE) international affairs. She holds an honorary title of NATO CCD COE Ambassador.

**Madan M. Oberoi** is an Indian Police Service officer of 1992 batch. He is presently on deputation as Director of Cyber Innovation and Outreach Directorate in the INTERPOL Global complex for innovation (IGCI), Singapore. He supervises two sub-directorates including 'Strategy & Outreach' sub-directorate and 'Research & Innovation' sub-directorate.

**Munish Sharma** is an Associate Fellow with the Cybersecurity Project at IDSA. He is an engineering graduate and holds masters in Geopolitics and International Relations. Prior to masters he worked as software engineer for four years with Accenture. His research areas are Cybersecurity, Critical Information Infrastructure Protection, Space Security, and Defence Technologies.

**Nandkumar Saravade** served as the Chief Executive Officer of the Data Security Council of India. He is a former officer of the Indian Police Service, who branched off to specialise in cyber security issues. Before taking voluntary retirement from IPS in 2008, he worked with National Association of Software and Service Companies (NASSCOM) as Director, Cyber Security and Compliance, on a three-year deputation. Post voluntary retirement, he led the security and crime prevention verticals at ICICI Bank and CitiBank, apart from being an advisor to Ernst & Young.

**Sanjeev Relia** was commissioned into the Corps of Signals of the Indian Army in 1986. He attended the Defence Services Staff College Course in Wellington. Presently serving as a Colonel in the army, he has been associated with modernization of IT and communication infrastructure and issues related to Cyber Security. During his study leave, he was associated with the research project on "Cyber Warfare: It's Implications on National Security" at The United Services Institution of India.

**Sico van der Meer** is a Research Fellow at the Netherlands Institute of International Affairs 'Clingendael'. His research is focused on non-conventional weapons like Weapons of Mass Destruction and cyber weapons from a strategic policy perspective. Before joining the Clingendael Institute he worked as a journalist and as a Fellow of a think tank on civil-military relations.

**Ted G. Lewis** was Professor of Computer Science and Executive Director of the Center for Homeland Defense and Security at the Naval Postgraduate School in Monterey, California. He has previously held a variety of positions within IEEE Computer Science and the industry as CEO of Daimler Chrysler Research and Technology NA and Senior Vice President of Eastman Kodak. Ted has advised the governments of Taiwan, Egypt, Mexico, and Italy in the areas of economic development and technology development parks and authored over 30 books and 100 papers on computing, critical infrastructure and complexity.

**Uchenna Jerome Orji** is an Attorney admitted to the Nigerian Bar as a Barrister and Solicitor of the Supreme Court of Nigeria. He is pursuing a Ph.D in law at the Nnamdi Azikiwe University in Nigeria, with a specialization in telecommunications regulation. He is also a Research Associate at the African Center for Cyber Law and Cybercrime Prevention (ACCP) located within the United Nations, African Institute for the Prevention of Crime and the Treatment of Offenders in Kampala, Uganda.

**Yasuaki Hashimoto** is the Head of Government and Law Division at The National Institute for Defense Studies (NIDS) and Lecturer (International Law) at Komazawa University, Japan. He also serves on the Committee on National Space Policy of Japan as *ad hoc* member. His areas of expertise are international law, space law, cyber law, international law of armed conflict and international humanitarian law. He has a career spanning over 25 years and has published articles in the space law field after becoming an International Institute of Space Law (IISL) member in 1987.

# INDEX